Metaheuristics
Progress in Complex Systems Optimization

T0189453

OPERATIONS RESEARCH/COMPUTER SCIENCE
INTERFACES SERIES

Professor Ramesh Sharda
Oklahoma State University

Prof. Dr. Stefan Voß
Universität Hamburg

Greenberg /*A Computer-Assisted Analysis System for Mathematical Programming Models and Solutions: A User's Guide for ANALYZE*

Greenberg / *Modeling by Object-Driven Linear Elemental Relations: A Users Guide for MODLER*

Brown & Scherer / *Intelligent Scheduling Systems*

Nash & Sofer / *The Impact of Emerging Technologies on Computer Science & Operations Research*

Barth / *Logic-Based 0-1 Constraint Programming*

Jones / *Visualization and Optimization*

Barr, Helgason & Kennington / *Interfaces in Computer Science & Operations Research: Advances in Metaheuristics, Optimization, & Stochastic Modeling Technologies*

Ellacott, Mason & Anderson / *Mathematics of Neural Networks: Models, Algorithms & Applications*

Woodruff / *Advances in Computational & Stochastic Optimization, Logic Programming, and Heuristic Search*

Klein / *Scheduling of Resource-Constrained Projects*

Bierwirth / *Adaptive Search and the Management of Logistics Systems*

Laguna & González-Velarde / *Computing Tools for Modeling, Optimization and Simulation*

Stilman / *Linguistic Geometry: From Search to Construction*

Sakawa / *Genetic Algorithms and Fuzzy Multiobjective Optimization*

Ribeiro & Hansen / *Essays and Surveys in Metaheuristics*

Holsapple, Jacob & Rao / *Business Modelling: Multidisciplinary Approaches — Economics, Operational and Information Systems Perspectives*

Sleezer, Wentling & Cude/*Human Resource Development And Information Technology: Making Global Connections*

Voß & Woodruff / *Optimization Software Class Libraries*

Upadhyaya et al / *Mobile Computing: Implementing Pervasive Information and Communications Technologies*

Reeves & Rowe / *Genetic Algorithms—Principles and Perspectives: A Guide to GA Theory*

Bhargava & Ye / *Computational Modeling And Problem Solving In The Networked World: Interfaces in Computer Science & Operations Research*

Woodruff / *Network Interdiction And Stochastic Integer Programming*

Anandalingam & Raghavan / *Telecommunications Network Design And Management*

Laguna & Martí / *Scatter Search: Methodology And Implementations In C*

Gosavi/ *Simulation-Based Optimization: Parametric Optimization Techniques and Reinforcement Learning*

Koutsoukis & Mitra / *Decision Modelling And Information Systems: The Information Value Chain*

Milano / *Constraint And Integer Programming: Toward a Unified Methodology*

Wilson & Nuzzolo / *Schedule-Based Dynamic Transit Modeling: Theory and Applications*

Golden, Raghavan & Wasil / *The Next Wave in Computing, Optimization, And Decision Technologies*

Rego & Alidaee/ *Metaheuristics Optimization via Memory and Evolution: Tabu Search and Scatter Search*

Kitamura & Kuwahara / *Simulation Approaches in Transportation Analysis: Recent Advances and Challenges*

Ibaraki, Nonobe & Yagiura / *Metaheuristics: Progress as Real Problem Solvers*

Golumbic & Hartman / *Graph Theory, Combinatorics, and Algorithms: Interdisciplinary Applications*

Raghavan & Anandalingam / *Telecommunications Planning: Innovations in Pricing, Network Design and Management*

Mattfeld / *The Management of Transshipment Terminals: Decision Support for Terminal Operations in Finished Vehicle Supply Chains*

Alba & Martí/ *Metaheuristic Procedures for Training Neural Networks*

Alt, Fu & Golden/ *Perspectives in Operations Research: Papers in honor of Saul Gass' 80th Birthday*

Baker et al/ *Extending the Horizons: Adv. In Computing, Optimization, and Dec. Technologies*

Zeimpekis et al/ *Dynamic Fleet Management: Concepts, Systems, Algorithms & Case Studies*

Metaheuristics
Progress in Complex Systems Optimization

edited by

Karl F. Doerner *(University of Vienna, Austria)*
Michel Gendreau *(CIRRELT, Montreal, Canada)*
Peter Greistorfer *(Karl-Franzens-Universität Graz, Austria)*
Walter J. Gutjahr *(University of Vienna, Austria)*
Richard F. Hartl *(University of Vienna, Austria)*
Marc Reimann *(ETH Zurich, Switzerland)*

 Springer

Karl F. Doerner
University of Vienna
Austria

Michel Gendreau
CIRRELT
Montréal, Canada

Peter Greistorfer
Karl-Franzens-Universität Graz
Austria

Walter Gutjahr
University of Vienna
Austria

Richard F. Hartl
University of Vienna
Austria

Marc Reimann
ETH Zurich
Switzerland

Series Editors:
Ramesh Sharda
Oklahoma State University
Stillwater, Oklahoma, USA

Stefan Voß
Universität Hamburg
Germany

ISBN-13: 978-1-4419-4421-4   e-ISBN-13: 978-0-387-71921-4

Preface

In the last decades the field of metaheuristics has grown considerably. Seen both from the technical point of view and from the application-oriented side, these optimization tools have established their value in a remarkable story of success. Researchers have demonstrated the ability of these methods to solve hard combinatorial problems of practical sizes within reasonable computational time. In this collection we highlight the recent developments made in the area of Simulated Annealing, Path Relinking, Scatter Search, Tabu Search, Variable Neighbourhood Search, Iterated Local Search, GRASP, Memetic Algorithms, evolutionary-inspired algorithms like Genetic Algorithms, Ant Colony Optimization or Swarm Intelligence, and several other paradigms for a variety of well-known application areas, like location problems, the travelling salesman and vehicle routing problems, timetabling problems and others. A specific part of this volume is also dedicated to papers addressing dynamic and stochastic problems, multi-objective optimization, parallel computation, as well as the discussion of general themes like the exploration of distance metrics for comparing solutions, cooperative learning and the use of statistical methods in metaheuristics' design.

The book is organized as follows. In the first four parts, metaheuristics applications to several combinatorial optimization problems are collected, where each part is dedicated to a particular solution technique. Part V treats problems with dynamic and stochastic characteristics, while Part VI addresses the design and application of distributed and parallel algorithms. The final part, Part VII, collects articles dealing with some ideas on algorithm tuning and design and the presentation of general, reusable software tools.

The first two papers in this volume deal with the application of Scatter Search to the multidemand multidimensional knapsack problem (Hvattum and Løkketangen) and for the fixed-charge multicommodity flow network design problem (Gendreau and Crainic). The main focus of both papers is the adaption of different algorithmic ideas and concepts of Scatter Search in respective application domains and the empirical analysis of these design decisions.

The next two topics cover Tabu Search and its use to solve large scale set covering problems and full truckload routing problems. Reflecting the maturity of Tabu Search, the paper by Caserta intertwines a Tabu Search based primal intensive scheme with a Lagrangian based dual intensive scheme to design a dynamic primal-dual algorithm that progressively reduces the gap between the upper and lower bound, while the paper by Hirsch and Gronalt presents a successful application of Tabu Search which solves a real world pickup and delivery problem of full truckloads in the timber industry.

Part III focuses on some recent bio-inspired methods. The paper of Aras et al. deals with the capacitated multi-facility Weber problem and among three nature-inspired methods developed and implemented for this problem, Simulated Annealing was found to outperform the competing approaches. In the paper of Schirrer et al., the reviewer assignment problem is solved by using a Memetic Algorithm. The algorithm developed is applied to the data gathered from the MIC 2001 and 2003 conferences and then used to solve the reviewer assignment problem for the MIC 2005.

A GRASP application to the TSP (Golbarg et al.) and a randomized iterative improvement algorithm for the university course timetabling problem (Abdullah et al.) are grouped in Part IV. In the former paper GRASP is hybridized with a path-relinking procedure, while the latter one uses a composite neighbourhood structure to further enhance the solution quality of the basic versions of the respective algorithms.

Uncertainty and/or dynamic problem formulations are the joint characteristics of the papers collected in Part V. which highlights the diversity of metaheuristic approaches and application domains.

Dejan Jovanović et al. present a new method for the probabilistic logic satisfiability problem, based on the Variable Neighborhood Search metaheuristic. The next paper by Mauro Birattari et al. introduces ACO/F-Race, an algorithm for tackling general combinatorial optimization problems under uncertainty, and addresses the TSP as an illustration. Abdunnaser Younes et al. present an idea of using diversity to guide evolutionary algorithms and investigate its merit on dynamic combinatorial optimization problems, exemplifying an implementation for the dynamic TSP. Joana Dias et al. develop a Memetic Algorithm for capacitated and uncapacitated dynamic location problems. Alba et al. compare different genetic algorithms applied to the non-stationary knapsack problem and study potentials and difficulties of applying GAs in dynamic contexts. Finally, Bartz-Beielstein and Blum present a Particle Swarm Optimization algorithm for problems in noisy environments. While the first five papers deal with uncertainty or dynamics with respect to some problem data, the last paper addresses the influence of noise in the evaluation of a solution on the convergence properties of a metaheuristic algorithm. It also combines the metaheuristic approach with noise reducing methods from statistics.

The application of metaheuristics to notoriously difficult (NP-hard) optimization problems has become a viable approach with the development of ever increasing computational power. However, as more and richer real world constraints are included into existing models with constantly increasing problem sizes, the inherent complexity asks for even more sophisticated computational methods, including parallel implementations of well-known metaheuristics, as well as the adaption of existing techniques for parallel architectures and the exploitation of parallelism within the algorithms. In this volume, two papers address these issues. Fischer and Merz propose a distributed version of the chained Lin-Kernighan algorithm for the Traveling Salesman Problem and show that – given an equivalent amount of computation time – the distributed version outperforms the original algorithm. Araújo et al. present four slightly differing strategies for the parallelization of an extended GRASP with iterated local search for the mirrored traveling tournament problem, with the objective of harnessing the benefits of grid computing. Computational grids are distributed high latency environments which offer significantly more computing power than traditional clusters. Experiments on such a dedicated cluster illustrate the effectiveness and the scalability of the proposed strategies.

The four papers grouped together in the last part of the book describe new methods with respect to algorithm tuning and design and reusable software tools for designing metaheuristics. First, Paquete et al. describe the usage of experimental design to analyze stochastic local search algorithms for multiobjective problems, particularly exemplified for the biobjective quadratic assignment problem. The goal of the paper is to enhance understanding of the influence of particular algorithm design decisions on the quality of the solutions and the dependance of this influence on problem instance features and characteristics, e.g. correlation between the objectives. Next, Kubiak introduces distance measures and a fitness-distance analysis for the capacitated vehicle routing problem based on a statistical analysis of the fitness landscape of problem instances. Halim and Lau present tuning strategies for tabu search via visual diagnosis, where the user and the computer can collaborate to diagnose the occurrence of negative incidents along the search trajectory on a set of training instances. Finally, Dorne et al. exhibit a software toolkit iOpt which provides reusable code to solve combinatorial optimization problems. A solution procedure for the vehicle routing problem is composed by using this toolkit. The authors explain in detail how to make use of the modeling and solving facilities available in iOpt to tackle this problem. At each step of this building process, they discuss the benefits of using iOpt rather than starting building a solution from scratch. The overall conclusion of this work is that the toolkit allows the user to maximize reuse of his code, significantly reduce the development time and focus attention on the design rather than the coding.

Given the range of potential design decisions and applications of meta-heuristics, the 20 papers presented here can only scratch the surface of this vast research field. We hope that this post conference volume will encourage further work in the area of metaheuristic search techniques.

Editing the post conference volume for MIC 2005 would not have been possible without the most valuable input of a large number of people. First of all, we wish to thank all the authors for their contributions. Furthermore we greatly appreciate the valuable help from the referees. Last but not least we are grateful to Monika Treipl for designing and implementing the online reviewing system and to Verena Schmid for editing the final version of the book.

Vienna, Montreal, Graz, Zurich

Karl F. Doerner
Michel Gendreau
Peter Greistorfer
Walter J. Gutjahr
Richard F. Hartl
Marc Reimann

Contents

Part I Scatter Search

**1 Experiments using Scatter Search for the Multidemand
Multidimensional Knapsack Problem**
Lars Magnus Hvattum, Arne Løkketangen 3

**2 A Scatter Search Heuristic for the Fixed-Charge
Multicommodity Flow Network Design Problem**
Teodor Gabriel Crainic, Michel Gendreau 25

Part II Tabu Search

**3 Tabu Search-Based Metaheuristic Algorithm for Large-scale
Set Covering Problems**
Marco Caserta 43

4 Log-truck scheduling with a tabu search strategy
Manfred Gronalt, Patrick Hirsch 65

Part III Nature-inspired methods

**5 Solving the Capacitated Multi-facility Weber Problem
by Simulated Annealing, Threshold Accepting and Genetic
Algorithms**
Necati Aras, Sadettin Yumuşak, Kuban Altınel 91

**6 A Memetic Algorithms for the Reviewer Assignment
Problem**
Alexander Schirrer, Karl F. Doerner, Richard F. Hartl 113

Part IV GRASP and Iterative Methods

7 GRASP with Path-relinking for the TSP
Elizabeth F. Gouvêa Goldbarg, Marco C. Goldbarg, João P. F. Farias ... 137

8 Using a randomised iterative improvement algorithm with composite neighbourhood structures for the university course timetabling problem
Salwani Abdullah, Edmund K. Burke, Barry McCollum 153

Part V Dynamic and Stochastic Problems

9 Variable Neighborhood Search for the Probabilistic Satisfiability Problem
Dejan Jovanović, Nenad Mladenović, Zoran Ognjanović 173

10 The *ACO/F-RACE* algorithm for Combinatorial Optimization Under Uncertainty
Mauro Birattari, Prasanna Balaprakash, Marco Dorigo 189

11 Adaptive Control of Genetic Parameters for Dynamic Combinatorial Problems
Abdunnaser Younes, Otman Basir, Paul Calamai 205

12 A memetic Algorithm for dynamic Location Problems
Joana Dias, M. Eugénia Captivo, João Clímaco 225

13 Panmictic versus Decentralized Genetic Algorithms for Non-Stationary Problems
Enrique Alba, Juan F. Saucedo Badia, Gabriel Luque 245

14 Particle Swarm Optimization and Sequential Sampling in Noisy Environments
Thomas Bartz-Beielstein, Daniel Blum 261

Part VI Distributed and Parallel Algorithms

15 Embedding a Chained Lin-Kernighan Algorithm into a Distributed Algorithm
Thomas Fischer, Peter Merz 277

16 Exploring Grid Implementations of Parallel Cooperative Metaheuristics
Aletéia P.F. Araújo, Cristina Boeres, Vinod E.F. Rebello, Celso C. Ribeiro, Sebastián Urrutia 297

Part VII Algorithm Tuning, Algorithm Design and Software Tools

17 Using Experimental Design to Analyze Stochastic Local Search Algorithms for Multiobjective Problems
Luís Paquete, Thomas Stützle, Manuel López-Ibáñez 325

18 Distance Measures and Fitness-Distance Analysis for the Capacitated Vehicle Routing Problem
Marek Kubiak 345

19 Tuning TABU Search Strategies via Visual Diagnosis
Steven Halim, Hoong Chuin Lau 365

20 Solving Vehicle Routing using iOpt
Raphael Dorne, Patrick Mills, Chris Voudouris 389

Committees

Program Steering Committee

Karl F. Doerner
University of Vienna, Austria (AT),

Michel Gendreau
CIRRELT, Montreal, Canada (CA),

Peter Greistorfer
Karl-Franzens-Universität Graz, Austria (AT),

Walter J. Gutjahr
University of Vienna, Austria (AT),

Richard F. Hartl (Chairman MIC2005Vienna)
University of Vienna, Austria (AT),

Marc Reimann
ETH Zurich, Switzerland (CH)

Local Organizing Committee, Vienna

Christian Almeder
Karl F. Doerner
Günther Füllerer
Walter J. Gutjahr
Richard F. Hartl
Vera Hemmelmayr

Barbara Kuglitsch
Pamela Nolz
Alexander Ostertag
Michael Polacek
Margaretha Preusser
Martin Romauch

Alexander Schirrer
Verena Schmid
Monika Treipl

Program Committee

Takao Asano (JP)
Jacek Blazewicz (PL)
Christian Blum (ES)
Peter Brucker (DE)
Edmund Burke (BG)
Jean-Francois Cordeau (CA)
Teodor Gabriel Crainic (CA)
Van-Dat Cung (FR)
Karl F. Doerner (AT)
Marco Dorigo (BE)
Andreas Fink (DE)
Nobuo Funabiki (JP)
Luca Maria Gambardella (CH)
Xavier Gandibleux (FR)
Michel Gendreau (CA)
Fred Glover (US)
Jens Gottlieb (DE)
Peter Greistorfer (AT)
Walter J. Gutjahr (AT)
Pierre Hansen (CA)
Jin-Kao Hao (FR)
Mark Harman (GB)
Richard F. Hartl (AT)
Alain Hertz (CA)
Toshihide Ibaraki (JP)
Sheldon H. Jacobson (US)
Andrzej Jaszkiewicz (PL)
Kengo Katayama (JP)
Graham Kendall (BG)
Gary Kochenberger (US)
Mikio Kubo (JP)
Andrea Lodi (IT)
Arne Løkketangen (NO)
Helena Lourenco (ES)

Vittorio Maniezzo (IT)
Silvano Martello (IT)
Rafael Martí (ES)
Martin Middendorf (DE)
Pablo Moscato (AU)
Ibrahim H. Osman (LB)
Erwin Pesch (DE)
Gerard Plateau (FR)
Chris Potts (BG)
Jean-Yves Potvin (CA)
Christian Prins (FR)
Günther Raidl (AT)
Cesar Rego (US)
Marc Reimann (CH)
Mauricio G. C. Resende (US)
Celso Ribeiro (BR)
Reuven Rubinstein (IL)
Frederic Semet (FR)
Marc Sevaux (FR)
Kenneth Sörensen (BE)
Thomas Stützle (DE)
Jorge Pinho de Sousa (PT)
El-Ghazali Talbi (FR)
Éric Taillard (CH)
Jacques Teghem (BE)
Paolo Toth (IT)
Michael Trick (US)
Vicente Valls (ES)
Stefan Voss (DE)
Andres Weintraub (CL)
Marino Widmer (CH)
David Woodruff (US)
Ming-Jong Yao (TW)

Part I

Scatter Search

Chapter 1

EXPERIMENTS USING SCATTER SEARCH FOR THE MULTIDEMAND MULTIDIMENSIONAL KNAPSACK PROBLEM

Lars Magnus Hvattum and Arne Løkketangen
Molde University College, Molde, Norway

Abstract: The evolutionary, population based metaheuristic called Scatter Search has been successfully applied to many combinatorial optimization problems. Within the Scatter Search framework, however, there are numerous alternatives for how to implement the different components of the search. In this paper we explore a variety of these alternatives in a Scatter Search for solving the demand constrained multidimensional knapsack problem. Our best Scatter Search implementations produce good results, compared both to previous heuristic work as well as to exact solvers.

Key words: Scatter Search, 0/1 Multidemand Multidimensional Knapsack Problem

1.   INTRODUCTION

Although the concepts of Scatter Search were first proposed in the 1970s, most of its applications (see, e.g., Glover, Laguna and Martí, 2002) are recent. Such applications have proved successful in producing good solutions for many different types of problems from combinatorial and non-linear optimization. As is normally the case for metaheuristics, one has to adapt the solution procedure to the problem at hand in order to achieve a well functioning solver. In this paper we focus on the application of Scatter Search to a formulation of the *0/1 Integer Programming Problem*, and, through experimenting with the different components of a Scatter Search implementation, try to discover how variations of the search affect the solution quality. The implementations are tested on instances of the *Multidemand Multidimensional Knapsack Problem*, as put forward by Cappanera and Trubian (2004).

After this brief introduction, Section 2 contains problem formulations for the *Multidemand Multidimensional Knapsack Problem* and the *0/1 Integer Programming Problem*. A very basic Scatter Search implementation is outlined in Section 3. Various extensions and improvements of this basic Scatter Search are described in Section 4. In Section 5 the computational results are reported, while conclusions and suggestions for future work are found in Section 6.

2.   PROBLEM FORMULATION

Before introducing the problem for which our heuristic solution methods are developed, we point out that our main interest is to study the behavior of the Scatter Search itself and to test different alternative implementations. Our treaty of the *Multidemand Multidimensional Knapsack Problem* (MDMKP) is therefore not extensive. For those interested in more information about the MDMKP and its applications, we recommend Cappanera (1999), Plastria (2001), and Romero-Morales, Carrizosa, and Conde (1997).

The original formulation of the MDMKP in Cappanera and Trubian (2004) is as follows,

$$\text{(MDMKP)} \quad \max \sum_{j=1}^{n} c_j x_j \tag{1}$$

$$\text{subject to} \quad \sum_{j=1}^{n} a_{ij} x_j \leq b_i \quad \forall i \in \{1,...,m\} \tag{2}$$

$$\sum_{j=1}^{n} a_{ij} x_j \geq b_i \quad \forall i \in \{m+1,...,m+q\} \tag{3}$$

$$x_j \in \{0,1\} \quad \forall j \in \{1,...,n\}, \tag{4}$$

where $b_i > 0$ and $a_{ij} \geq 0$ $\forall i \in \{1,...,m+q\}$, $\forall j \in \{1,...,n\}$. The m constraints of family (2) are called *knapsack constraints*, while the q constraints (3) are referred to as *demand constraints*.

The solution methods presented in this paper, however, solve the more general *0/1 Integer Programming Problem* (0/1 IP), where constraints (2) and (3) above can be replaced by

$$\sum_{j=1}^{n} a_{ij}^{*} x_j \leq b_i^{*} \qquad \forall i \in \{1,...,m+q\}, \qquad (5)$$

where b_i^{*} and a_{ij}^{*} can take any value The transformation from the MDMKP to the 0/1 IP is straightforward. The reader may notice that, due to having constraints of both type (2) and (3), obtaining feasible solutions is not trivial, and any heuristic solution method must take this into consideration.

3.   BASIC SCATTER SEARCH

As a basis for further investigation of the Scatter Search paradigm, this section contains a description of a basic Scatter Search implementation. The Scatter Search is usually described through the following five components (see, e.g., Laguna and Martí, 2003, or Martí, Laguna, and Glover, 2006):

1. Diversification Generation Method
2. Improvement Method
3. Reference Set Update Method
4. Subset Generation Method
5. Solution Combination Method

The Scatter Search procedure combines these five components in hope of finding good solutions. In our implementation, the Diversification Generation Method and the Improvement Method are first used to produce a pool of solutions. Each solution is encoded as a 0/1 vector. The Reference Set Update Method then chooses among the available solutions to build a reference set. Subsequently, the Subset Generation Method selects a family of subsets of the reference set, which are input to the Solution Combination Method. The output of the Solution Combination Method is then a set of solutions called trial solutions, which after being subjected to the Improvement Method are fed to the Reference Set Update Method. At this point one can repeat from the application of the Subset Generation Method if the Reference Set Update Method has altered the reference set, or alternatively stop or restart from scratch. In our basic Scatter Search implementation, the search continues, by restarting if necessary, until a time limit has been exceeded. Next follows a description of the five components as chosen for our basic Scatter Search implementation.

3.1   The Diversification Generation Method

The purpose of the Diversification Generation Method is to produce a set of diverse solutions. Although using systematic and deterministic procedures or controlled randomization is advocated in the Scatter Search methodology, the method introduced in our basic Scatter Search implementation is totally randomized. Each solution vector x is initialized component by component, by assigning, with equal probability, either $x_j = 0$ or $x_j = 1$. New solutions are subjected to the Improvement Method and inserted in an initial pool of solutions until p unique solutions have been generated.

3.2   The Improvement Method

Solutions generated randomly, as in the Diversification Generation Method described above, are likely to be of very poor quality, and not necessarily feasible with respect to the set of constraints. The Improvement Method chosen here is a simple steepest ascent local search, using a flip-neighborhood (that is, the neighborhood consists of all solutions of Hamming distance one from the current solution) and moving to the best solution in the current neighborhood until no improving move is found. To handle infeasible solutions we evaluate solution vectors using two measures:

$$z(x) \;\;=\;\; \sum_{j=1}^{n} c_j x_j$$

$$v(x) \;\;=\;\; \sum_{i \in S_x} (\sum_{j=1}^{n} a_{ij}^* x_j - b_i^*) \;\;,\;\; S_x = \left\{ i : \sum_{j=1}^{n} a_{ij}^* x_j > b_i^* \right\}.$$

Here $z(x)$ gives the objective function value of x, whereas $v(x)$ is the sum of violations in the set S of violated constraints. We use these two values to compare two solutions and say that x^1 is better than x^2 if either $v(x^1) < v(x^2)$ or both $v(x^1) = v(x^2)$ and $z(x^1) > z(x^2)$. Note that a solution x^1 is feasible if and only if $v(x^1) = 0$.

In Section 4.1 we discuss other alternatives for the Improvement Method.

3.3   The Reference Set Update Method

From a large number of solutions produced by the Diversification Generation Method, the task of the Reference Set Update Method is to select a smaller set of interesting solutions, called the reference set, which will be the input to the Subset Generation Method. The Reference Set Update Method will also be used to update the reference set with trial solutions generated by the Solution Combination Method.

Typically, the reference set consists of b_1 solutions of high quality and b_2 solutions that are diverse (with respect to the high quality solutions and each other), for a total of $b = b_1 + b_2$ solutions. Thus, the reference set is said to have two tiers, one with good solutions and one with diverse solutions. In our implementation, the reference set is (re-)built by the following steps:

1. Choose the b_1 best solutions available from the pool generated by the Diversification Generation Method, from the previous reference set, and from the set of trial solutions.
2. Choose the b_2 most diverse solutions available from the pool generated by the Diversification Generation Method and from the previous reference set.

For every solution that is accepted into the reference set, a diverseness measure is updated for all solutions that are candidates to be selected during step 2. The diverseness measure is the minimum Hamming distance from a solution to any solution selected for the reference set. High values of the diverseness measure thus indicate that a solution is diverse with respect to all solutions in the reference set. Note that in this implementation the trial solutions are considered for inclusion based only on quality, and not on diversity.

3.4   The Subset Generation Method

This method generates subsets of the reference set, after which each subset is used to create trial solutions by the Solution Combination Method. The basic implementation simply generates all subsets consisting of exactly two solutions where at least one of the solutions was added to the reference set since the last execution of the Subset Generation Method.

3.5   The Solution Combination Method

There are numerous alternatives for the Solution Combination Method, several of which will be discussed further in Section 4.4. The method selected for our basic Scatter Search, however, produces five trial solutions for each pair of solutions selected by the Subset Generation Method. Letting x^1 and x^2 be two reference solutions, they will be combined in the following way (as suggested in Laguna and Martí, 2003):

1. x^a is such that $x_j^a = x_j^1 x_j^2$ (i.e., the intersection of ones).
2. x^b is such that $x_j^b = x_j^1 + x_j^2 - x_j^a$ (i.e., the union of ones).
3. x^c is such that $x_j^c = x_j^1(1 - x_j^2)$ (i.e., the ones that belong to x^1 but not x^2).
4. x^d is such that $x_j^d = x_j^2(1 - x_j^1)$ (i.e., the ones that belong to x^2 but not x^1).
5. x^e is such that $x_j^e = x_j^c + x_j^d$ (i.e., the symmetric difference of x^1 and x^2).

Each of the five resulting solutions is subjected to the Improvement Method before being added to the set of trial solutions.

3.6   Parameter Testing

There are three parameters in the above description for which good values are difficult to ascertain a priori, namely p, b_1 and b_2. Typically, the reference set has 20 solutions or less, while the size of the initial pool, p, is suggested to be no more than 100 (see Glover, Laguna and Martí, 2003). In order to perform empirical tests on parameter settings, four test instances were chosen from a larger set of 836 instances (see description in Section 5): 100-100-1-1, 100-10-1-0-0, 250-30-30-1-0, and 500-5-2-0-0. They have from 100 to 500 variables and from 7 to 101 constraints (knapsack constraints and demand constraints counted together).

We first tested different values of b_1 and b_2 while setting $p = 100$. Figure 1-1 illustrates average best value found on ten runs of 240 seconds each and with different random seeds for instance 100-10-1-0-0. The test instance 100-10-1-0-0 has 100 variables, 10 knapsack constraints and 1 demand constraint, and the optimal solution is known to be 28504. The reference set tiers were varied in size, with b_1 from zero to thirty-five and b_2 from zero to twenty-five, both with a step length of five. It is clear from the tests that the size of the reference set should not be too small in our basic Scatter Search implementation, thus slightly contradicting the rule of thumb stated in the previous paragraph. Good results were often found when using b_2 equal either to zero or five, and with b_1 close to thirty. Similar results were found on the other test instances, based on which a choice was made to continue with $b_1 = 25$ and $b_2 = 5$.

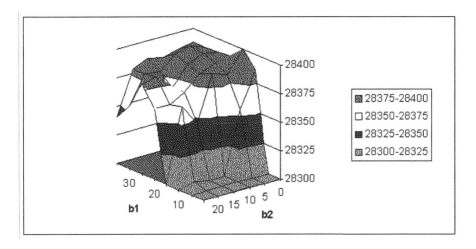

Figure 1-1. Average best values for different sizes of the reference set

Having decided the size (and partitioning) of the reference set, the effect of the size of the initial pool of solutions was examined. Different values for p was tried, from 30 to 300. Preliminary testing indicated a best value for the size of the initial pool to be 200 different solutions.

Table 1-1 in Section 5 contains results for an entire test set of 836 instances using the basic Scatter Search implementation just described with the parameter settings outlined above, and limited to 240 seconds per instance. Evidently, even though the selected components seemed fairly sensible, its performance is poor in comparison to previous heuristic methods developed for the MDMKP, as well as to exact methods. However, it does succeed at finding feasible solutions to all but three instances, while other methods fail to find feasible solutions for as many as 28 (Cappanera and Trubian) to 72 (CPLEX) instances. The results presented are from a single run, but since they are summarized by class, and each class contains 45 instances (except one class containing 26 instances), some indication of the performance of the solution methods can be inferred.

4. EXPERIMENTS WITH DIFFERENT SCATTER SEARCH COMPONENTS

This section of the paper examines several improvements/alternatives to the components in the basic Scatter Search, and attempts to provide an insight into how the components can interact to create high quality solutions.

4.1   **Alternatives for the Improvement Method**

We now examine different strategies for improving the solutions encountered during the search, which shows to have a significant impact on the final solution. The first alternative is a slightly modified steepest ascent local search. In the regular steepest ascent the stop criterion ensures that the next solution visited must be better than the current solution. Here we change the stop criterion so that the search continues as long as the next solution is better than the previous solution. This allows, in situations where there are many local optimas – several of which are close to each other with respect to a given neighborhood operator, the search to continue from one local optimum to the next (plotting the objective function value for each iteration will yield a jagged curve, hence we label it "jagged steepest ascent"). There is no danger of looping since solutions may not be repeated; as the neighborhood is symmetric and we move to the best neighbor, the worst we can do is to go back to the previous solution – which is disallowed since a solution is never better than itself.

Rather than relying only on neighborhoods in which the members differ from the current solution in one variable only, one can inspect neighborhoods that flips two or more variables. Cappanera and Trubian (2004) used a swap move, limited to search in feasible space, where a zero variable and a one variable interchange values. Since the 0/1-IP model allows arbitrary values for the a_{ij}^{*} variables one must also consider the flip of two zero variables or two one variables (i.e., this gives a double flip neighborhood, rather than a swap neighborhood), and since we have ordered the infeasible solutions (through $v(x)$), the neighborhood may be used also in infeasible parts of the search space.

Cappanera and Trubian also occasionally apply a double swap move, where two zero variables and two one variables interchange values. Similar to the previous neighborhood, swaps are replaced by flips, yielding a quadruple-flip neighborhood rather than a double swap neighborhood. In the approach tested here the neighborhood is only examined when the current solution is feasible. Then the neighborhood is explored by looking at variables that have the better influence on the objective function value first, and the first improving move found is accepted.

We now summarize the available improvement methods, presented in increasing order of merit when tested in a 300 seconds multi-start local search:

- **NONE** – do not apply any improvement method.
- **SA** – a steepest ascent local search, using single-flip moves only.

- **JSA** – a jagged steepest ascent local search, which extends **SA** by allowing the search to continue if the next solution is better than the previous.
- **2SA** – a steepest ascent local search which, at each iteration, combines the single-flip neighborhood with the double-flip neighborhood described above.
- **J2SA** – same as **2SA**, but the search is allowed to continue as long as the next solution is better than the previous.
- **4SA** – similar to **2SA**, but if the current solution is feasible and locally optimal with respect to **2SA**, then the quadruple-flip neighborhood described above is searched.
- **J4SA** – similar to the **J2SA**, but the quadruple-flip neighborhood is searched when no improved solution can be reached otherwise. That is, quadruple-flip moves will not be considered, unless two consecutive moves by the combined single- and double-flip neighborhoods do not improve the solution.

4.2   A Systematic Diversification Generation Method

We also tried a diversification generator that does not rely on randomization, but which attempts to produce diverse solutions in a more systematic way. The approach is similar to one described by Glover (1998).

Let x^{seed} be a solution vector used to seed the generator, and let o (offset), s (step size) and c (cluster size) be parameters that decide how the next solution is generated. Let the length of the solution vectors be n, and let the separate variables in this case be denoted $x_0, x_1, ..., x_{n-1}$. For given values of the parameters two solutions x^1 and x^2 are generated as follows:

1. Set $x^1 = x^{seed}$.
2. For $j = o, o + sc, o + 2, sc..., o + o^* sc$, where o^* is the largest integer such that $o + o^* sc < n$ and for $i = 0, ..., (c-1)$, with $k = (i + j) \bmod n$, set $x_k^1 = 1 - x_k^{seed}$.
3. Let x^2 be the complement of x^1.

We produce solutions by setting $c = 1, 2, ..., n/2$ with $s = 1, 2, ..., n/5c$ and $o = 0, 1, ..., sc - 1$. Note that when $n = s$ there is no need of generating the complement solution x^2. Empirically, this scheme is found to allow more than $n^2/10$ unique solutions to be generated from each seed, which is sufficient for large n. For small n the generator may be restarted using a different seed. In our implementation the first seed is the solution vector with only 0's and if another seed is needed then this is generated randomly.

Using the systematic diversification generator, one achieves an increase in the number of restarts per second, indicating that it is either quicker to produce starting solutions without calls to the pseudorandom number generator, or that each starting solution may be, on average, closer to a local optimum. With the exception of **NONE** and **JSA** the results are better, although often found later. In fact, if the runs were stopped after 60 seconds it seemed that generating all starting solutions randomly led to the better solutions. Thus, this systematic diversification generator may need long time before being useful vis-à-vis a purely randomized scheme. One should note, though, that the goal of this diversification generator is not to generate better solutions when used in a multi-start local search, but to generate diverse solutions for a population based metaheuristic.

4.3   Elaborations of the Reference Set Update Method

The Reference Set Update Method used in the implementation described in Section 3 has a two-tier based selection process, where one tier is based on solution quality and the other is based on solution diversity. A third tier, as suggested in Laguna and Martí (2003), could be based on including solutions that have contributed to high quality solutions when used in the Solution Combination Method. Thus, an alternative update of the reference set could be as follows.

1. Choose the b_1 best solutions available from the pool generated by the Diversification Generation Method, from the previous reference set and from the set of trial solutions.
2. Choose the b_3 solutions that have contributed to the best solutions after being input to the Solution Combination Method. Only solutions that have previously been selected for the reference set need to be considered.
3. Choose the b_2 most diverse solutions available from the pool generated by the Diversification Generation Method and from the previous reference set.

In some preliminary testing, with $b_1 = 10$, $b_2 = 10$ and otherwise using the basic Scatter Search implementation outlined in Section 3, we varied b_3 from 0 to 21 in steps of three, in order to assess the value of including such a tier in the reference set. The results were not unanimous, but in all cases using small non-zero values for b_3 was better than not including the tier at all.

One way of changing the Reference Set Update Method that conflicts slightly with the approach usually considered in Scatter Search is to rebuild the reference set using trial solutions only (akin to the generational approach in Genetic Algorithms, see Reeves, 2003). Since solutions that are in the current reference set are not permitted among the trial solutions, the reference set will change completely between iterations. Using $b_1 = 25$ and $b_2 = 5$ this approach gave good results on the four instances selected for testing, with the best solution from ten runs being better than the best solution using the regular update mechanism on all four instances and the average results being better on three out of the four instances. The improved results can be explained by the avoidance of restarts, as restarting (which occurs when the reference set is not updated after the generation of trial solutions) will squander the information about good solutions that have been gained thus far, whereas the generational approach will to some extent carry over such information.

4.4 Variations of the Subset Generation and Solution Combination Methods

Since the Subset Generation Method and the Solution Combination Method are tightly connected, we examine these together. As the deterministic manner in which pairs of solutions were combined in the Scatter Search of Section 3.5 seems rather limiting, we examine different approaches. However, the focus here is mainly on methods that combine only two solutions at a time, although combinations of three or more solutions are encouraged in the literature (e.g., Glover, Laguna and Martí, 2003).

Only one approach for combining three solutions has been tested. It is similar to the approach described in Section 3.5, and creates the following combinations from x^1, x^2 and x^3:

1. x^a, in which $x_j^a = 1$ iff at least one of x_j^1, x_j^2 and x_j^3 is one
2. x^b, in which $x_j^b = 1$ iff exactly one of x_j^1, x_j^2 and x_j^3 is one
3. x^c, in which $x_j^c = 1$ iff exactly two of x_j^1, x_j^2 and x_j^3 are one
4. x^d, in which $x_j^d = 1$ iff two or three of x_j^1, x_j^2 and x_j^3 are one
5. x^e, in which $x_j^e = 1$ iff exactly three of x_j^1, x_j^2 and x_j^3 are one

For combining two solutions, a few other approaches have been tried. An option is to use the concept of path relinking, where the idea is to create a path between two solutions, x^1 and x^2, consisting of moves as defined through a neighborhood operator (see Glover, 1998, for more on path relinking). One of the solutions encountered on this path is considered as output from the Solution Combination Method, either the best solution found (having a certain minimum distance from the combined solutions, as both of these are locally optimal with respect to the choice of Improvement Method and one does not want to return to the same local optimum when applying the Improvement Method on the resulting solution), or the solution found in the middle of the path, i.e. equally far from x^1 and x^2. We consider both starting the path in x^1 as well as in x^2, since the two directions are likely to yield different paths. In addition there is the option of simultaneous path relinking, where the path is built starting from both solutions, merging somewhere in between after extending each path segment in turn towards the other. The paths are generated by using the single-flip neighborhood described in Section 3.2. For more about path relinking and its relation to Scatter Search, see Martí, Laguna, and Glover (2006).

All combination methods presented so far have been completely deterministic. Since the type of solutions they can produce may seem somewhat limited, we have also tested two combination methods that rely on randomized choice. The first method is inspired from the one-point cross-over operator used in Genetic Algorithms (see, e.g., Reeves, 2003), where a cross-over point is selected, dividing two parent solutions in two parts, and two offspring solutions are generated by combining one part from each parent. Since two given solutions are combined only once in the Scatter Search paradigm, it may be promising to repeat the procedure with different cross-over points. Thus, in the approach implemented here we first preprocess the parent solutions, finding all possible distinct cross-over points, and then choose a number n_{cp} of these cross-over points, resulting in a total of $2 \cdot n_{cp}$ trial solutions from each pair of reference set solutions that are combined.

The second combination method using randomization is inspired by a description in Laguna and Martí (2003), page 63. Suppose the solutions x^1 and x^2 are to be combined. For each variable calculate

$$score(i) = \frac{z(x^1)x_i^1 + z(x^2)x_i^2}{z(x^1) + z(x^2)},$$

and then, using a randomly drawn number $r_i \in [0,1]$, let

$$x_i = \begin{cases} 1 & \textit{if } r_i \leq score(i) \\ 0 & \textit{if } r_i > score(i) \end{cases}.$$

Thus, good solutions have a greater influence than poor solutions. However, two weaknesses need to be mended. Firstly, if both x^1 and x^2 are infeasible the score will be misleading as to which solution is better. In such cases a better choice for computation of score is

$$score(i) = \frac{v(x^1)x_i^2 + v(x^2)x_i^1}{v(x^1) + v(x^2)},$$

where the infeasibility level is used instead of the objective function value. Secondly, the method is not able to create solutions with variables different from both x^1 and x^2; if both x_i^1 and x_i^2 are one (or zero), then the new solution will have x_i equal to one (or zero). The following formula has proven useful to generate n_{sb} solutions from each pair of reference set solutions, allowing variables that are different from both solutions:

$$score(i) = \frac{z(x^1)x_i^1 + z(x^2)x_i^2}{z(x^1) + z(x^2) + offset_j},$$

where $offset_j = offset_{sb}(j - n_{sb}/2)$ for $j = 0,...,n_{sb} - 1$. A similar change is made in the formula for when x^1 and x^2 are both infeasible.

The different combination methods can be summarized thus:

- **D2S** – Combine two solutions deterministically using five different formulas (see Section 3.5).
- **D3S** – Combine three solutions deterministically using five different formulas.
- **D2/3S** – Apply both **D2S** and **D3S**.
- **PRL-B** – Combine two solutions using path relinking (both ways, output best solution).
- **PRL-M** – Combine two solutions using path relinking (both ways, output middle solution).
- **SPRL** – Combine two solutions using simultaneous path relinking (i.e., by extending paths from both solutions, which merge into one path somewhere in between).
- **COO** – Combine two solutions using a cross-over operator.
- **SB** – Combine two solutions using a score based scheme.

Otherwise using the basic Scatter Search with the parameters described in Section 3, these methods have been tested on our selection of four test instances. Since the methods may yield quite different results on each test instance, we report the average gap, over ten runs of 240 seconds for each instance, to the best upper bound found by running the exact solver CPLEX 9.0 (see Section 5). Figure 1-2 shows the gap for the eight methods on the list above, as well as the gap when no combination method is used (**NONE**, which corresponds to just generating random solutions and subjecting them to the improvement method, without combining them). The corresponding gap for CPLEX is 0.066, and four of the methods tested gave better gaps than CPLEX on the selected set of test instances. Methods based on path relinking seems to perform best here, but the poor results of the two methods that combine three solutions (**D3S** and **D2/3S**) is apparently caused by using a too large reference set for the largest instances.

For the cross-over based method (**COO**), one need to decide the maximum number of solutions to generate from each pair of reference set solutions, n_{cp}. Different values from zero to twenty were tested, and it turned out that all values from three to twenty worked well. A decision was made to use $n_{cp} = 5$, which seemed quite robust for all problem sizes.

The score based method (**SB**) was found to be rather robust with respect to the two parameters n_{sb} and *offset*$_{sb}$, with the best combination on the four selected test instances being $n_{sb} = 4$ and *offset*$_{sb} = 0.1$. Both methods that incorporate random choices perform better than all the deterministic methods, but the difference is small compared to the best path relinking based combination method.

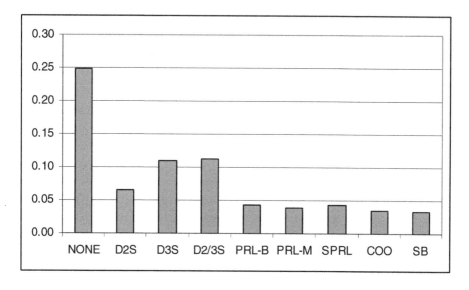

Figure 1-2. Average gap to CPLEX upper bound for different combination methods

4.5   Restarts

Deciding what to do when the search has converged, i.e., does not produce trial solutions that are included in the reference set, may be important, depending on how the different components are selected. In our case, when using a relatively large reference set, the search does not converge quickly for the larger problem instances. In our basic Scatter Search the decision was simply to restart from scratch by disregarding all solutions found thus far, and in Section 4.3 we mentioned a Reference Set Update Method that removed the need for restarts by replacing the entire reference set every iteration. A third option is to carry over a number, n_{co}, of the best solutions of the current reference set to be included in the pool when restarting. This will preserve some information about good solutions, making the convergence more rapid for the following reference set.

Based on preliminary testing using the basic SS settings, it seems that an advantage can be gained from using $n_{co} = 2$.

4.6   Combining the different components into a complete Scatter Search

Having established that there are many viable alternatives to each of the 5 components of the basic Scatter Search implementation, we can try to combine them into a well functioning Scatter Search. Optimally we would like to test the different parameters for each of the alternative components together, but since this represents a too heavy computational effort we instead select some good values for the parameters, as found in Sections 4.1-4.5, and create a few combinations of the components to inspect further. These are then assembled into complete Scatter Search implementations and tested on the full set of test instances (see Section 5). The following complete implementations are considered, here summarized with respect to the choice of methods: 1) Diversification Generation, 2) Improvement, 3) Reference Set Update, 4) Subset Generation, and 5) Solution Combination.

- **BASIC SS** – using only very basic and straightforward components.
  1. Starting solutions are generated using pure random choice.
  2. Every solution encountered is improved using SA.
  3. The reference set is updated as explained in Section 3.3, with $b_1=25$ and $b_2=5$.
  4. All subsets of size two are generated from the reference set, with the restriction that at least one of the solutions in each subset was included in the reference set during the previous iteration.
  5. The solutions are combined using deterministic formulas, yielding five different new solutions for every pair of reference set solutions combined.
- **DET SS** – combining the deterministic components with best test results in Section 4.
  1. Using the strategic diversification generator, with the only randomization appearing on the rare occasion that the generator needs a new seed.
  2. Every solution encountered is improved using J2SA.
  3. The reference set is updated as explained in Section 4.3, with $b_1=20$, $b_2=5$, and $b_3=3$.
  4. Same as for BASIC SS.
  5. The solutions are combined using path relinking, and the solution subjected to improvement is the solution that is encountered on the path equidistant from the combined solutions (PRL-M). This

potentially produces two new solutions, as a path is generated starting from each of the two solutions in the subset.

- **SB SS** – based around the score based combination method.
  1. Start solutions are generated using pure random choice.
  2. 3. and 4. same as for DET SS.
  5. Solutions are combined using score based scheme, with $n_{sb}=4$ and $offset_{sb}=0.1$.
- **GA SS** – based on concepts from Genetic Algorithms
  1. and 2. same as for SB SS.
  3. In this approach the reference set is rebuilt using trial solutions only (cf. the generational approach). There is no need for the third tier, so we use $b_1=20$, $b_2=5$, and $b_3=0$.
  4. All subsets of size two is generated at each iteration, yielding 300 different subsets when using the specified reference set size of 25.
  5. Solutions are combined using the cross-over based method, with $n_{cp}=5$.
- **D2/3S SS** – based on combining sets of both two and three solutions
  1. Using the strategic diversification generator, with the only randomization appearing on the rare occasion that the generator needs a new seed.
  2. Every solution encountered is improved using J2SA.
  3. Since we assume the poor performance of the Solution Combination Method D2/3S during testing was due to the large reference set used, we alter the reference set size to the more standard size, with with $b_1=5$, $b_2=4$, and $b_3=1$.
  4. All subsets of size two and three are generated from the reference set, with the restriction that at least one of the solutions in each subset was included in the reference set during the previous iteration.
  5. The solutions are combined using deterministic formulas, yielding five different new solutions for every pair and every triple of reference set solutions combined.

Each of these four implementations are limited to run for 240 seconds per instance, although they are only aborted after the Solution Combination Method has been completed, which on the larger test instances may cause a slight violation of the time limit to incur. A summary of results is reported in Section 5.

5. COMPUTATIONAL RESULTS

There are in total 836 test instances, which in this presentation are divided in 19 classes. One class, called Obnoxious, consists of 26 instances whose structure is based on the problem of simultaneously locating obnoxious facilities and routing obnoxious materials. These instances all have 100 variables and one demand constraint, with either 50 or 100 knapsack constraints. The other 18 classes have 45 instances each, and are randomly generated based on modifications of Multidimensional Knapsack Problems (see Cappanera and Trubian, 2004, for more information regarding the generation of test instances). These classes are named n-m-x, where n is the number of variables, m is the number of knapsack constraints and x is 0 if all cost coefficients are positive or 1 if there exists negative cost coefficients. Each class have fifteen instances with $q = 1$, $q = m/2$ and $q = m$ demand constraints respectively. The instances can thus be described using the notation n-m-q-x-t, where t is the instance number, except for the instances of class Obnoxious, which are labelled n-m-q-t. All problem instances, except for the class Obnoxious, are, at the time of writing, publicly available from Beasley (1995). The class Obnoxious can be obtained from the authors of Cappanera and Trubian (2004).

All the instances have been attempted solved by the commercially available, exact solver CPLEX 9.0, running for one hour on a standard 2.6GHz Pentium 4. These runs produce upper bounds for the optimal values, and the different methods are compared by calculating average gap in % from the upper bound given by CPLEX 9.0 to the best solution found (**G**). Also reported are the number of instances for which no feasible solution was found (**F**) and the average time spent before the best solution was found (**TB**). In some cases the average total time spent (**TT**) is reported, rather than the time to best. Note that the gap is calculated as the average over the instances for which the method finds feasible solutions, and that the gap can be quite large depending on ability of CPLEX to find good upper bounds.

Table 1-1 contains results by the exact solver **CPLEX,** as well as previous heuristic work by Cappanera and Trubian (2004). Their heuristic **NT** (Nested Tabu search), was run on a 600 MHz Pentium III for a given number of iterations. For **CPLEX** and **NT** we report both total running time and time to best solution. Cappanera and Trubian presented results by CPLEX 7.0 for comparisons, but these results are neglected, since they are very much similar to the results of **CPLEX** above, only slightly worse. Results for the **BASIC SS** is also presented in Table 1-1, and note that total running time is not stated, since this was limited to 240 seconds per run. The final row (**Average***) of the table shows average values over all the classes, except that solution values are only averaged over instance classes where **CPLEX** finds feasible solutions.

Table 1-1. Comparison of overall results for the basic Scatter Search implementation

Class	CPLEX				NT				BASIC SS		
	G	F	TB	TT	G	F	TB	TT	G	F	TB
Obnoxious	0.41	0	363	1130	0.50	0	73	105	1.15	0	115
100- 5-0	0.00	0	14	61	0.07	0	19	39	0.56	0	100
100- 5-1	0.00	0	16	103	0.25	0	20	37	2.16	0	103
100-10-0	0.13	0	316	1678	0.47	0	25	49	1.48	0	123
100-10-1	0.16	0	248	1004	0.64	0	27	47	2.67	0	118
100-30-0	4.44	18	1455	3600	4.37	14	94	207	8.15	1	121
100-30-1	8.92	22	912	2262	8.23	14	84	267	18.67	2	131
250- 5-0	0.04	0	404	2697	0.22	0	140	217	1.16	0	135
250- 5-1	0.20	0	704	2843	1.04	0	152	198	4.51	0	123
250-10-0	0.41	0	1685	3600	0.92	0	146	230	2.65	0	157
250-10-1	0.79	0	984	3593	1.68	0	147	212	4.50	0	156
250-30-0	4.38	10	2195	3600	2.72	0	385	627	6.31	0	187
250-30-1	13.15	11	1927	3600	5.55	0	372	588	10.51	0	179
500- 5-0	0.05	0	1232	3591	0.24	0	606	875	2.37	0	195
500- 5-1	0.17	0	1364	3599	0.97	0	615	791	7.20	0	197
500-10-0	0.21	0	1512	3600	0.64	0	627	867	4.40	0	224
500-10-1	0.54	0	1662	3600	1.48	0	616	814	6.92	0	218
500-30-0	3.56	5	1695	3600	1.71	0	1176	1836	11.05	0	220
500-30-1	7.28	6	1801	3600	2.91	0	946	1361	13.13	0	229
Average*	0.24	3.8	1078	2703	0.70	1.5	330	493	3.21	0.2	160

Results for the four Scatter Search implementations based on the different components tested in Section 4 are reported in Table 1-2. These runs, with headings **DET SS**, **SB SS**, **GA SS**, and **D2/3S SS**, were also limited to 240 seconds per instance. The best overall results seems to be produced by **DET SS**, being better than **NT** on 17 of the 19 classes but better than **CPLEX** on only 6 classes. The running time of the Scatter Search implementations are much shorter than those of **CPLEX**, though, whereas it is more difficult to compare running times with **NT**. The only method that finds a feasible solution to all problems in one run is **SB SS**. For the other Scatter Search implementations it is two particular problem instances that are most difficult, one in class 100-30-0 and one in 100-30-1. Both **CPLEX** and **NT** fail to find feasible solutions to many problems (14-22 instances per class) in these classes. It is interesting to note that each of the other implementations improve upon the **BASIC SS** with quite a large margin, even when considering the enhancements in the Improvement Method and the Solution Combination Method.

Table 1-2. Overall results for different Scatter Search implementations

Class	DET SS			SB SS			GA SS			D2/3S SS		
	G	F	TB	G	F	TB	G	F	TB	G	F	TB
Obnoxious	0.41	0	20	0.41	0	11	0.49	0	16	0.42	0	21
100- 5-0	0.03	0	21	0.01	0	42	0.04	0	46	0.03	0	64
100- 5-1	0.06	0	17	0.02	0	20	0.08	0	23	0.20	0	55
100-10-0	0.29	0	42	0.23	0	57	0.27	0	58	0.30	0	79
100-10-1	0.37	0	34	0.23	0	40	0.28	0	37	0.43	0	79
100-30-0	4.95	1	60	4.92	0	77	5.48	1	96	5.76	1	112
100-30-1	13.09	0	55	13.05	0	88	13.36	1	95	14.57	1	106
250- 5-0	0.09	0	108	0.15	0	123	0.15	0	186	0.27	0	140
250- 5-1	0.44	0	101	0.52	0	127	0.47	0	179	1.01	0	129
250-10-0	0.62	0	126	0.70	0	138	0.76	0	197	0.84	0	150
250-10-1	1.15	0	110	1.25	0	133	1.14	0	177	1.63	0	135
250-30-0	2.10	0	152	2.77	0	213	2.77	0	224	2.60	0	197
250-30-1	4.27	0	160	5.71	0	207	5.84	0	220	5.70	0	187
500- 5-0	0.13	0	164	1.12	0	228	1.86	0	221	0.51	0	221
500- 5-1	0.48	0	178	1.28	0	228	1.92	0	224	1.79	0	199
500-10-0	0.45	0	201	1.52	0	229	1.86	0	225	1.10	0	200
500-10-1	1.01	0	224	2.18	0	227	2.63	0	215	2.31	0	203
500-30-0	2.18	0	217	3.14	0	196	3.24	0	187	2.91	0	184
500-30-1	3.84	0	215	5.11	0	199	5.30	0	191	5.21	0	205
Average*	0.43	0.1	121	0.77	0.0	143	0.96	0.1	156	0.87	0.1	147

We have also tested the same four Scatter Search implementations, but where the stopping criterion is not time, but rather that no new solutions have been included in the reference set during the last iteration. The results on the smaller instances are inferior, as they no longer benefit from the restarts. They get better results on the larger instances, but using almost as much time as CPLEX on the largest ones, as can be seen in Table 1-3.

The best of these implementations, **DET(1)SS**, gets better results than the nested tabu search method (**NT**) of Cappanera and Trubian on 15 of 19 problem classes, and, due to the stopping criterion, uses on average only 2-4 seconds per instance on the four problem classes on which it performs worse than **NT**.

Table 1-3. Overall results for different Scatter Search implementations, using convergence of reference set as stopping criterion.

Class	DET(1)SS			SB(1)SS			GA(1)SS			D2/D3(1)SS		
	G	F	TT	G	F	TT	G	F	TT	G	F	TT
Obnoxious	0.45	0	10	0.41	0	17	0.42	0	27	0.47	0	22
100- 5-0	0.12	0	2	0.13	0	6	0.17	0	11	0.25	0	5
100- 5-1	0.27	0	3	0.27	0	5	0.32	0	9	0.85	0	5
100-10-0	0.52	0	4	0.43	0	10	0.53	0	17	0.66	0	8
100-10-1	0.77	0	4	0.68	0	8	0.78	0	14	1.17	0	8
100-30-0	5.68	1	15	5.37	1	39	5.83	1	49	6.49	1	33
100-30-1	14.6	1	14	12.78	2	36	14.36	1	47	16.33	1	31
250- 5-0	0.14	0	27	0.17	0	97	0.22	0	162	0.31	0	87
250- 5-1	0.56	0	32	0.61	0	82	0.67	0	130	1.07	0	84
250-10-0	0.69	0	43	0.73	0	134	0.82	0	212	0.86	0	141
250-10-1	1.26	0	46	1.29	0	113	1.33	0	166	1.74	0	108
250-30-0	2.15	0	116	2.14	0	409	2.27	0	455	2.42	0	284
250-30-1	4.37	0	125	4.36	0	366	4.46	0	436	5.1	0	279
500- 5-0	0.13	0	188	0.15	0	698	0.19	0	1350	0.25	0	829
500- 5-1	0.45	0	229	0.49	0	575	0.54	0	1075	0.9	0	736
500-10-0	0.41	0	283	0.46	0	995	0.56	0	1629	0.56	0	1119
500-10-1	0.88	0	314	1.01	0	827	1.07	0	1236	1.32	0	854
500-30-0	1.27	0	798	1.36	0	2310	1.42	0	3012	1.46	0	2422
500-30-1	1.96	0	819	2.09	0	2209	2.14	0	2681	2.56	0	1926
Average*	0.51	0.1	162	0.53	0.2	470	0.59	0.1	669	0.80	0.1	473

6. CONCLUSIONS AND FUTURE WORK

Through experimenting with the different components of the Scatter Search methodology, this work has shown that the Scatter Search is quite robust with respect to the choices made for each component. Four quite different implementations, based on good deterministic components (**DET SS, D2/3S SS**), based on randomized components (**SB SS**), and based on components inspired by Genetic Algorithms (**GA SS**) have all produced competitive results on problem instances of the *Multidemand Multidimensional Knapsack Problem*. Even though our methods have been designed to solve the more general *0/1 Integer Programming Problem*, and previous heuristic work (**NT**) has been designed to take advantage of problem specific features, feasible solutions are found on most instances with small gaps to the optimal solutions.

One could point out that the parameters to neither of the Scatter Search implementations have been fine tuned, and that the parameter search was conducted from the perspective of the basic Scatter Search, rather than in combination with the particular implementation in which the parameters were later incorporated. Thus one might expect that each method could perform better if important parameters were retuned.

Since the empirical parameter testing showed that a rather large (as compared to the recommended figures) reference set was useful, one of the drawbacks of the different implementations of Scatter Search was the quite high associated computational effort. However, there is probably a connection between the size of the reference set and the ability to find good solutions on problems with many constraints (and many local optima). A small reference set would probably find a quite good solution relatively fast, whereas a larger reference set increases the possibility of finding better solutions, albeit at a later time. As the initial parameter tests were allotted 240 seconds per run, this may have favored the latter strategy.

For several components one has the option of either relying on randomized choices (random initial solution, making combinations of solutions with random variations) or deterministic choices (strategically generated initial solutions, combining solutions deterministically, etc.). Although the best overall Scatter Search results were obtained by a pure (with disregard to the rare event of restarting the algorithm for generating initial solutions) deterministic method (**DET SS**), making randomized decisions does seem to have some merit, and on some problem classes the methods incorporating randomized choices performs well: the **SB SS** being the only method to find feasible solutions to all instances in one run, and in general being the best heuristic method for the classes with small (100 variables) problems.

The particular choice of Improvement Method seems to have an impact on solution quality that is easily recognizable, though the cooperation with other mechanisms in the Scatter Search framework will improve the solution quality. Relying blindly on the most simple steepest ascent (or descent) local search may be dangerous, and very simple extensions (extending neighborhoods or changing the stopping criterion) can be helpful.

For future work, it may be interesting to delve into the effects of changing the size of the reference set. Preliminary experimental results indicated that large reference sets were more appropriate than the sizes usually recommended in the literature. There seems to be a trade-off between obtaining good results quickly and getting very good results in the long run. This suggests an idea that the size of the reference set should be dynamic, starting out small and growing as the solution quality improves. Bearing in mind that the Reference Set Update Method which was based on the generational approach from Genetic Algorithms also performed well, a combination of this strategy and a growing reference set, could lead to more robust methods of maintaining the reference set.

Although this work has focused mainly on Combination Methods based on combining pairs of solutions, the possibility of combining several solutions simultaneously should not be neglected. However, there seems to be a conflict between using large reference sets and allowing combinations of more than two solutions, since the number of possible combinations grows rapidly. Whether or not larger reference sets, combinations of more than two solutions at a time, both, or neither of these choices are to be preferred, is ostensibly a problem specific issue, and may need further investigation based on this perspective.

REFERENCES

Beasley, J. (1995). OR Library. (http:\\www.ms.ic.ac.uk/info.html).

Cappanera, P. (1999). "Discrete Facility Location and Routing of Obnoxious Facilities". Ph.D. thesis, University of Milano.

Cappanera, P. and M. Trubian. (2004). "A Local Search Based Heuristic for the Demand Constrained Multidimensional Knapsack Problem", *INFORMS Journal of Computing*, to appear.

Glover, F. (1998). "A Template for Scatter Search and Path Relinking", In: *Artificial Evolution, Lecture Notes in Computer Science 1363*, Springer, eds.: J.-K. Hao, E. Lutton, E. Ronald, M. Schoenauer and D. Snyers, pp. 13-54.

Glover, F., M. Laguna and R. Martí. (2000). "Fundamentals of Scatter Search and Path Relinking", *Control and Cybernetics* 39, pp. 653-684.

Glover, F., M. Laguna and R. Martí. (2002). "New Ideas and Applications of Scatter Search and Path Relinking". In: *New Optimization Techniques in Engineering*, Springer, ed.: G. Onwubolu, to appear.

Glover, F., M. Laguna and R. Martí. (2003). "Scatter Search". In: *Advances in Evolutionary Computation: Theory And Applications*, Springer, eds.: A. Ghosh and S. Tsutsui, pp. 519-537.

Laguna, M. and R. Martí. (2003). *Scatter Search: Methodology and Implementations in C*, Kluwer Academic Publishers.

Martí, R., M. Laguna, and F. Glover. (2006). "Principles of Scatter Search". *European Journal of Operations Research* 169, pp. 359-372.

Plastria, F. (2001). "Static Competitive Facility Location: an overview of Optimization Approaches". *European Journal of Operational Research* 129, pp 461-470.

Reeves, C. (2003). "Genetic Algorithms". In: *Handbook of Metaheuristics*, Kluwer Academic Publishers, Boston, eds.: F. Glover and G. Kochenberger, pp. 55-82.

Romero-Morales,D., E. Carrizosa and E. Conde. (1997). "Semi-Obnoxious Location Models: a Global Optimization Approach". *European Journal of Operational Research* 102, pp. 295-301.

Chapter 2

A SCATTER SEARCH HEURISTIC FOR THE FIXED-CHARGE CAPACITATED NETWORK DESIGN PROBLEM

Teodor Gabriel Crainic[1,2], Michel Gendreau[1,3]

[1] *Centre interuniversitaire de recherche sur les réseaux d'entreprise,*
la logistique et le transport (CIRRELT), Université de Montréal
C.P. 6128, Succursale Centre-ville, Montreal, H3C 3J7, Canada
{theo, michelg}@crt.umontreal.ca

[2] *Département de management et technologie, Université du Québec à Montréal*
C.P. 8888, Succursale Centre-ville, Montreal, H3P 3P8, Canada

[3] *Département d'informatique et recherche opérationnelle, Université de Montréal*
C.P. 6128, Succursale Centre-ville, Montreal, H3C 3J7, Canada

Abstract   The Fixed-Charge Capacitated Multi-commodity Network Design (CMND) problem consists in finding the optimal configuration, i.e., the arcs to include in the final design, of a network on which the flows of several products ("commodities") must be routed to satisfy given demands between origin-destination pairs. Each of the arcs that can possibly be included in the design is characterized by its capacity (the maximum amount of flow of all commodities it can support), a fixed cost to be incurred if the arc is selected, and a variable cost for each unit of flow that uses the arc. The objective of the problem is to minimize the total system cost (the sum of the fixed costs of selected arcs and routing costs), while respecting capacity limits.

In this paper, we report on an extensive investigation of different variants of a new metaheuristic, based on the Scatter Search concept originally proposed by Glover, for the CMND. Computational results on a set of small and medium size benchmark instances show that while scatter search is not yet able to match the results of the best existing metaheuristics for the problem, all variants are successful in finding better solutions on some instances.

Keywords: scatter search; network design; multi-commodity flows; fixed-charge
          problems.

1.   Introduction

The fixed-charge capacitated multicommodity network design problem (CMND) is a generic model that covers a wide range of network planning problems in the areas of transportation, logistics, telecommunication, and production management [2, 14, 15]. The problem consists in finding the optimal configuration, i.e., the arcs to include in the final design, of a network on which the flows of several products (or "commodities") must be routed to satisfy given demands between origin-destination pairs. Each of the arcs that can possibly be included in the design is characterized by its capacity (the maximum amount of flow of all commodities it can support), a fixed cost to be incurred if the arc is selected, and a variable cost for each unit of flow that uses the arc. The objective of the problem is to minimize the total system cost, computed as the sum of the fixed costs of selected arcs and routing costs, while respecting capacity limits.

The CMND problem is usually modeled as a 0–1 mixed integer programming problem and it has been shown to be NP-hard in the strong sense. Not surprisingly, even though significant efforts have been devoted to the development of exact methods for this problem (see, e.g., [3, 5, 12]), heuristics must be resorted to when dealing with large instances with several commodities.

Over the last few years, two main approaches based on metaheuristics have been proposed for the general CMND model. The first, which was developed by Crainic, Gendreau, and Farvolden [4], is a tabu search heuristic for the path-based formulation of the problem. It exploits the fixed-charge nature of the problem by exploring the space of the continuous path-flow variables using pivot-like moves in a column generation environment. This method produces impressive results compared to simple heuristics, but its efficiency remains limited by the fact that each move considers the impact of changing the flow of only one commodity (a pivot from one path to another), thus making it difficult to properly account for the multi-commodity nature of the problem. To overcome this limitation, Ghamlouche, Crainic, and Gendreau [6] proposed another tabu search heuristic, but for the arc-based formulation of the problem. The key element of this heuristic was a new neighbourhood structure for the CMND, the so-called "cycle-based neighbourhood", that allows changing the flow pattern of several commodities simultaneously, as well as opening and closing several arcs. In spite of the fact that the global search strategy used in this paper amounted to a fairly basic tabu search scheme, it was at the time the best approximate solution method for the CMND in terms of robust performance, solution quality, and computing efficiency. The authors extended their approach in a second paper [7] in which tabu search was combined with Path Relinking [9–11] to yield an enhanced and more effective search strategy. As far as we know, this is currently the most effective method for tackling large CMND instances. However, for some of the larger and more complex instances, the relative gaps observed between the lower bounds computed using Lagrangian relaxation [3, 5] and the best solutions obtained by the combined tabu search-path relinking hybrid heuristic can be as large as 20.9%, thus suggesting that there might exist significantly better solutions than these. It is therefore relevant to pursue examining other heuristic approaches for the CMND.

The purpose of this paper is to report on an extensive investigation of different variants of a new metaheuristic for the CMND. This new heuristic is based on the Scatter Search concept originally introduced by Glover [8, 9].

The remainder of the paper is organized as follows. In section 2, we recall the arc-based formulation of the CMND, as well as some of its basic properties that will be exploited in our scatter search heuristic. Following a brief outline of the scatter search methodology, section 3 details our scatter search implementation for the CMND. Computational results are reported and analyzed in section 4; we compare, in particular, the scatter search results to those obtained by the path relinking hybrid. Section 5 concludes the paper and suggests future research directions for the application of scatter search to network design problems.

2.   Formulation and basic properties

Let $\mathcal{G} = (\mathcal{N}, \mathcal{A})$ be a network with set of nodes $\mathcal{N}$ and set of directed arcs $\mathcal{A}$. Let $\mathcal{P}$ denote the set of commodities to move using this network and for each $p \in \mathcal{P}$, let d^p denote the required amount of flow of commodity p to be shipped from its origin $o(p)$ to its destination $s(p)$. The total flow on each arc $(i, j) \in \mathcal{A}$ is limited by the capacity u_{ij}. There are two costs involved in the network. The unit cost of moving commodity $p \in \mathcal{P}$ through the arc (i, j), denoted c_{ij}^p, and the fixed cost of including arc (i, j) in the design of the network, denoted f_{ij}. The problem consists in minimizing the sum of all costs while satisfying the demand of transportation.

The arc-based formulation of the CMND can then be written as

$$\min z(x,y) = \sum_{(i,j)\in\mathcal{A}} f_{ij}y_{ij} + \sum_{p\in\mathcal{P}}\sum_{(i,j)\in\mathcal{A}} c_{ij}^p x_{ij}^p \tag{1}$$

subject to

$$\sum_{j\in\mathcal{N}^+(i)} x_{ij}^p - \sum_{j\in\mathcal{N}^-(i)} x_{ji}^p = \begin{cases} d^p & if\ i = o(p) \\ -d^p & if\ i = s(p) \\ 0 & \text{otherwise.} \end{cases} \quad \forall i\in\mathcal{N}, \forall p\in\mathcal{P}, \tag{2}$$

$$\sum_{p\in\mathcal{P}} x_{ij}^p \leq u_{ij}y_{ij} \quad \forall(i,j)\in\mathcal{A}, \tag{3}$$

$$x_{ij}^p \geq 0 \quad \forall(i,j)\in\mathcal{A}, \forall p\in\mathcal{P}, \tag{4}$$

$$y_{ij} \in \{0,1\} \quad \forall(i,j)\in\mathcal{A} \tag{5}$$

where y_{ij}, $(i,j) \in \mathcal{A}$, represent the design variables that equal 1 if arc (i,j) is selected in the final design (and 0 otherwise), x_{ij}^p stand for the flow distribution decision variables indicating the amount of flow of commodity $p \in \mathcal{P}$ on arc (i,j), and $\mathcal{N}^+(i)/\mathcal{N}^-(i)$ denotes the set of outward/inward neighbours of node i.

The objective function (1) accounts for the total system cost, the fixed cost of arcs included in a given design plus the cost of routing the demand of all commodities, and aims to select the minimum cost design. Constraints (2) represent the network flow conservation relations, while constraints (3) state that for each arc, the total flow of all commodities cannot exceed its capacity if the arc is opened ($y_{ij} = 1$) and must be 0 if the arc is closed ($y_{ij} = 0$). Relations (4) and (5) are the usual non-negativity and integrality constraints for decision variables.

For a given design vector $\bar{y}$, the arc-based formulation of the CMND becomes a capacitated multicommodity minimum cost flow problem (CMCF)

$$\min z(x(\bar{y})) = \sum_{p\in\mathcal{P}}\sum_{(i,j)\in\mathcal{A}(\bar{y})} c_{ij}^p x_{ij}^p \tag{6}$$

subject to (2) plus

$$\sum_{p\in\mathcal{P}} x_{ij}^p \leq u_{ij}\bar{y}_{ij} \quad \forall(i,j)\in\mathcal{A}(\bar{y}),$$

$$x_{ij}^p \geq 0 \quad \forall(i,j)\in\mathcal{A}(\bar{y}), \forall p\in\mathcal{P},$$

where $\mathcal{A}(\bar{y})$ stands for the set of arcs corresponding to the design $\bar{y}$. A solution to the CMND may thus be viewed as an assignment $\bar{y}$ of 0 or 1 to each design variable, plus the optimal flow pattern of the corresponding multicommodity minimum cost flow problem $x^*(\bar{y})$. Similarly, the objective function value associated with a solution $(\bar{y}, x^*(\bar{y}))$ is the sum of the fixed cost of the open arcs in $\bar{y}$ and the objective function value of the CMCF associated with $\bar{y}$

$$z(\bar{y}, x^*(\bar{y})) = \sum_{(i,j) \in \mathcal{A}(\bar{y})} f_{ij}\bar{y}_{ij} + z(x^*(\bar{y})). \tag{7}$$

3. Scatter search

Scatter search is a population-based search heuristic that was originally introduced by Glover in 1977 [8]. The basic idea of the method is to create new (hopefully) *interesting* solutions to a problem by combining values from *elite* solutions previously obtained either by some other (meta-)heuristic or by the method itself. As in genetic algorithms, the key idea behind the method is that the best solutions to a problem must share some common attributes. Scatter search is particularly well-suited to problems with continuous decision variables, since elite solutions can then be combined through linear intra- or extrapolation. However, it can also be applied to combinatorial problems, but this requires some ingenuity in the definition of the procedures used to combine elite solutions, as we shall see in the following. The basic template for scatter search as proposed by Glover in 1998 [9] is as follows:

Scatter search template

1 Generate an initial population of *good* and *diverse* solutions.

2 Select a subset of the population to form the *Reference Set* (RS).

3 Extract N solutions from RS to create the *Candidate Set* (CS).

4 Create new solutions by combining the solutions from CS.

5 If necessary, repair these solutions to make them feasible.

6 Improve the new solutions.

7 Update the reference set and go back to step 3.

Interested readers will find more details on scatter search in the original papers by Glover [8, 9] or in the book by Laguna and Martí [13].

Our implementation of scatter search for the CMND focuses on the y_{ij} *design* variables; the continuous flow variables are computed by solving the associated CMCF subproblem. The initial population of solutions is obtained by applying the cycle-based tabu search of Ghamlouche,

Crainic, and Gendreau [6] until it has performed a pre-set number of iterations without improvement (20, in the current implementation). The
reference set is made up of S local optima extracted from the initial
population. If more than S local optima are identified, we keep the S
best ones; if there are less than S local optima, we complete RS with the
best solutions encountered by tabu search that were not local optima.
At each iteration, the candidate set is created by selecting the best solution in RS, along with the one that is farthest from it; if $N > 2$, CS is
completed by randomly chosen solutions in RS.

We create a single new solution from the candidate set. To combine
solutions $y^l, 1, ..., N$, of CS, we first compute for each arc (i, j) a *desirability factor*, $0 \leq m_{ij} \leq 1$, :

$$m_{ij} = \frac{\sum_l w_l y_{ij}^l}{\sum_l w_l}, \forall (i, j),$$

where w_l represents the *weight* of solution y^l. Three alternate variants
were examined to define the weights:

- Voting (V): $w_l = 1, \forall l$;

- Cost (C): $w_l = 1/($cost difference between solution l and the best
  solution$), \forall l$;

- Distance (H): $w_l = 1/($Hamming distance between solution l and
  the best solution$), \forall l$.

In a fourth variant (Frequency – F), desirability factors were weighted
with respect to the frequency of appearance of each arc in the best solutions encountered so far. Note that in variants (C) and (H), if solution
l is the best solution, its weight is set to 100.

The desirability factor of each arc is then assessed on a *desirability
scale* ranging from 0 (do not open the arc) to 1 (absolutely open the
arc). We define two thresholds $t_c \leq t_o$ on this scale and perform the
following comparisons:

- if $0 \leq m_{ij} < t_c$, the arc is closed in the new solution;

- if $t_c \leq m_{ij} \leq t_o$, the arc is undecided;

- if $t_o < m_{ij} \leq 1$, the arc is open in the new solution.

To complete this solution, we first solve a modified CMCF problem
using CPLEX. In this modified CMCF, the capacity of closed arcs is set
to 0, while the variable cost for undecided arcs is set to $(f_{ij}/u_{ij}) + c_{ij}$,

i.e., the cost of arcs in the linear relaxation of CMND. The optimal solution of this problem is then used to create a feasible solution for CMNF by opening all arcs on which there is flow. The cycle-based tabu search is launched from that solution in the hope of finding an improved solution. In a variant of the algorithm, an *intensification* phase is applied to the best solution obtained by TS. This intensification phase is similar to the one described in the original tabu search procedure of Ghamlouche, Crainic, and Gendreau [6]; it involves iteratively modifying the flow distribution of a single commodity at the time and only accepting improving moves. If the best solution found by the above procedure is better than the worst solution in the reference set, it replaces it.

4. Computational results

Computational experiments were performed to first identify good parameter values, and then to evaluate the performance of the method by comparing the results that it produces with those of the path relinking heuristic.

For these experiments, we used one of the original data sets used by Ghamlouche, Crainic, and Gendreau to test the tabu search heuristic [6] and the path relinking method [7]. The 43 problems in this set are general transshipment networks with no parallel arcs. Each commodity corresponds to a single origin-destination pair. On each arc, routing costs are the same for all commodities. Problem instances have been generated to offer for each network size (20 to 100 nodes, 100 to 700 arcs, 10 to 400 commodities), a variety of fixed cost to routing cost ratios and capacity to demand ratios. Each instance is thus uniquely identified by a label consisting of five entries:

1 Number of nodes,

2 Number of arcs,

3 Number of commodities,

4 A letter indicating if variable (V) or fixed (F) costs are dominant in the objective,

5 A letter indicating if capacity constraints are tight (T) or loose (L).

Instance difficulty is largely driven by the number of commodities involved. For a given number of commodities, instance difficulty increases in general with the number of nodes and arcs. For a given network size and number of commodities, problems with tight capacities (T) and where fixed costs dominate (F) are the most difficult. A detailed description of problem instances is given in Crainic, Frangioni, and Gendron [3]. The problem generators as well as the problem instances can be obtained from the authors.

The computer code is written in C++. The exact evaluation of the CMNF problems is done using the LP solver of CPLEX 7.5. All tests were conducted on one 400MHz processor of a 64-processor Sun Enterprise 10000 with 64 Gigabyte of RAM, operating under Solaris 8.

Preliminary testing was performed to find good parameter values on a subset of 10 representative instances. In particular, two values were tested for (t_c, t_o): (0.25, 0.75) and (0.4, 0.6). These tests indicated that the second combination performed significantly better and all further tests were conducted with $(t_c, t_o) = (0.4, 0.6)$. Extensive computational experiments were performed for several values of N using the four variants for combinations, with or without intensification. Different versions of the basic method involving different ways of building the reference set were also considered and tested. The analysis of the results of these various runs, some of which will not be reported in detail here, allowed several conclusions to be drawn:

1 The use of intensification does not significantly improve the results obtained.

2 The best values of N range between 3 and 5; in particular, using $N = 2$ produces markedly inferior results.

3 It is important to use a sufficiently large reference set and to ensure that it is full at initialization step.

4 The combination rule based on frequencies (F) is clearly inferior to the three other ones.

On the basis of these conclusions, we now report on the most interesting variants and combinations. In these, intensification is not used and the size of the reference set is fixed to 20, a value that provides sufficient diversity in the set. Furthermore, as indicated in section 3, at initialization, if less than 20 local optima have been identified by tabu search, the reference set is filled with the best solutions encountered by tabu search that were not local optima, since preliminary testing showed that this had a positive impact on results. In Table 2.1, we report the percentage gaps observed between the solutions obtained with scatter search for $N = $ 3, 4, 5 and for combination rules (V), (C) and (H) with those produced by path relinking [7] for the 43 instances tested. These gaps are computed as the difference between the value of the scatter search solutions and the average value of the path relinking solutions divided by this path relinking value; negative gaps indicate that scatter search produced a better solution than path relinking. Summary statistics for all 43 runs are provided in Table 2.2.

These tables show that the performance of the scatter search heuristic varies quite significantly from one instance to another (with gaps ranging from -3.51 to 10.07%) and sometimes between variants for a given instance (see, e.g. instance (100,400,30,F,L) for which gaps range from -2.11 to 5.42%). In general, scatter search does fairly well, but no single variant outperforms path relinking on average. It is interesting to note that, with an average gap of 0.47%, the variant that displays the best overall performance is $(N = 3, V)$, which is the "simplest" one, since it requires less solutions than others and combines them in the most straightforward way.

More detailed statistics by problem class (see Table 2.3) and by problem size (see Table 2.4) confirm the slight superiority, on average, of the Voting combination rule over the others, but also highlight the fact that there are problem classes and problem sizes where combining more than 3 solutions pays off. Furthermore, one may notice that there are indeed problem classes ((F, L) instances for $(N = 4, V)$) and problem sizes (100–200 commodities for $(N = 3, V$ or $C)$) for which scatter search can do better on average than path relinking.

A more interesting conclusion can be drawn, however, by considering the minimum gap observed over all 9 runs reported here for a given instance: on average, it is equal to -0.37%. By considering multiple runs of scatter search, it is thus possible to obtain better results on average than with path relinking. We pushed this analysis further to determine what could be said about the best solutions obtained by considering the three combinations for a given value of N or the three values of N for any combination rule. In all cases, we observed, unfortunately, positive average gaps. Therefore, if we consider independent runs of scatter search, we probably need to combine all 9 variants to outperform path relinking. There is, however, a good reason for not accepting this as a final answer in the search for methods capable of doing better than path relinking and this is running times. These running times (in CPU seconds) are reported for the case $N = 3$ in Table 2.5; similar, but in general somewhat longer times were observed for $N = 4$ or 5. As one may easily remark, running times are reasonable for the smaller instance sizes, but they grow quite rapidly. Moreover, one must also note that the running times for scatter search are, except for a very few cases, consistently higher (often 3 to 5 times higher) than the times required by path relinking.

Table 2.1. Percentage gaps between the scatter search and the path relinking heuristics

Problem	$N = 3$			$N = 4$			$N = 5$		
	(V)	(C)	(H)	(V)	(C)	(H)	(V)	(C)	(H)
25,100,10,V,L	0.76	0.69	0.00	0.00	0.69	0.00	0.69	0.76	0.00
25,100,10,F,L	0.75	0.75	-0.54	0.75	0.75	-0.54	1.65	-1.81	-0.54
25,100,10,F,T	1.20	2.89	-0.33	2.02	2.65	0.36	3.79	0.81	2.76
100,400,10,V,L	0.07	0.07	0.07	0.07	0.07	0.07	0.07	0.07	0.07
100,400,10,F,L	1.65	1.88	1.83	-1.21	0.07	-1.21	-0.22	-0.16	-0.22
100,400,10,F,T	7.54	9.55	10.07	7.01	8.58	9.21	9.75	6.40	7.22
25,100,30,F,L	-1.81	0.86	-0.09	-0.66	0.24	0.80	0.23	1.69	2.90
25,100,30,V,T	0.00	0.00	0.00	0.00	0.00	0.00	0.00	0.00	0.00
25,100,30,F,T	2.62	2.02	2.02	0.79	1.35	1.21	1.29	1.45	1.82
100,400,30,F,L	0.94	1.03	1.52	1.76	-1.60	2.42	-0.84	5.42	-2.11
100,400,30,V,T	-0.03	-0.03	-0.03	-0.03	-0.02	-0.03	-0.02	-0.03	-0.02
100,400,30,F,T	1.04	1.06	1.01	0.51	0.73	0.33	0.33	1.29	0.41
20,230,40,V,L	0.23	0.33	0.23	0.24	0.33	0.43	0.23	0.33	0.23
20,230,40,V,T	0.07	0.07	0.07	0.06	0.07	0.07	0.06	0.07	-0.01
20,230,40,F,T	0.62	0.66	0.80	0.70	0.70	0.70	0.80	0.82	0.82
20,300,40,V,L	0.01	0.19	0.10	0.16	0.10	0.09	0.05	0.17	0.00
20,300,40,F,L	0.27	0.99	0.87	0.74	0.99	0.99	0.99	0.82	0.99
20,300,40,V,T	0.18	0.17	0.17	0.17	0.17	0.18	0.18	0.21	0.17
20,300,40,F,T	0.38	0.47	0.56	0.25	0.64	0.92	0.80	-0.03	0.24
30,520,100,V,L	1.52	2.01	2.70	2.73	2.20	2.62	2.87	3.04	2.72
30,520,100,F,L	-0.11	-0.11	0.13	0.24	-0.35	1.02	0.84	-1.81	0.86
30,520,100,V,T	1.54	1.89	1.64	1.79	1.80	1.81	1.54	1.88	1.66
30,520,100,F,T	4.68	4.92	4.24	1.54	4.68	3.72	3.39	4.98	5.79
30,700,100,V,L	1.22	1.22	1.22	1.22	1.22	1.22	1.42	1.22	1.22
30,700,100,F,L	1.80	1.58	1.89	0.29	3.06	1.86	0.46	1.55	2.36
30,700,100,V,T	1.39	1.39	2.08	1.84	1.87	1.89	1.66	2.05	1.91
30,700,100,F,T	2.04	1.60	1.21	1.81	1.84	2.06	2.55	1.76	2.55
20,230,200,V,L	-2.81	-1.28	-1.40	-1.03	-0.77	0.75	-0.54	0.06	1.00
20,230,200,F,L	-1.84	-3.26	-1.82	-1.72	-2.08	-1.46	-1.81	-0.84	-1.81
20,230,200,V,T	-1.53	-2.09	1.05	-0.26	-0.65	0.75	-0.77	-0.96	1.30
20,230,200,F,T	-0.61	-3.26	-0.97	-1.09	-1.05	-0.54	-0.86	-0.21	-0.53
20,300,200,V,L	-2.80	-1.67	-0.47	-1.26	-1.75	-1.42	-0.21	-0.67	-1.31
20,300,200,F,L	-1.04	-0.69	0.12	-1.44	-1.14	-0.92	1.18	-2.21	2.01
20,300,200,V,T	-2.23	-1.69	1.05	0.42	-1.64	0.42	0.86	1.18	-0.97
20,300,200,F,T	-3.10	-2.90	-1.94	-2.88	-2.50	-3.41	-1.89	-3.51	-2.31
30,520,400,V,L	1.38	0.52	-0.84	0.38	0.26	1.45	0.97	0.57	0.10
30,520,400,F,L	-0.40	-0.92	-0.38	-1.71	-0.80	-1.91	-1.48	-0.73	0.68
30,520,400,V,T	0.92	0.84	0.93	1.02	0.89	1.21	1.30	1.04	0.81
30,520,400,F,T	-0.32	-0.28	1.16	0.88	0.88	0.82	-0.20	-0.20	0.54
30,700,400,V,L	1.55	1.40	4.51	3.15	1.70	2.90	2.21	2.34	3.97
30,700,400,F,L	0.62	-1.02	3.39	-0.88	1.72	0.62	-1.06	0.60	0.21
30,700,400,V,T	1.78	1.47	-0.08	1.59	1.64	1.18	1.46	1.44	1.37
30,700,400,F,T	0.27	0.08	0.94	1.60	0.84	1.68	-0.20	1.76	1.04

Table 2.2. Percentage gaps between the scatter search and the path relinking heuristics – Global statistics

	N = 3			N = 4			N = 5		
	(V)	(C)	(H)	(V)	(C)	(H)	(V)	(C)	(H)
Average	0.47	0.54	0.90	0.50	0.66	0.80	0.78	0.76	0.93
Std. dev.	1.88	2.12	1.99	1.61	1.88	1.87	1.89	1.85	1.83
Minimum	-3.10	-3.26	-1.94	-2.88	-2.50	-3.41	-1.89	-3.51	-2.31
Maximum	7.54	9.55	10.07	7.01	8.58	9.21	9.75	6.40	7.22

Table 2.3. Average percentage gaps by problem class between the scatter search and the path relinking heuristics

Problem class	#	N = 3			N = 4			N = 5		
		(V)	(C)	(H)	(V)	(C)	(H)	(V)	(C)	(H)
V,L	10	0.11	0.35	0.61	0.57	0.41	0.81	0.78	0.79	0.80
F,L	11	0.08	0.10	0.63	-0.35	0.08	0.15	-0.01	0.23	0.48
V,T	10	0.21	0.20	0.69	0.66	0.41	0.75	0.63	0.69	0.62
F,T	12	1.36	1.40	1.56	1.10	1.61	1.42	1.63	1.28	1.70
All	43	0.47	0.54	0.90	0.50	0.66	0.80	0.78	0.76	0.93

Table 2.4. Average percentage gaps by problem size between the scatter search and the path relinking heuristics

Number of commodities	#	N = 3			N = 4			N = 5		
		(V)	(C)	(H)	(V)	(C)	(H)	(V)	(C)	(H)
10 – 40	19	0.87	1.24	0.96	0.70	0.87	0.84	1.04	0.96	0.78
100 – 200	16	-0.12	-0.15	0.67	0.14	0.30	0.65	0.67	0.47	1.03
400	8	0.73	0.26	1.20	0.75	0.89	0.99	0.38	0.85	1.09
All	43	0.47	0.54	0.90	0.50	0.66	0.80	0.78	0.76	0.93

Table 2.5. Running times (in seconds) for the scatter search ($N = 3$) and the path relinking heuristics

Problem	Scatter search			Path
	(V)	(C)	(H)	Relinking
25,100,10,V,L	30	16	50	13
25,100,10,F,L	84	91	79	16
25,100,10,F,T	107	43	152	25
100,400,10,V,L	347	1044	419	99
100,400,10,F,L	796	641	710	112
100,400,10,F,T	2,475	2,351	1,413	201
25,100,30,F,L	431	282	349	79
25,100,30,V,T	69	42	124	93
25,100,30,F,T	90	58	211	100
100,400,30,F,L	2,959	4,279	2,137	301
100,400,30,V,T	927	2,279	3,500	451
100,400,30,F,T	3,886	2,623	2,581	579
20,230,40,V,L	258	148	507	132
20,230,40,V,T	385	418	514	149
20,230,40,F,T	198	440	305	146
20,300,40,V,L	447	784	719	247
20,300,40,F,L	464	117	913	241
20,300,40,V,T	218	620	965	246
20,300,40,F,T	534	733	751	138
30,520,100,V,L	13,840	3,057	13,187	1,351
30,520,100,F,L	9,932	10,400	19,370	1,843
30,520,100,V,T	3,817	6,090	5,902	1,423
30,520,100,F,T	10,256	13,336	12,196	1,371
30,700,100,V,L	6,453	6,860	14,220	1,899
30,700,100,F,L	8,707	18,718	16,131	2,190
30,700,100,V,T	7,489	7,933	13,755	1,674
30,700,100,F,T	4,948	5,244	7,211	1,765
20,230,200,V,L	7,943	7,413	10,119	2,035
20,230,200,F,L	3,080	7,011	6,090	2,508
20,230,200,V,T	4,412	6,116	5,814	1,946
20,230,200,F,T	6,390	10,082	7,919	2,954
20,300,200,V,L	11,017	11,583	11,238	3,561
20,300,200,F,L	8,276	10,416	8,026	3,913
20,300,200,V,T	11,088	12,352	18,681	3,860
20,300,200,F,T	7,365	9,832	9,318	4,001
30,520,400,V,L	29,581	35,185	101,605	31,546
30,520,400,F,L	104,151	90,483	113,216	35,671
30,520,400,V,T	15,674	38,438	39,859	23,546
30,520,400,F,T	67,333	94,994	83,548	60,123
30,700,400,V,L	100,944	112,100	74,263	19,433
30,700,400,F,L	107,977	131,285	120,494	58,762
30,700,400,V,T	31,122	88,147	115,992	32,450
30,700,400,F,T	111,778	93,911	106,186	51,235

Table 2.6. Number of iterations, new solutions, and reference set updates ($N = 3$)

Problem	(V)			(C)			(H)		
	#iter	new	upd.	#iter	new	upd.	#iter	new	upd.
100,400,10,F,L	68	57	11	49	42	8	47	41	9
100,400,10,F,T	72	64	13	51	48	10	35	34	6
100,400,10,V,L	22	19	3	56	45	16	23	23	4
100,400,30,F,L	66	56	13	102	92	24	43	37	12
100,400,30,F,T	51	50	10	32	29	7	33	30	7
100,400,30,V,T	12	12	2	27	26	7	39	34	9
25,100,10,F,L	73	60	13	73	60	13	61	51	11
25,100,10,F,T	58	50	12	20	19	4	78	71	15
25,100,10,V,L	35	25	6	16	13	3	47	39	9
25,100,30,F,L	150	143	32	85	83	18	105	103	20
25,100,30,F,T	24	23	5	15	14	2	49	47	16
25,100,30,V,T	19	15	3	10	10	3	29	28	8
20,230,40,V,L	20	20	6	11	11	2	35	35	7
20,230,40,V,T	28	28	7	28	28	7	33	33	8
20,230,40,F,T	15	15	4	29	28	6	19	19	5
20,230,200,V,L	116	113	25	95	92	31	115	114	28
20,230,200,F,L	43	43	11	96	92	20	71	70	15
20,230,200,V,T	66	66	13	85	81	20	70	68	14
20,230,200,F,T	90	89	20	126	121	35	93	87	17
20,300,40,V,L	23	22	5	35	32	11	32	30	6
20,300,40,F,L	22	20	4	6	6	0	38	35	8
20,300,40,V,T	11	11	2	25	25	7	36	35	9
20,300,40,F,T	25	23	5	32	30	7	31	30	6
20,300,200,V,L	114	113	26	115	109	26	94	93	19
20,300,200,F,L	85	84	18	100	94	26	66	63	15
20,300,200,V,T	114	111	24	113	107	29	166	163	36
20,300,200,F,T	83	81	15	97	96	21	82	81	16
30,520,100,V,L	94	84	26	21	20	5	82	74	18
30,520,100,F,L	73	72	14	73	72	14	134	134	23
30,520,100,V,T	29	28	8	44	41	11	39	38	10
30,520,100,F,T	81	79	22	99	93	26	81	79	18
30,520,400,V,L	49	49	15	49	42	15	124	117	30
30,520,400,F,L	183	183	38	130	128	26	137	136	27
30,520,400,V,T	27	25	6	54	47	12	48	47	11
30,520,400,F,T	106	106	24	125	115	27	88	87	19
30,700,100,V,L	30	29	9	31	30	7	57	51	14
30,700,100,F,L	42	37	11	85	77	21	70	68	12
30,700,100,V,T	32	32	8	32	32	8	53	53	10
30,700,100,F,T	24	21	5	25	21	5	31	27	6
30,700,400,V,L	119	118	25	134	128	29	61	58	14
30,700,400,F,L	125	124	29	142	138	30	109	108	20
30,700,400,V,T	35	35	10	96	90	21	120	116	23
30,700,400,F,T	148	142	36	92	90	17	89	87	19

Regarding running times, it is interesting to examine the breakdown of CPU times with respect to the different steps of the proposed algorithm. Detailed statistics were collected for a representative subset of 10 instances using profiling software. These statistics clearly show that the tabu search step is the most time-consuming one: on average, it requires 80% of the CPU time. This figure somewhat varies among instances and it can be observed that the instances that are solved faster than comparable-size ones (e.g., (30,520,400,V,T)) have a smaller fraction of the CPU time used devoted to tabu search. This simply reflects the fact that these instances are require less CPU time, **because** the tabu search step is, in general, shorter. The initialization phase of the algorithm (i.e., steps 1 and 2) takes on average 5% of the total running time and the remainder of the procedure (steps 3, 4, 5, and 7) the last 15%.

Another critical issue with respect to the performance of a scatter search procedure is its ability to explore new, meaningful portions of the solution space, since it is in these portions that one hopes to find better solutions. Table 2.6 provides detailed statistics on the search performed when $N = 3$ for combination rules (V), (C) and (H) on all 43 instances of the benchmark. For each instance, we report the number of iterations performed, the number of new solutions (i.e., solutions not present in the reference set) generated, and the number of times the reference set was updated (because the current solution was better than the worst solution in the reference set). On average, more than 60 iterations of the procedure are performed and more than 90% of these iterations yield "new" solutions, while the reference set is updated in more than 20% of iterations. This confirms that the procedure is consistently able to yield new solutions of high quality.

5.   Conclusion and future research directions

In this paper, we have proposed a scatter search heuristic for the fixed-charge capacitated multicommodity network design problem (CMND). As far as we know, this is the first time that scatter search is applied to the generic CMND formulation. Our heuristic, which is based upon the integer design variables, allows for several variants that use different rules for combining previously obtained solutions. Extensive computational experiments performed on a fairly large set of benchmark problems have shown that, on average, the most effective variants of the scatter search heuristic do not perform better than the best existing method for the problem (a path relinking approach), but that they do come very close.

Further analysis of the computational results has highlighted the fact that multiple runs of scatter search could lead to better results than path relinking, but at the price of very high computational requirements. We believe, however, that a more refined parallel scatter search approach, involving for instance parallel search threads which exchange meaningful information, could better exploit the full potential of scatter search and thus prove much more effective (see, e.g., [1]). We intend to start investigating such an approach in the very near future.

Another possibility for further work would be to go in the completely opposite direction and to use a simpler local search scheme than the cycle-based tabu search to improve solutions. This would allow one to perform more iterations for a given allotment of CPU time and perhaps to explore a wider range of potentially interesting solutions.

Acknowledgments

Funding for this project has been provided by the Natural Sciences and Engineering Council of Canada through its Research Grant program. This support is gratefully acknowledged by the authors. We also want to thank Ms. Geneviève Hernu for the care and attention she displayed in the programming and testing of the scatter search algorithm.

References

[1] Crainic, T. G. and M. Toulouse (2003), "Parallel Strategies for Meta-heuristics", in F. Glover and G. Kochenberger (eds.), *Handbook of Metaheuristics*, Kluwer Academic Publishers, Boston, USA, 475–513.

[2] Balakrishnan, A., T.L. Magnanti, and P. Mirchandani (1997), "Network Design", in M. Dell'Amico, F. Maffioli, and S. Martello (eds.), *Annotated Bibliographies in Combinatorial Optimization*, Wiley, New York, USA, 311–334.

[3] Crainic, T.G., A. Frangioni, and B. Gendron (2001), "Bundle-Based Relaxation Methods for Multicommodity Capacitated Network Design", *Discrete Applied Mathematics* **112**, 73–99.

[4] Crainic, T.G., M. Gendreau, and J.M. Farvolden (2000), "A Simplex-Based Tabu Search Method for Capacitated Network Design", *INFORMS Journal on Computing* **12**, 223–236.

[5] Gendron, B., T.G. Crainic, and A. Frangioni (1998), "Multicommodity Capacitated Network Design", in B. Sansó and P. Soriano (eds.), *Telecommunications Network Planning*, Kluwer Academic Publishers, Boston, USA, 1–19.

[6] Ghamlouche, I., T.G. Crainic, and M. Gendreau (2003), "Cycle-based Neighbourhoods for Fixed-Charge Capacitated Multicommodity Network Design", *Operations Research* **51**, 655–667.

[7] Ghamlouche, I., T.G. Crainic, and M. Gendreau (2004), "Path Relinking, Cycle-based Neighbourhoods and Capacitated Multi-commodity Network Design", *Annals of Operations Research* **131**, 109–133.

[8] Glover, F. (1977), "Heuristics for Integer Programming using Surrogate Constraints", *Decision Sciences* **8**, 156–166.

[9] Glover, F. (1998), "A Template for Scatter Search and Path Relinking", in J. Hao, E. Lutton, E. Ronald, M. Schoenauer, and D. Snyers (eds.), *Artificial Evolution*, (Lecture Notes in Computer Science, Vol. 1363), Springer, Berlin, Germany, 13–54.

[10] Glover, F., and M. Laguna (1997), *Tabu Search*, Kluwer Academic Publishers, Boston, USA.

[11] Glover, F., M. Laguna, and R. Martí (2000), "Fundamentals of Scatter Search and Path Relinking", *Control and Cybernetics* **39** (3), 653–684.

[12] Holmberg, K. and D. Yuan (2000), "A Lagrangean Heuristic Based Branch-and-Bound Approach for the Capacitated Network Design Problem", *Operations Research* **48** (3), 461–481.

[13] Laguna, M. and R. Martí (2003), *Scatter Search: Methodology and Implementations in C*, Kluwer Academic Publishers, Boston, USA.

[14] Magnanti, T.L. and R.T. Wong (1986), "Network Design and Transportation Planning: Models and Algorithms", *Transportation Science* **18** (1), 1–55.

[15] Minoux, M. (1986), "Network Synthesis and Optimum Network Design Problems: Models, Solution Methods and Applications", *Networks* **19**, 313–360.

Part II

Tabu Search

Chapter 3

TABU SEARCH-BASED METAHEURISTIC ALGORITHM FOR LARGE-SCALE SET COVERING PROBLEMS

Marco Caserta

Instituto Tecnológico de Monterrey
Calle del Puente, 222
Col. Ejidos de Huipulco
Del. Tlalpan, México DF, 14380
México

marco.caserta@itesm.mx

Abstract   This paper presents an algorithm for the Set Covering Problem whose centerpiece is a new primal-to-dual scheme aimed at linking any primal solution to the dual feasible vector that best reflects the quality of the primal solution. This new mechanism is used to intertwine a tabu search based primal intensive scheme with a Lagrangian based dual intensive scheme to design a dynamic primal-dual algorithm that progressively reduces the gap between upper and lower bound. The algorithm has been tested on benchmark problems from the literature: the gap between upper and lower bound in 6 instances of problems whose optimal solution is not known has been further reduced, 4 of them via improvements in the lower bound, and 4 by producing a solution that is better than the best solution provided by other procedures.

Keywords: Set Covering; Tabu Search; Metaheuristic; Primal-to-Dual.

1.   Introduction

Our interest toward the set covering problem (SCP) is motivated by its use in the minimization of the number of patterns required to discriminate observations from a given population. Having an effective SCP algorithm, designed to tackle very large instances of SCP, is vital in order to define a pattern generation and pattern minimization scheme with high classification power. This paper is devoted to the development of a tabu search-based metaheuristic algorithm for very large scale set covering instances. We design a dynamic primal-and-dual scheme especially suited for large instances of SCP that are typical in the classification of data from massive data sets.

The set covering problem is a $0 - 1$ integer problem with m rows in $M = \{1, \ldots, m\}$, and n columns in $N = \{1, \ldots, n\}$. A mathematical formulation for SCP is

$$(\text{SCP}) : \min \{z = \mathbf{cx} : \mathbf{Ax} \geq \mathbf{1}, \mathbf{x} \in \mathbb{B}^n\},$$

where $\mathbf{c} \in \mathbf{Z}_+^n$ and $\mathbf{A}$ is a matrix of 0's and 1's. In the following, we call *cover* a binary vector $\mathbf{x} \in \mathbb{B}^n$ that is a feasible solution of SCP, while a *prime cover* is a cover with no redundant columns. Also, let $J_i = \{j \in N : a_{ij} = 1\}$ be the index set of columns covering row i, and $I_j = \{i \in M : a_{ij} = 1\}$ the index set of rows covered by column j.

Many real-world applications can be formulated as SCP, including traditional delivery and routing problems, as well as scheduling and location problems. More recent applications of SCP are found in probe selection in hybridization experiments for DNA sequencing (*e.g.*, Borneman et al., 2001) and feature selection and pattern construction in LAD, the logical analysis of numerical data (*e.g.*, Boros et al., 1996).

SCP is $\mathcal{NP}$-complete (Garey and Johnson, 1979), hence exact solution procedures are doomed to fail in solving practical SCP problems. Furthermore, it is parameterized intractable, which is, $W[2]-$complete with respect to the parameter "solution size" (Downey and Fellows, 1999; Niedermeier, 2006; Dom et al., 2006). Supported by its applicability, the need for solution procedures that can efficiently handle large-scale instances of SCP has attracted a vast amount of interest in the optimization community in the past four decades and a great deal of effort has been directed, especially in the past two decades, toward the development of approximate algorithms for SCP. As a result, some algorithms are capable of solving SCPs with thousands of rows and millions of columns (*e.g.*, Ceria et al., 1998, Caprara et al., 1999).

To summarize, most approximate solution procedures for SCP are dual heuristic procedures based upon the solution of the Lagrangian relaxations of SCP via subgradient optimization (*e.g.* Caprara et al., 1999, Ceria et al., 1998, Balas and Carrera, 1996, Fisher and Kedia, 1990, Vaasko and Wilson, 1984, and Balas and Ho, 1980). As the dual procedures require greedy-type primal heuristics in order to build a primal cover, they can also be viewed as primal-and-dual algorithms with "dual-to-primal" mechanisms. In addition, more "advanced" dual procedures for SCP typically feature some forms of probing and variable fixing schemes that dynamically update primal and dual information of SCP and aid in finding more effective solutions of SCP (*e.g.*, Caprara et al., 1999, Ceria et al., 1998, Balas and Carrera, 1996, and Beasley, 1990). Most algorithms designed to tackle very large scale instances work on a subset of variables, called "core problem" or "kernel problem". An interesting approach aimed at identifying the kernel problem has been proposed by Weihe, 1998, whose paper presents an effective data reduction technique that has been tested on very large railway problems. The objective is to select the minimum set of stations needed to cover a given set of trains. Real-world instances from the German and European Railroad network have been successfully solved by the author. The proposed scheme can be divided into two phases: first, the irreducible core problem is identified via dominance and equivalence relations; next, the core instance is solved via brute-force, when possible, or via heuristic scheme, when the dimensions of the core make an exhaustive search still too expensive. The author proposes an interesting approach, since he suggests that, when dealing with large scale instances, one should first work on the preprocessing scheme and, afterwards, design the routine that will work on the core instances, since it is only then that the characteristics of the core instances are known.

A major contribution of the paper is the development of a "primal-to-dual" (p2d) mechanism that, for any given primal solution, constructs a *feasible* dual vector that minimizes the gap between the upper bound of SCP given by *the* cover and the lower bound given by a feasible dual solution with respect to the sufficient optimality conditions presented in Theorem 3.1. The benefit of the primal-to-dual mechanism is twofold: (i) if the current cover is optimal to SCP, it verifies the optimality and the search process can be terminated; (ii) otherwise, it constructs a dual vector **u** that serves as a new starting vector for subgradient optimization. If different prime covers are provided, the primal-to-dual scheme constructs different **u**'s, allowing subgradient optimization to explore different regions of the dual solution space. This, in turn, allows greedy-type dual-to-primal heuristics to construct different prime covers for SCP.

In this paper we integrate effective dual-to-primal mechanisms from the literature and a specialization of the novel primal-to-dual mechanism provided in Caserta and Ryoo, 2001 for SCP. We develop a primal-intensive, "dynamic" primal-and-dual metaheuristic for large-scale SCP. Computational experiments with the proposed metaheuristic on 94 benchmarks from Caprara et al., 1999, Balas and Carrera, 1996, and Wedelin, 1995 indicate that the proposed algorithm advances the state-of-the-art in SCP quite substantially. Out of 94 benchmark problems, 21 of them have not been solved to optimality. For 6 of these 21 problems, our algorithm reduces the gap between best lower and upper bounds: new best solutions to 4 problems are found and the lower bounds of 4 problems have been improved. For the 73 benchmarks solved to optimality, the proposed algorithm finds the optimal solutions.

The proposed algorithm is made up of metaheuristic components that contribute to the efficiency and efficacy of the proposed algorithm. We first present an overview of the overall algorithm in Section 2. Subsequently, we present the metaheuristic components of the proposed algorithm in Sections 3-7. Computational experiments with 94 SCP benchmark problems are summarized in Section 8 and concluding remarks are provided in Section 9.

2.   Overall Algorithm

In this section we present the overall algorithm, while the remaining sections will clarify each step of the proposed scheme. The basic idea of the proposed scheme is related to the development of a mechanism that connects the search in the primal space with the exploration of the dual space. This scheme, called (p2d), is thoroughly presented in Section 6 and is what makes the algorithm quite effective. The reason why (p2d) sensibly improves the performance of the algorithm is that it allows to create "synergies" between the primal phase, based upon the Tabu Search paradigm, and the dual phase, based upon the Lagrangian Relaxation technique coupled with subgradient optimization.

The pseudocode of Algorithm PD_SCA() along with Figure 3.1 provide a first overview of the general algorithm.

3.   Tabu Search Metaheuristic

The tabu search metaheuristic of the proposed algorithm is the result of a specialization of the meta-strategy provided in Caserta and Ryoo, 2001 for SCP. For reasons of space, we provide details for those components that are problem-specific in nature for SCP. The proposed scheme is aimed at thoroughly exploring the feasible space along with a portion of the infeasible space. Furthermore, by introducing random and memory-based mechanisms, it aims at striking the balance between diversification and intensification.

The overall tabu search metaheuristic procedure is summarized in Procedure Tabu_Search_Metaheuristic(), while the remainder of the section is devoted to explaining the different ingredients of such procedure.

Algorithm PD_SCA();
  initialize **u** via (3.3)
  call Define_Core_Problem()   {Section 7}
  $K = \left\lceil \frac{|N|}{|N_C|} \right\rceil$ {total number of core problems examined}
  $k = 0$ {cycles counter}
  while $k < K$ **do**
    call Tabu_Search_Metaheuristic()   {Section 3}
    solve (p2d)   {Section 6}
    call Define_Core_Problem()   {Section 7}
    call Lagrangian_Optimization()   {Section 5}
    call Fixing_to_Zero()   {Section 7}
    call Fixing_to_One()   {Section 7}
    $k \leftarrow k + 1$ {increase cycles counter}
  end while

Procedure Tabu_Search_Metaheuristic();
Input: $\mathbf{x}^*$, UB, $\mathbf{x}^0$ (initial cover), $\mathcal{TL}$, (core) problem instance
Output: $\mathbf{x}^*$, UB, $\mathcal{TL}$
  for $phase \in regular, intensification, diversification$ **do**
    $k \leftarrow 0$ {# excursions into allowed infeasible region}
    $\bullet = -$ {start with the releasing phase}
    $t \leftarrow 0$ {tabu search counter}
    while $k < 2$ **do**
      call Composite_Move_Assignment()
      if $\mathbf{x}^{t+1} \in X$ **then**
        if $\mathbf{c}\mathbf{x}^{t+1} < UB$ **then**
          $\mathbf{x}^* \leftarrow \mathbf{x}^{t+1}; \; UB \leftarrow \mathbf{c}\mathbf{x}^{t+1}$ {update primal information}
        end if
        if $(\mathbf{x}^t \in X)$ and $(\mathbf{x}^{t+1} \in \overline{X})$ **then**
          $k \leftarrow k + 1$ {end tabu iteration}
        end if
        if $(\mathbf{x}^{t+1} \in X)$ and $(\mathbf{x}^t \in \overline{X})$ **then**
          solve (p2d)   {see Section 6}
          partial pricing   {see Section 7}
          call Lagrangian_Optimization()   {see Section 5}
        end if
      end if
    end while
  end for

Let $\mathbf{x}^k$ denote the current prime cover. Let us denote by B the index set of columns that take value 1 in $\mathbf{x}^k$. Let $M^0 = \{i \in M : J_i \cap B = \emptyset\}$ denote the set of rows that are uncovered in $\mathbf{x}^k$. With this notation, the feasible space of SCP can be defined as $X := \{\mathbf{x} \in \{0,1\}^n : |M^0| = 0\}$. In contrast to X, let us define $\overline{X} := \{\mathbf{x} \in \{0,1\}^n : |M^0| \leq \alpha m\} \setminus X$ as the "allowed infeasible space" of SCP, where α is a predetermined parameter chosen in $[0, 1)$. A key feature of the proposed tabu search metaheuristic is its ability to escape from a locally optimal solution via an excursion into the allowed infeasible space. Owing to the monotone decreasing property of the objective function in $\mathbf{x}$, solutions in $\overline{X}$ are, usually, more attractive than the feasible solutions. Thus, even if $\mathbf{x}^k$ is a locally dominant prime cover, the search path will be able to escape from it to a remote, different prime cover $\mathbf{x}^{k+1}$ through a sequence of 1-neighborhood moves in $X \cup \overline{X}$.

Each composite move, from $\mathbf{x}^k$ to $\mathbf{x}^{k+1}$, is comprised of a sequence of a finite number of 1-neighborhood moves, selected in such a way that a monotonic property in the search path is preserved with respect to $|M^0|$, a measurement of the amount of infeasibility associated with $\mathbf{x}^k$ (see Figure 3.1-(a).) Furthermore, let us indicate with $I_j^- = \{i \in I_j : |J_i \cap B| = 1\}$ the set of rows uniquely covered, in the current solution $\mathbf{x}^k$, by a column $j \in B$, and with $I_j^+ = I_j \cap M^0$ the set of rows currently uncovered that would be covered by adding a column $j \in N \setminus B$ to the partial cover $\mathbf{x}^k$. Finally, let us indicate with $\mathcal{TL}$ the tabu list. The primal phase is made up by two sub-phases, which allow to implement a strategic oscillation mechanism around the boundaries of the feasible region:

1. ascending sub-phase: columns are constantly 'released' (set to zero), in such a way that, on the one hand, the objective function value monotonically improves and, on the other hand, the infeasibility level monotonically increases. During this phase, at each iteration, a non-tabu move ($j \notin \mathcal{TL}$) is chosen as $j_1 \in \Gamma^-$, where:

$$\Gamma^- := \left\{ j_l \in B, I_{j_l}^- \neq \emptyset, j_l \notin \mathcal{TL} : c_{j_l}|I_{j_l}^-| \geq c_{j_l+1}|I_{j_l+1}^-| \right\}$$

2. descending sub-phase: columns are constantly 'added' (set to one), such that the infeasibility level monotonically decreases, eventually reaching a prime cover. During this phase, at each iteration, a non-tabu move ($j \notin \mathcal{TL}$) is chosen as $j_1 \in \Gamma^+$, where:

$$\Gamma^+ := \left\{ j_l \in N \setminus B, I_{j_l}^+ \neq \emptyset, j_l \notin \mathcal{TL} : \overline{r}_{j_l} \leq \overline{r}_{j_l+1} \right\}$$

with $\overline{r}_j := c_j - \sum_{i \in I_j^+} u_i$.

We switch from one sub-phase to the other when the corresponding $\Gamma^\bullet$ is empty. To allow the search path to deviate from following a predetermined trajectory given by the use of the greedy merit functions, at each iteration we select j_1 probabilistically, as indicated by the scheme `Select_First_Move()`. Each move is, in turn, classified as either a regular, diversified or intensified move depending upon the way j_1 is selected.

Let $\mathbf{x}^*$ denote the best solution found so far and let $0 < \gamma_1 < \gamma_2 < 1$. Let $j\bullet$ indicate a 1-neighborhood move that sets the j−th component of $\mathbf{x}^t$ to 1 if $\bullet = +$ (a set covering move) and to 0 if $\bullet = -$ (a set releasing move) and let $\bar{\bullet}$ denote the move in the opposite direction of $\bullet$. Then, each composite move from $\mathbf{x}^t$ to $\mathbf{x}^{t+1}$ is comprised of a sequence of a finite number of 1-neighborhood moves, and the choice of the first move plays a critical role in the proposed meta-strategy.

Procedure `Select_First_Move()`;
Input: $\bullet$ ($= +$ or $-$), $\mathbf{x}^*$, $\mathbf{x}^t$, $\mathcal{TL}$
Output: j_1
    generate a random number γ in $[0, 1]$
    if $\gamma \in [0, \gamma_1]$ **then**
        select a move in $\Gamma^\bullet$ {normal scheme}
    else if $\gamma \in (\gamma_1, \gamma_2]$ **then**
        randomly select j_1 among $j \in N$, $j\bullet \notin \mathcal{TL}$ {random scheme}
    else
        $I_{d\bullet} := \left\{ j \in N, j\bullet \notin \mathcal{TL} : x_j^t \neq x_j^* \right\}$
        if $I_{d\bullet} = \emptyset$ **then**
            select a move in $\Gamma^\bullet$ {memory-based scheme}
        else
            randomly select j_1 from $I_{d\bullet}$
        end if
    end if

Remark. In order to allow for a more rigorous search of the solution space, we recur to three different strategies that define three search phases of the algorithm, namely the regular, diversification, and intensification phases. During the regular phase, we use $\gamma_1 = 0.8$ and $\gamma_2 = 0.9$ for the procedure `Select_First_Move()`, in such a way that the normal scheme is privileged above the random and memory-based scheme. For the diversification phase, we increase the probability of selecting a random move by using $\gamma_1 = 0.6$ and $\gamma_2 = 0.9$. Likewise, for the intensification phase, we use $\gamma_1 = 0.6$ and $\gamma_2 = 0.7$, thus granting a higher chance to the selection of a memory-based move.

Finally, the overall definition of a composite move is illustrated in Procedure `Composite_Move_Assignment()`. Denote by $\bullet e_l$ a unit vector whose l−th component is -1 if $\bullet = -$ and $+1$ otherwise. We first identify the portion of the search space that is being explored. It the boundary of the "allowed infeasible region" has been reached or if feasibility has been restored (lines 2,3), then the sub-phase is inverted and the process is restarted. On the other hand, if the algorithm is currently in an ascending or descending phase, mainly within the $\overline{X}$ region, the first (ascending or descending) move is executed (line 5). The next steps are aimed at identifying a set of moves that go in the opposite direction of the first move. For example, if the algorithm is ascending into the "allowed infeasible space", we want to identify a set of descending moves in such a way that the net effect is still to uncover rows. To illustrate, if the first ascending move is such that column $j_1 \in B$ is set to 0, with the consequence that $|I_j^-|$ rows will be uncovered, a set of descending moves ζ^+ will be identified in such a way that the number of rows uncovered by j_1 is higher than the number of rows covered by all the moves in ζ^+. This is accomplished as illustrated in lines 6–10. Finally, lines 10 and 11 show how the composite move is executed and how the tabu list is updated.

Procedure `Composite_Move_Assignment()`;
Input: $\bullet$, $\mathbf{x}^t$, $\mathcal{TL}$, (core) problem instance
Output: $\bullet$, $\mathbf{x}^{t+1}$, $\mathcal{TL}$

1: call `Select_First_Move()`   {identify j_1 in $\Gamma^\bullet$}
2: **if** $(\mathbf{x}^t + \bullet \mathbf{e}_{j_1} \notin X \cup \overline{X})$ or $(\mathbf{x}^t \in X)$ **then**
3:    $\bullet \leftarrow \overline{\bullet}$, go to line 1   {invert search direction}
4: **end if**
5: $\mathcal{TL} \leftarrow \mathcal{TL} \cup \{\overline{j}_1\}$ {set first move as tabu}
6: **if** $|I_{j_1^t}^\bullet| \geq \sum_{j \in \Gamma^{\overline{\bullet}}} |I_j^{\overline{\bullet}}|$ **then**
7:    $\zeta^{\overline{\bullet}} := \Gamma^{\overline{\bullet}}$ {identify set of opposite moves}
8: **else**
9:    $\zeta^{\overline{\bullet}} := \left\{ j_1, \ldots, j_i \in \Gamma^{\overline{\bullet}} : |I_{j_1^t}^\bullet| \geq \sum_{j=j_1}^{j_i} |I_j^{\overline{\bullet}}|,\ |I_{j_1^t}^\bullet| < \sum_{j=j_1}^{j_i+1} |I_j^{\overline{\bullet}}| \right\}$
10: **end if**
11: $\mathbf{x}^{t+1} \leftarrow \mathbf{x}^t + \bullet \mathbf{e}_{j_1} + \sum_{j \in \zeta^{\overline{\bullet}}} \overline{\bullet} \mathbf{e}_j$ {execute composite move}
12: $\mathcal{TL} \leftarrow \mathcal{TL} \cup \{\overline{j}_l : j_l \in \zeta^{\overline{\bullet}}\}$ {update tabu status}

Remark. Note in the above that each move is selected in such a way that a monotonic property in the search path is preserved with respect to $|M^0|$, a measurement of the amount of infeasibility associated with $\mathbf{x}^t$.

Remark. Since the algorithm is especially designed to handle large-scale instances of SCP, we always work on a subset of columns $N_C \subset N$ and we employ pricing techniques to add or remove columns to and from N_C (See section 7). Consequently, each occurrence of N in the definition of neighborhoods must be replaced by N_C.

4.   Lagrangian Relaxation & Greedy Heuristics

The best known primal heuristic is the greedy one, which uses the reduced cost information provided by the dual phase to construct a prime cover. Balas and Ho, 1980, presented a list of scores based upon the column cost per row covered to create a prime cover. Vaasko and Wilson, 1984, selected a column to be added to the partial cover according to the value of a score function, randomly chosen among a pool of functions based upon the column cost per row covered. At every iteration the primal heuristic is run 30 times with randomly chosen score functions. Beasley, 1990, proposed a Lagrangian based primal heuristic scheme that extended the partial cover of the Lagrangian problem to a prime cover. A score based upon the column cost per row covered is used to rank the columns. Fisher and Kedia, 1990, proposed as score the reduced cost computed using only the multipliers of rows left uncovered, rather than the actual reduced cost. Bricker and Techapicjetvanich, 1993, studied the effectiveness of five different primal heuristic scores, based upon the column cost per row covered and the reduced cost per row covered, both the real and the modified reduced cost. Balas and Carrera, 1996, coupled the approach of Vaasko and Wilson, 1984, with a primal scheme that creates a prime cover extending the partial cover of the Lagrangian phase by choosing columns based upon their reduced cost. The primal scheme, as a byproduct, produces an improved dual vector.

Let M^0 denote the set of rows left uncovered by $\mathbf{x}$, and B denotes the set of columns fixed to 1 in the current (partial) cover $\mathbf{x}$. Let $|I_j \cap M^0|$ be the number of rows currently uncovered that would be covered by setting x_j to 1. During the dual Lagrangian phase we use the score

$$
s(j) = \frac{c_j - \displaystyle\sum_{i \in I_j \cap M^0} u_i}{|I_j \cap M^0|},
$$

as in Fisher and Kedia, 1990, within the simple heuristic described in Procedure `Greedy_Heuristic()`.

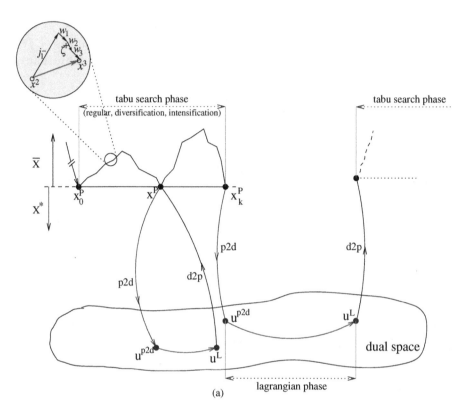

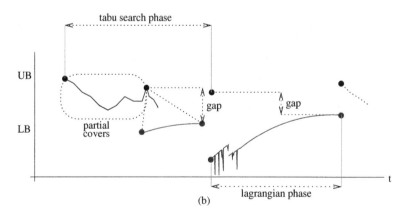

Figure 3.1.   Dynamic Primal-and-Dual scheme: (a) from the feasibility point of view; (b) from the objective function value point of view.

Procedure Greedy_Heuristic();

Input: u

Output: x

$\quad M^0 \leftarrow M; \quad \mathbf{x} \leftarrow 0; \quad B \leftarrow \emptyset$ $\hspace{4cm}$ {initialization}

$\quad$ **while** $M^0 \neq \emptyset$ **do**

$\qquad j \leftarrow \underset{j \in N_C \setminus B}{\operatorname{argmin}} \{s(j)\}$ $\hspace{4cm}$ {make a cover}

$\qquad x_j \leftarrow 1; \quad M^0 \leftarrow M^0 \setminus I_j; \quad B \leftarrow B \cup \{j\}$ $\hspace{1.5cm}$ {updating}

$\quad$ **end while**

$\quad$ remove redundancy in **x** $\hspace{3cm}$ {prime cover is obtained}

5. Subgradient Optimization

The Lagrangian relaxation of SCP is defined as

$$L(\mathbf{x}, \mathbf{u}) = \min_{\mathbf{x} \in \{0,1\}^n} \sum_{j=1}^{n} r_j x_j + \sum_{i=1}^{m} u_i,$$

where r_j (the reduced cost for j, $j = 1, \ldots, n$) is defined as $c_j - \sum_{i \in I_j} u_i$ and requires **u** such that an optimal vector $\mathbf{x}^L$ minimizing the Lagrangian function can be computed by a standard technique:

$$x_j^L = \begin{cases} 1, & \text{if } c_j - \sum_{i \in I_j} u_i < 0 \\ 0, & \text{otherwise} \end{cases} , \; j \in N \qquad (3.1)$$

It is worth noting that, since vector $\mathbf{x}^L$ is optimal to the Lagrangian problem, $L(\mathbf{u})$ provides a valid lower bound for SCP. For this reason, we are interested in finding the vector **u** that solves the Lagrangian dual problem, which is

$$L_D(\mathbf{u}) = \max_{\mathbf{u} \in \mathbb{R}_+^m} L(\mathbf{x}, \mathbf{u}).$$

Most successful approaches for SCP in the literature solve a series of Lagrangian relaxations of SCP and use the subgradient optimization technique to generate a near-optimal vector **u** for $L_D(\mathbf{u})$. For subgradient optimization, we use the formula of Held and Karp, 1971:

$$u_i^{k+1} = \max \left\{ u_i^k + \lambda \frac{UB - LB}{\|s(u^k)\|^2} s_i(u^k), 0 \right\}, \; i \in M, \qquad (3.2)$$

where UB and LB are the upper and lower bounds of the optimum of SCP, λ is the step size parameter, and $s_i(\mathbf{x}^k) = 1 - \sum_{j \in J_i} x_j$ is the component i of the subgradient. As in Caprara et al., 1999, $\mathbf{u}^0$ is initialized as

$$u_i^0 = \min_{j \in J_i} \frac{c_j}{|I_j|}, \; i \in M \qquad (3.3)$$

and λ is updated after every $p = 20$ iterations, utilizing the best and worst lower bounds information obtained during the last p iterations. In addition, if the lower bound improvement in the last $4p$ iterations is below the threshold limit of 1%, we apply a "perturbation scheme" based upon the primal-to-dual scheme of Section 6 to enforce a drastic modification of the vector $\mathbf{u}$. We summarize the steps of the Lagrangian optimization phase in Procedure `Lagrangian_Optimization()`.

6. Primal-to-Dual Scheme

Let $\mathbf{x}^P$ and $\mathbf{x}^L$ denote a prime cover for SCP and a Lagrangian solution for a given vector $\mathbf{u} \in \mathbb{R}_+^m$, respectively. Denote by $z(\bullet)$ and $L(\bullet, \mathbf{u})$ the objective value of SCP and the value of the Lagrangian function evaluated at $\bullet$, respectively. Let $B^P = \{j \in N : x_j^P = 1\}$, $B^L = \{j \in N : x_j^L = 1\}$, $B^{LP} = B^L \setminus B^P$ and $B^{PL} = B^P \setminus B^L$.

Lema 3.1. *Suppose that* $\mathbf{x}^P \in \{0,1\}^n$, $\mathbf{x}^L \in \{0,1\}^n$, *and* $\mathbf{u} \in \mathbb{R}_+^m$ *satisfy* $u_i(\sum_{j \in J_i} x_j^P - 1) = 0$, *for all* $i \in M$. *Then,* $z(\mathbf{x}^P) - L(\mathbf{x}^L, \mathbf{u}) = \sum_{j \in B^{PL}} r_j - \sum_{j \in B^{LP}} r_j$.

Proof. We have

$$
\begin{aligned}
L(\mathbf{x}^L, \mathbf{u}) &= \sum_{i \in M} u_i + \sum_{j \in B^L} \left(c_j - \sum_{i \in I_j} u_i \right) \\
&= \sum_{i \in M} u_i \sum_{j \in J_i} x_j^P + \sum_{j \in B^L} c_j - \sum_{j \in B^L} \sum_{i \in I_j} u_i \\
&= \sum_{j \in B^L} c_j + \sum_{j \in B^P} \sum_{i \in I_j} u_i - \sum_{j \in B^L} \sum_{i \in I_j} u_i \\
&= \sum_{j \in B^L} c_j + \sum_{j \in B^{PL}} \sum_{i \in I_j} u_i - \sum_{j \in B^{LP}} \sum_{i \in I_j} u_i,
\end{aligned}
$$

Procedure `Lagrangian_Optimization();`
Input: $LB, UB, \mathbf{x}^*, \mathbf{u}^0$
Output: $LB, UB, \mathbf{x}^*$
   $k \leftarrow 0$ `{Lagrangian iteration counter}`
   $w \leftarrow 0$ `{perturbation scheme counters}`
   $\mathbf{x}^{old} = \mathbf{0}, \lambda^0 = 0.1$
   while lower bound termination tolerance is not met **do**
      if $(k \bmod 20) = 0$ **then**
         if $LB_{best} - LB_{worst} > 0.01 LB_{best}$ **then**
            $\lambda^k \leftarrow 0.5\lambda^k$ `{modify step size}`
            $w \leftarrow 0$ `{reset perturbation scheme counter}`
         else
            $w \leftarrow w + 1$ `{increase perturbation scheme counter}`
            if $w < 4$ **then**
               $\lambda^k \leftarrow 1.5\lambda^k$
            else
               $\lambda^k = 0.1$ `{apply perturbation scheme}`
               call $p2d(\mathbf{x}^k, \mathbf{u}^k)$ `{see Section 6}`
               if $\mathbf{x}^{old} = \mathbf{x}^k$ **then**
                 $\delta = \mathrm{random}(0, 0.1 u_{\max})$, where $u_{\max} = \min_{i \in M}\{u_i\}$
                 $u_i \leftarrow \delta u_i$ for randomly chosen 10% of $\mathbf{u}$
               else
                 $\mathbf{x}^{old} = \mathbf{x}^k$
               end if
            end if
         end if
         $LB_{best} = LB_{worst} = L(\mathbf{x}^L, \mathbf{u})$ `{set best and worst LB}`
      end if   `{new step size available}`
      $k \leftarrow k + 1$ `{increase Lagrangian iteration counter}`
      update $\mathbf{u}$ via (3.2)   `{perform` k^{th} `Lagrangian iteration}`
      solve Lagrangian relaxation via (3.1)   `{`$\mathbf{x}^L$ `is obtained}`
      if $L(\mathbf{x}^L, \mathbf{u}) > LB$ **then**
         $LB \leftarrow L(\mathbf{x}^L, \mathbf{u})$ `{update lower bound on SCP}`
      end if
      if $L(\mathbf{x}^L, \mathbf{u}) > LB_{best}$ **then**
         $LB_{best} \leftarrow L(\mathbf{x}^L, \mathbf{u})$ `{update best lower bound}`
      else if $L(\mathbf{x}^L, \mathbf{u}) < LB_{worst}$ **then**
         $LB_{worst} \leftarrow L(\mathbf{x}^L, \mathbf{u})$ `{update worst lower bound}`
      end if
   end while   `{lower bound termination tolerance met}`

where the second equality is obtained via $u_i(\sum_{j \in J_i} x_j^P - 1) = 0, \forall i \in M$.
Now, we have

$$
\begin{aligned}
z(\mathbf{x}^P) - L(\mathbf{x}^L, \mathbf{u}) &= \sum_{j \in B^P} c_j - \sum_{j \in B^L} c_j - \sum_{j \in B^{PL}} \sum_{i \in I_j} u_i + \sum_{j \in B^{LP}} \sum_{i \in I_j} u_i \\
&= \sum_{j \in B^{PL}} c_j - \sum_{j \in B^{LP}} c_j - \sum_{j \in B^{PL}} \sum_{i \in I_j} u_i + \sum_{j \in B^{LP}} \sum_{i \in I_j} u_i \\
&= \sum_{j \in B^{PL}} \left(c_j - \sum_{i \in I_j} u_i \right) - \sum_{j \in B^{LP}} \left(c_j - \sum_{i \in I_j} u_i \right) \\
&= \sum_{j \in B^{PL}} r_j - \sum_{j \in B^{LP}} r_j.
\end{aligned}
$$

$\square$

Theorem 3.1 (Sufficient Conditions). *Suppose that* $\mathbf{x}^P \in \{0,1\}^n$, $\mathbf{x}^L \in \{0,1\}^n$, *and* $\mathbf{u} \in \mathbb{R}_+^m$ *satisfy:*

$$
\begin{aligned}
&(i)   &&u_i \left(\sum_{j \in J_i} x_j^P - 1 \right) = 0,   &&\forall i \in M \\
&(ii)   &&r_j = c_j - \sum_{i \in I_j} u_i = 0,   &&\forall j \in B^P \\
&(iii) &&r_j = c_j - \sum_{i \in I_j} u_i \geq 0,   &&\forall j \in N \setminus B^P
\end{aligned}
$$

Then, $\mathbf{x}^P$ *solves SCP to optimality.*

Proof. We need to show that both feasibility and optimality are ensured. Feasibility of $\mathbf{x}^P$ is enforced via conditions *(ii)* and *(iii)*, while optimality is ensured by conditions *(i)* and *(ii)*, along with $\mathbf{x}^* \in \{0,1\}^n$. $\square$

The sufficient optimality conditions of $\mathbf{x}^*$ for SCP in Theorem 3.1 can be exploited in the derivation of a mechanism that constructs a "feasible" dual solution $\mathbf{u}$ that properly reflects the importance of each constraint of SCP with respect to the characteristics of $\mathbf{x}^P$.

First, note that Conditions (ii) and (iii) of Theorem 3.1, along with the requirement $\mathbf{u} \in \mathbb{R}_+^m$ give the feasibility of $\mathbf{u}$ to the dual linear program of the linearized SCP. Conditions (i) and (ii), along with $\mathbf{x}^* \in \{0,1\}^n$ ensure that the primal and dual solutions are optimal to their respective programs. Let $M^1 := \{i \in M : \sum_{j \in J_i} x_j^P = 1\}$ be the set of rows covered only once by a given solution $\mathbf{x}^P$. Furthermore, let $N_C \subseteq N$ be the set of columns in the current core problem, with $|N_C| \ll |N|$. Consider the following linear program:

$$(\text{p2d}): \quad \begin{vmatrix} \min & g = \sum_{j \in B^P \setminus B^L} \left(c_j - \sum_{i \in I_j} u_i \right) \\ \text{s.t.} & c_j - \sum_{i \in I_j \cap M^1} u_i = 0, \quad j \in B^L \cap B^P \\ & c_j - \sum_{i \in I_j \cap M^1} u_i \geq 0, \quad j \in N_C \setminus B^P \\ & u_i \geq 0, \quad\quad\quad\quad i \in M^1 \end{vmatrix}$$

It is worth noting that (p2d) is a LP with $|(N_C \setminus B^P) \cup (B^P \cap B^L)|$ rows and $|M^1|$ columns. Also, note that the two non-trivial constraints of (p2d) set $u_i = 0$ for all $i \in M \setminus M^1$, and, through the minimization process, (p2d) modifies the remaining components of the vector $\mathbf{u}$ feasible to the dual of the linearized SCP that satisfies the sufficiency conditions of Theorem 3.1 "as much as possible" to yield $\mathbf{u}$ that reflects the characteristics of $\mathbf{x}^P$. It is easy to see that, if (p2d) has a feasible solution, such solution is dual feasible and, consequently, $\sum_i u_i$ provides a valid lower bound for SCP.

The following is an obvious consequence of (p2d) and Theorem 3.1:

Corollary 3.1. *If the optimum of (p2d) is equal to zero, then* $\mathbf{x}^P$ *solves SCP.*

The following also holds true:

Theorem 3.2. *Of all dual feasible* $\mathbf{u} \in \mathbb{R}_+^m$, $\mathbf{u}^*$ *obtained from solving (p2d) minimizes the gap* $z(\mathbf{x}^P) - L(\mathbf{x}^L, \mathbf{u}^*)$ *with respect to* $\mathbf{x}^P$ *and* $\mathbf{x}^L$ *and Condition (i) of Theorem 3.1.*

Proof. The dual feasibility of $\mathbf{u}^*$ is immediate. The formulation of (p2d) and Theorem 3.1 easily show that $z(\mathbf{x}^P) - L(\mathbf{x}^L, \mathbf{u}^*)$ is minimized by the $\mathbf{x}^P$ and $\mathbf{u}^*$ pair. □

7. Variable Fixing, Pricing and Core Problem Generation

In this section we present the variables fixing schemes for SCP. When probing x_j at 1, not only the Lagrangian multipliers of all rows $i \in I_j$ must be set to 0 but also all r_q, $q \in J_i$ for every $i \in I_j$, must be reduced to properly update the importance of the columns after setting $x_j = 1$. Let:

$$\delta_j^+ := \sum_{i \in I_j} u_i \times (|q \in J_i : r_q \leq 0| - 1)$$

The proposed score is embedded in the `Fixing_to_Zero()` scheme as in Balas and Carrera, 1996:

Procedure `Fixing_to_Zero`();

Input: N, N_C, LB, UB, **u**

Output: N, N_C

    for $j \in N_C \setminus B^P$ **do**

      if $\lceil LB + r_j + \delta_j^+ \rceil \geq UB$ **then**

        $N_C \leftarrow N_C \setminus \{j\}$ {eliminate column from core problem}

        $N \leftarrow N \setminus \{j\}$ {permanently eliminate column}

      end if

    end for

Remark. Ceria et al., 1998, fixed a variable to zero if its reduced cost is greater than the gap between upper and lower bound. Balas and Carrera, 1996, computed a factor Δ_j for every column $j \in N \setminus B$ defined as the improvement in the value of the vector **u** obtained by fixing x_j to one. Subsequently, one column j is fixed to zero if $\lceil LB + r_j + \Delta_j \rceil \geq UB$.

To fix a column $j \in B^P$ permanently at 1 compute, for each $i \in I_j^-$, the variation of u_i required in order for at least another column $q \in J_i$, $q \neq j$ to have a non-positive reduced cost. This amount of modification required by u_i is

$$\delta_j^- = \sum_{i \in I_j^-} \min_{q \in J_i} \{r_q\},$$

and x_j, $j \in B^P$, can be permanently fixed to 1 if $\lceil LB - r_j + \delta_j^- \rceil \geq UB$. Procedure `Fixing_to_One`() summarizes the scheme used.

Procedure `Fixing_to_One`();

Input: N, N_C, LB, UB, F, **u**

Output: N, N_C, F

    for $j \in B^P$ **do**

      if $\lceil LB - r_j + \delta_j^- \rceil \geq UB$ **then**

        $x_j \leftarrow 1$

        $N_C \leftarrow N_C \setminus \{j\}$ {eliminate column from core problem}

        $F \leftarrow F \cup \{j\}$ {include column in fixed columns set F}

      end if

    end for

To define core problems $N_C \subseteq N$, we employ a pricing scheme that resembles the one presented in Caprara et al., 1999. We first add to the core problem N_C all the columns whose reduced cost is less than 0.1. Subsequently, whenever possible, for each row $i \in M$, we add enough columns $j \in J_i$ to N_C in such a way that each row is covered by at least

5 columns in the core problem. These columns are added according to the reduced cost value.

8.   Computational Results on SCP Benchmarks

In this section we present the results obtained by testing the algorithm on benchmark problems. The algorithm was implemented in C++ and compiled with the GNU C++ compiler with the -O2 option. The (p2d) problem is solved using the linear programming solver Clp() of COIN-OR Library (Lougee-Heimer, 2003). The computing platform used is a Linux workstation with Intel Pentium 4 1.1GHz processor and 256 Mb of RAM memory.

The parameters value for the tabu search metaheuristic are: $\theta = 2$ (number of excursion into the infeasible region for each TS phase), $\alpha = 0.1$ (maximum infeasibility allowed) and $\tau \in [\tau_{min}, \tau_{max}]$ (tabu tenure), where:

$$\tau_{max} = \alpha \times |\mathbf{x}|, \quad \tau_{min} = 0.1\tau_{max}$$

The value of τ is set to τ_{min} every time a new best solution is found and increased every time dominated solutions are visited. The rationale behind such a choice is that, on the one hand, we want to thoroughly explore promising regions, in which "good" feasible solutions are found, while, on the other hand, we aim at escaping from unattractive regions by increasing the tabu tenure, thus forcing the algorithm to move toward a different region.

Computational results for Beasley's OR Library (Beasley, 1990) are not reported because the algorithm always finds the optimal, or the best known, solution. We only report, in Table 3.1, the results of Beasley's OR Library RAIL problems. The table shows that to 4 out of 7 instances the gap between upper and lower bound has been further reduced. In addition, for the two biggest instances a new best result is found, which indicates that the algorithm is especially suited for very large scale problems. Finally, Table 3.2 reports the results on the instances appeared in Wedelin, 1995. Out of 6 instances, for 4 of them the algorithm finds the optimal solution, and for the last two it finds a solution that is better than any other solution found so far.

9.   Conclusions

We have presented a new dynamic scheme for large scale set covering problems. The backbone of the algorithm is a new primal-to-dual mechanism that, given any prime cover, constructs the dual feasible vector that better reflects the quality of the prime cover. Using this new mechanism, the algorithm updates the status of the search in the dual space any time a new prime cover is found and vice versa, dynamically linking the primal intensive phase with the dual intensive phase.

When tested on benchmark problems, the algorithm improved the best known results on 6 instances, 2 of them by providing a better lower bound, 2 by finding a solution that is better that any other solution found so far, and 2 by improving both upper and lower bound.

Owing to the intensive use of primal-based schemes, the algorithm is especially suited for those instances of SCP with a number of rows much larger than the number of columns. Considering a classification problem, where a set of observations is partitioned into true and false, one wants to classify future observations based upon the value of certain attributes. The problem of selecting the smallest support set of attributes needed to classify a population can be formulated as SCP. However, if we indicate with m the number of observations, equally divided between positive and negative observations, the number of rows of SCP is of the order of $\mathcal{O}(m^2)$, leading to SCPs with $m \gg n$. For this reason, some new applications of SCP, such as probe selection problem for hybridization experiment as well as attributes identification and patterns selection in logical analysis of data, can be better tackled with a primal intensive approach rather that via the traditional Lagrangian based approach. This approach could be fostered by the design of a parallel algorithm for very large instances of SCP and, hence, applied to large problems in data mining and genetics.

Finally, it is also worth noting that the technique proposed in Section 3 of this paper, dealing with the swap of columns within and outside of the current solution is a generalization of oscillation mechanisms as well as k-flip mechanisms, such as the ones of Glover and Kochenberger, 1996, Chu and Beasley, 1998, Caserta et al., 2006 or Yagiura et al., 2006. The results obtained on SCPs by these authors along with the promising results of the proposed scheme endorse the idea that oscillating mechanisms (continually crossing the boundaries of the feasible region) are very powerful ingredients of a metaheuristic scheme when it comes to solving large scale combinatorial optimization problems.

References

Balas, E. and Carrera, M. C. (1996). A Dynamic Subgradient-Based Branch-and-Bound Procedure for Set Covering Problem. *Operations Research*, 44(6):875–890.

Balas, E. and Ho, A. (1980). Set Covering Algorithms Using Cutting Planes, Heuristics and Subgradient Optimization: a Computational Study. *Mathematical Programming Study*, 112:37–60.

Table 3.1. Results on the RAIL test instances from Beasley's OR-Library

Name	Size	Best in Literature			PD-SCP		
		LB	UB	Time	LB	UB	Time
RAIL582	582×55,515	210	211	570[#]	210	211	131
RAIL507	507×63,009	173	174	817[†]	173	174	139
RAIL516	516×47,311	182	182	3000[#]	182	182	217
RAIL2536	2536×1,081,841	685	691	10000[§]	**687**	691	338
RAIL2586	2,586×920,683	937	948	1183[§]	**939**	948	399
RAIL4284	4284 ×1,092,610	1051	1065	10000[§]	**1055**	**1063**	1022
RAIL4872	4,872×968,672	1,509	1,534	4566[§]	**1514**	**1532**	1166

[#]: Caprara et al. (1999) - time in PC486/33 CPU seconds.
[§]: Caprara et al. (1999) - time in HP735/125 CPU seconds.
[†]: Ceria et al. (1998) - time in PC486/33 CPU seconds.

Table 3.2. Results on instances from Wedelin (1995)

Name	Range	Best in Literature		PD-SCP	
		UB	Time	UB	Time
b727scratch	29×157	94,400 [†,§]	0.3	94,400	0.1
alitalia	118×1,165	27,258,300[†,§]	6.2	27,258,300	2.1
a320	199×6,931	12,620,100[†,§]	79.5	12,620,100	37.3
a320coc	235×18,753	14,495,500[†]	1,023.7	14,495,500	228.1
sasjump	742×10,370	7,339,537 [§]	396.3	**7,339,521**	221.7
sasd9imp2	1,366×25,032	5,262,190 [†]	1,579.7	**5,262,140**	1,066.3

[§]: Caprara et al. (1999) - time given in DECstation 5000/240 CPU seconds.
[†]: Wedelin (1995) - time given in DECstation 5000/240 CPU seconds.

Beasley, J. E. (1990). A Lagrangian Heuristic for Set Covering Problems. *Naval Research Logistics*, 37:151–164.

Borneman, J., Chrobak, M., Della Vedova, G., Figueroa, A., and Jiang, T. (2001). Probe Selection Algorithms with Applications in the Analysis of Microbial Community. *Bioinformatics - Discovery Notes*, 1(1):1–9.

Boros, E., Hammer, P. L., Ibaraki, T., Kogan, A., Mayoraz, E., and Muchnik, I. (1996). An Implementation of Logical Analysis of Data. Technical report, Rutgers University, New Brunswick, NJ - 08903-5062.

Bricker, D. and Techapicjetvanich, K. (1993). Investigation of Lagrangian Heuristics for Set Covering Problems. Technical Report Iowa City, IA, 52242, University of Iowa.

Caprara, A., Fischetti, M., and Toth, P. (1999). A Heuristic Method for the Set Covering Problem. *Operations Research*, 47(5):730–743.

Caserta, M., Quiñonez Rico, E., and Márquez Uribe, A. (2006). A Cross Entropy Algorithm for the Knapsack Problem with Setups. *Computers and Operations Research*.

Caserta, M. and Ryoo, H. S. (2001). Efficient Tabu Search-Based Procedure for Optimal Redundancy Allocation in Complex System Reliability. In *Proc. 5th Intl Conference on Optimization: Techniques and Applications*, pages 592–599.

Ceria, S., Nobili, P., and Sassan, A. (1998). A Lagrangian-based Heuristic for Large-scale Set Covering Problems. *Mathematical Programming Ser B*, 81:215–228.

Chu, P. C. and Beasley, J. E. (1998). A Genetic Algorithm for the Multidimensional Knapsack Problem. *Journal of Heuristics*, 4(1):63–86.

Dom, M., Guo, J., Niedermeier, R., and Wernicke, S. (2006). Minimum Membership Set Covering and the Consecutive Ones Property. In Arge, L. and Freivalds, R., editors, *SWAT 2006 - 10th Scandinavian Workshop on Algorithm Theory*, volume 4059 of *Springer LNCS*, page 337.

Downey, R. G. and Fellows, M. R. (1999). *Parameterized Complexity*. Monographs in Computer Science. Springer.

Fisher, M. L. and Kedia, P. (1990). Optimal Solutions of Set Covering/Partitioning Problems using Dual Heuristics. *Management Science*, 36:674–688.

Garey, M. R. and Johnson, D. S. (1979). *Computers and Intractability: a Guide to the Theory of NP-Completeness*. Freeman.

Glover, F. and Kochenberger, G. A. (1996). *Critical Event Tabu Search for Multidimensional Knapsack Problems*, chapter In Meta-Heuristics:

Theory and Applications – I.H. Osman and J.P. Kelly, pages 407–427. Kluwer Academic Publishers.

Held, M. and Karp, R. M. (1971). The Traveling Salesman Problem and Minimum Spanning Tree: Part II. *Mathematical Programming*, 1:6–25.

Lougee-Heimer, R. (2003). The Common Optimization INterface for Operations Research. *IBM Journal of Research and Developmentg*, 47:57–66.

Niedermeier, R. (2006). *Invitation to Fixed-Parameter Algorithms*. Oxforf University Press.

Vaasko, F. J and Wilson, G. R. (1984). An Efficient Heuristic for Large Set Covering Problems. *Naval Research Logistics*, 31:163–171.

Wedelin, D. (1995). An Algorithm for Large Scale 0-1 Integer Programming with Applications to Airline Crew Scheduling. *Annals of Operational Research*, 57:283–301.

Weihe, K. (1998). Covering Trains by Stations or the Power of Data Reduction. In Battiti, R. and Bertosi, A. A., editors, *OnLine Proceedings of the 1st Workshop on Algorithms and Experiments (ALEX'98) - Trento, Italy*, pages 1–8.

Yagiura, M., Kishida, M., and Ibaraki, T. (2006). A 3-Flip Neighborhood Local Search for the Set Covering Problem. *European Journal of Operational Research*, 172:472–499.

Chapter 4

LOG-TRUCK SCHEDULING WITH A TABU SEARCH STRATEGY

Manfred Gronalt and Patrick Hirsch
Institute of Production Economics and Logistics, BOKU – University of Natural Resources and Applied Life Sciences, Vienna, Feistmantelstrasse 4, A-1180 Wien

Abstract: The Austrian forest sector has experienced extensive development in recent years. In 2003, approximately 27.9 million cubic meters of logs were processed in Austria. In order to enable a stable supply, an efficient and economical operation for round timber transport is necessary. In this paper, we present a Tabu Search based solution method for log-truck scheduling. A fleet of m log-trucks that are situated at the respective homes of the truck drivers must fulfill n transports of round timber between various wood storage locations and industrial sites. All of the transports are carried out as full truckloads. Since the full truck movements are known, our objective is to minimize the overall duration of empty truck movements. In addition to the standard VRP, we have to take into consideration weight constraints on the road network, multi-depots, and time windows at the industrial sites and homes of the truck drivers. We applied the Unified Tabu Search method and modified it by an oscillating change of the neighborhood size in some selected iteration steps. Our heuristics are verified with extensive numerical studies. The Tabu Search based heuristics are able to solve real-life problems within a reasonable timeframe by providing good solution quality.

Keywords: Log-truck scheduling; Timber Transport Vehicle Routing Problem; Tabu Search

1.   INTRODUCTION AND PROBLEM DESCRIPTION

The Austrian forest sector has experienced extensive development in recent years. With respect to the Austrian economy, forest based industry is in second place in terms of exporting goods and services. In 2003, Austria had 1,400 sawmills, 30 pulp mills, ground wood pulp mills, and paper mills, and 39 chipboard factories that processed approximately 27.9 million cubic meters of logs (Schwarzbauer, 2005). In order to enable a stable supply, an efficient and economical operation for round timber transport is necessary. A number of natural restrictions, such as regional topology, storms, and heavy snow may hinder a steady supply for industrial recipients. In this regard, the availability of the forest road network is of great importance. A number of research efforts mainly focus on information supply and the provision of GIS based applications for supporting truck drivers in finding storage locations and to further support wood transfer. Moreover, there exist numerous proposals for the efficient use of wood transportation systems. In order to sustain the competitiveness of the Austrian forest based industry improvements in transportation logistics are often considered an essential starting point. Our work focuses on the log-truck scheduling problem, which typically has many sources and few recipients. At the beginning of a planning period, the transportation orders are given. Here, we are considering full truckloads when it is that the truck moves from the wood storage location to a particular industrial site. When starting at the home location, and after unloading at the mill, we have to decide where the trucks should collect a new load in order to minimize the overall empty truck movements.

The problem we are discussing here is relevant for large forestry companies that serve a number of different mills; it describes the challenge of reducing the mileage of log-trucks. In our background forestry application, 10 trucks and approximately 30 trips are scheduled daily. In Austria, forest owners usually need to organize the log transport by employing forwarding companies. These forwarders usually serve several forest owners per day and aim to minimize their transportation costs. With this respect, the presented scheduling problem is highly relevant for log transport companies.

The emerging vehicle routing problem is denoted as a Timber Transport Vehicle Routing Problem (TTVRP) (see also Karanta et al., 2000 and Weintraub et al., 1996). It can be characterized as follows: a heterogeneous fleet of m log-trucks that are situated at the respective homes of truck drivers must fulfill n transports of logs between various wood storage locations and industrial sites, such as pulp mills and sawmills, during a specified timeframe. All of the transports are carried out as full truckloads; the vehicle is loaded at the wood storage location and unloaded at the industrial site. Each route commences at the home of the truck driver who leaves with an empty truck for loading round timber. Subsequently, he drives to the designated industrial site and completes the transport. The truck driver can now finish his tour and return back home or start a new delivery. Due to the transportation orders, each wood storage location and industrial site can be visited more than once during the planning horizon.

Log transport has a few specific constraints to consider such as the fact that some parts of the forest road networks are unsuitable for larger trucks, as their weight occasionally damages the road. Therefore, some wood storage locations can only be reached by trucks with a certain capacity. We denote this as the route weight limits. Due to industry operating hours, time windows for unloading wood must be considered. Time windows also occur at the truck starting points since truck drivers are only on duty at certain times. Additionally, we have to observe tour length constraints and capacity constraints. According to the given transportation orders, the objective is to minimize empty truck movements.

In Figure 1, we present a small example to illustrate the planning problem. Two log-trucks have to perform eight transportation orders (*1*, ..., *8*). The log-trucks are situated at the home-locations *A* and *B*, respectively. Wood is provided at six different wood storage locations (*P1*, ..., *P6*) and must be transported to three industrial sites (*I1*, ..., *I3*). The number of rectangles and triangles provides the number of visits at the respective location. *I1* receives three loads: two from *P1* and one from *P2*. Figure 1a) shows the required transports and demonstrates the problem of linking these transports in a cost-efficient manner, taking into account the above-mentioned constraints. Figure 1b) shows the cost-optimal solution for this problem. *A1* is the first trip taken from the log-truck situated at *A*, *A2* is the second one, etc.; the same is true for *B1* to *B11*. Altogether, we have scheduled 8 transports and 10 empty truck movements.

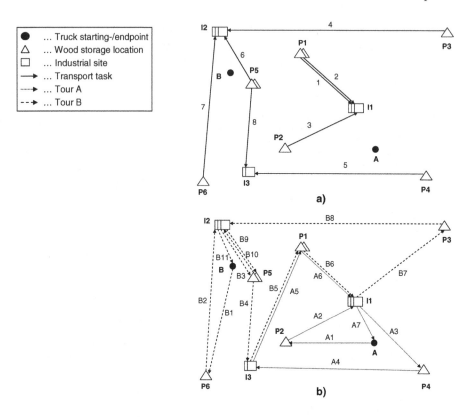

Figure 4-1. Conceptual formulation and solution for a TTVRP

The TTVRP is a special application of the full truckload vehicle routing problem (Gronalt et al., 2003). Murphy (2003) presents an approach that attempts to reduce the number of log-trucks that are used to perform transports of round timber. He developed a MIP model that minimizes the total transport costs. His approach does not take into account, however, time windows at industrial sites, availability times of the drivers, or route weight limits. He uses standard solver software to solve his problems but he only provides the best found solution after a certain computing time and not the global optimal solution. The approach of Palmgren et al. (2003) unites tactical and operational planning in wood transport. They provide a model formulation for the Log Truck Scheduling Problem (LTSP) and present a column generation based solution approach.

According to the established notation on VRPs, the TTVRP is related to the Multi Depot Vehicle Routing Problem with Pickup and Delivery, and Time Windows (MDVRPPDTW); supplementarily, one has to deal with specific route weight limits and full truckloads. An overview of the Vehicle Routing Problems can be found for example in Toth and Vigo (2002). The transport activities of the TTVRP have a similar structure to the Stacker

Crane Problem (SCP) (see Righini et al., 1999). Coja-Oghlan et al. (2004) provide an example of a SCP, which describes the scheduling of a delivery truck. Glover and Laguna (1997) provide a general introduction to the Tabu Search metaheuristic. For solving the TTVRP the Unified Tabu Search (Cordeau et al., 2001) is adapted and modified.

This present paper is organized in the following way: In Section 2, we present a model formulation of the TTVRP. The heuristic solution approach is outlined in the third section. We have developed three variants of the Tabu Search in order to obtain solutions for the TTVRP. Section 4 describes our numerical studies and the generation of test data. The results of the numerical experiments are provided in Section 5. We use different parameter sets for our heuristics and compare the three variants of the Tabu Search with each other and the best found feasible solution obtained with solver software. Finally, our conclusions are drawn in Section 6, in which an outlook on our future research is also provided.

2.   MODEL FORMULATION

The transportation orders are predefined and can therefore be considered as tasks that must be fulfilled in order to obtain a feasible solution. A feasible solution must include all of the tasks that are represented by arcs. Figure 2 demonstrates the same problem as Figure 1, which is transformed into a special case of the SCP. We have two kinds of tasks, so-called artificial tasks (A', B') and transport tasks ($1, …, 8$). The artificial tasks are introduced in order to connect the starting point and endpoint of a cycle. The direct connection between two vertices is always the shortest one. It is impossible to transport directly from one wood storage location to another or from one industrial site to another. This is because we have to deal with full truckloads. In Figure 2a) the tasks and vertices are displayed. Figure 2b) shows the corresponding optimal solution, using the same notation as in Figure 1b).

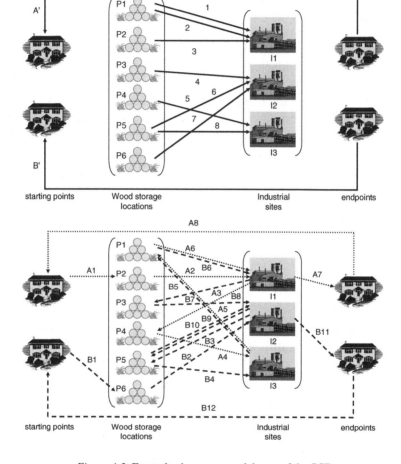

Figure 4-2. Example shown as special case of the SCP

In order to facilitate a further description the following notations are used:

- *n*-element set of transport tasks W,
- *m*-element set of artificial tasks V,
- and *m*-element set of trucks R.

The notation of the elements in V and R is identical. Truck $r \in R$ has a maximum capacity Q_r and a duration limit T_r. The availability time of a truck driver starts at e_r and ends at l_r. A specific route of a truck is named after this truck r.

Each transport task $i \in W$ has the following attributes:

- loading time a_i at the wood storage location,
- route weight limit k_i given in units of weight,
- order quantity q_i,

- unloading time s_i at the industrial site,
- time window $[e_i, l_i]$ at the industrial site,
- and traveling time u_i.

Each truck is allowed to arrive at an industrial site at a time $0 \leq b_i \leq l_i$; if the truck arrives at a time $b_i < e_i$ it must wait for the period $w_i = e_i - b_i$. t_{ij} represents the time that is needed to move from the endpoint of task i to the starting point of task j; this is the time needed for the empty truck movement.

The following binary decision variables are defined:

- $x_{ijr} = 1$, if task j is visited directly after task i with truck r; 0 otherwise.
- $y_{ir} = 1$, if task i is visited with truck r; 0 otherwise.

The set presented in (1) includes all of the tasks.

$$\tilde{W} = W \cup V \tag{1}$$

The objective function (2) of the model minimizes the duration of empty truck movements.

$$\min \sum_{r \in R} \sum_{i \in \tilde{W}} \sum_{j \in \tilde{W}} t_{ij} \cdot x_{ijr} \tag{2}$$

The following constraints have to be fulfilled:

$$\sum_{i \in \tilde{W}} x_{ihr} - \sum_{j \in \tilde{W}} x_{hjr} = 0 \quad \dots \forall h \in \tilde{W}, r \in R \tag{3}$$

$$\sum_{r \in R} \sum_{j \in \tilde{W}} x_{ijr} = 1 \quad \dots \forall i \in \tilde{W} \tag{4}$$

$$\sum_{j \in W} x_{ijr} + x_{iir} = 1 \quad \dots \forall i \in V, r \in R \quad (i = r) \tag{5}$$

$$\sum_{i \in W} x_{ijr} + x_{jjr} = 1 \quad \dots \forall j \in V, r \in R \quad (j = r) \tag{6}$$

$$y_{ir} = \sum_{j \in \tilde{W}} x_{ijr} \quad \dots \forall i \in \tilde{W}, r \in R \tag{7}$$

$$q_i \cdot y_{ir} \leq Q_r \quad \dots \forall i \in W, r \in R \tag{8}$$

$$Q_r \cdot y_{ir} \leq k_i \quad \dots \forall i \in W, r \in R \tag{9}$$

$$\sum_{i \in \tilde{W}} \sum_{j \in \tilde{W}} t_{ij} \cdot x_{ijr} + \sum_{i \in W} u_i \cdot y_{ir} \leq T_r \quad \dots \forall r \in R \tag{10}$$

$$b_i + w_i + s_i + t_{ij} + a_j + u_j - M \cdot (1 - x_{ijr}) \leq b_j \quad \dots \forall i \in \tilde{W}, j \in W, r \in R \tag{11}$$

$$b_i + w_i \geq e_i \quad \dots \forall i \in \tilde{W} \tag{12}$$

$$b_i \leq l_i \quad \dots \forall i \in \tilde{W} \tag{13}$$

$$b_i + w_i + s_i + t_{ij} - M \cdot (1 - x_{ijr}) \leq l_r \quad \dots \forall i \in \tilde{W}, j \in V, r \in R \quad (j = r) \tag{14}$$

$$x_{ijr} \in \{0,1\} \quad \dots \forall i \in \tilde{W}, j \in \tilde{W}, r \in R \tag{15}$$

$$y_{ir} \in \{0,1\} \quad \dots \forall i \in \tilde{W}, r \in R \tag{16}$$

$$w_i \geq 0 \quad \dots \forall i \in \tilde{W} \tag{17}$$

$$b_i \geq 0 \quad \dots \forall i \in \tilde{W} \tag{18}$$

Constraints (3) guarantee a tour, (4) to (6) define the predecessor and successor relationships, and (7) links the binary variables. Constraints (8) to (10) guarantee the observance of the truck capacity, route weight limits, and maximum travel times. (11) to (14) deal with the time windows at the industrial sites and truck starting points. (15) and (16) define the binary variables. (17) and (18) are non-negativity constraints. We validated our model for small instances, using Xpress-MP software. For real-life problems, it is necessary to develop a customized heuristic.

3.   SOLUTION APPROACH

The solution approach consists of the following steps:
1. Restrict the solution space.
2. Find an initial solution with a greedy heuristic.
3. Find an improved solution by applying one of the following Tabu Search procedures:
    a. Standard Tabu Search
    b. Tabu Search with a limited neighborhood
    c. Tabu Search with an alternating strategy
4. Apply a post-optimization heuristic based on 2opt.

The Unified Tabu Search heuristic served as a starting point for our solution procedures. Three variants that differ with respect to the size of their solution space in each iteration step are developed and subsequently discussed.

3.1 Tabu Search

3.1.1 Solution space and initial solution

The overall heuristic commences with a reduction of the solution space. Looking at the problem characteristic, we see that some transport tasks can only be executed by certain truck types. On the one hand, this is because it is impossible to split transport tasks; therefore a truck r with capacity Q_r cannot perform a transport task with an order quantity q_i if $Q_r < q_i$. On the other hand, we have wood storage locations that cannot be reached by each truck type because of the route weight limits. A truck r with capacity Q_r cannot perform a transport task with a route weight limit k_i if $Q_r > k_i$. In the first step, it is guaranteed that a truck r is only assigned to a task that can be handled by this truck with respect to the truck capacity and the route weight limit.

To construct an initial solution we use a regret-heuristic. The gained solution may violate the duration- and time window constraints. The regret-heuristic works in the following way:

- Initialization:
  - For every artificial task $i \in V$: find the closest transport task j and the second-closest transport task z.
  - Calculate a regret-value $REG_i = t_{iz} - t_{ij}$.
  - Sort the regret-values in descending order.
  - Allocate the closest transport tasks to the artificial tasks according to this order; if a transport task is the closest to two or more artificial tasks, it is assigned to the one with the highest regret value.
- Continue with the same procedure until all of the transport tasks are assigned to a tour. Always find the closest and second-closest transport task to the last included task.

3.1.2 Parameter setup

Based on the initial solution a rank indicator B_{ir} with $i \in W$ and $r \in R$ is defined. If $B_{ir} = 0$ this means that transport task i is not on tour (of truck) r. If for example $B_{ir} = 3$ this means that transport task i is ranked third on tour r. While traversing the solution space we apply different notations for marking the solutions: current solution s, a neighbor solution $s°$, the best neighbor solution s', and the best found feasible solution $s*$. The costs associated with a solution are given by $c(s)$ and are equal to the total travel time of the empty trucks. The Tabu Search permits infeasible intermediate solutions. The total violation of tour duration constraints and time windows is denoted by $d(s)$

and $h(s)$, respectively. The variables α and β are used to weight the total violation of constraints. Their values are updated in each iteration step with the help of a parameter δ. α and β are used in order to guide the search process. If we are gaining feasible solutions for a number of iterations, these variables encourage the search process to move to areas with infeasible solutions. If the search process stays in an area with infeasible solutions for a longer time, the search process is driven to areas with feasible solutions. The parameter λ is used to weight the penalizing factor for deteriorating neighbor solutions.

The array ρ_{ir} is used to store how often a transport task i was part of a tour r in a solution s. The tabu status is stored in the array τ_{ir}. We save the information up to which iteration step a task i may not be part of a tour r. An aspiration criterion is used to permit the bypassing of the tabu status. We use fixed tabu durations that are dependent on the number of log-trucks and transport tasks, in which the tabu duration is given by θ. The array σ_{ir} saves the value of the best found feasible solution, in which transport task i was part of tour r. The parameter η gives the number of iteration steps. The function $f(s)$ is equal to the cost function $c(s)$ plus the weighted violations of constraints. The decision function $g(s)$ is used to determine which neighbor solution is chosen; it is equal to $f(s)$ plus a possible penalty function $p(s)$.

3.1.3   Search procedure

The Tabu Search algorithm works as follows:
- Initialization
  - If the initial solution s is feasible set $s^* := s$ and $c(s^*) := c(s)$; else set $s^* := \{\ \}$ and $c(s^*) := \infty$.
  - Initialize α and β.
  - For all attributes (i,r):
    - Set $\tau_{ir} := 0$ and $\rho_{ir} := 0$.
    - If the initial solution s is feasible and $B_{ir} > 0$ then set $\sigma_{ir} := c(s)$; else set $\sigma_{ir} := \infty$.
  - Set the parameters δ and λ. We use the following values for these parameters:
    - $\delta \in [0.1, 0.9]$
    - $\lambda \in [0.010, 0.025]$
- For $\kappa = 1$ To η do
  - Determine all neighbor solutions s° of s and their costs $c(s^\circ)$. A neighbor solution s° is generated by moving a transport task i from a tour r to a tour o (move-operator).

- Each transport task i is taken out of its current tour r and tentatively inserted into all of the other tours that fulfill the capacity- and route weight limits.
- If a transport task i is eliminated in a tour r, the direct predecessor and direct successor of i are connected. In its new tour o transport task i is inserted at the position with the least additional costs.
- For each attribute (i,r) that is part of a neighbor solution $s°$, but was not part of solution s, the procedure checks if τ_{ir} is smaller than κ. This means that the attribute is checked as to whether it is tabu or not. If the attribute (i,r) is tabu the algorithm checks if $s°$ is a feasible solution and $c(s°) < \sigma_{ir}$. In this case, the aspiration criterion is fulfilled, in which it is permitted to use this neighbor solution $s°$ despite its tabu status. If the tabu status remains, the value of the decision function to choose a neighbor solution must be set to $g(s°) := \infty$.
- For all neighbor solutions $s°$ that are not tabu or meet the aspiration criterion, the algorithm computes $f(s°)$ and $g(s°)$. If $f(s°) < f(s)$, then set $g(s°) := f(s°)$; otherwise set $g(s°) := f(s°) + p(s°)$. Equation (19) shows the calculation of $f(s°)$; (20) shows the computation of the penalty function $p(s°)$.

$$f(s°) = c(s°) + \alpha \cdot d(s°) + \beta \cdot h(s°) \tag{19}$$

$$p(s°) = \lambda \cdot c(s°) \cdot \sqrt{n \cdot m} \cdot \sum_{i \in W} \sum_{r \in R} \rho_{ir} \quad ...\forall (i,r) \in s° \tag{20}$$

The penalty function $p(s°)$ penalizes the neighbor solutions $s°$ for having the same or a higher function value $f(s°)$ as the current solution s. The parameter λ is predefined; n is the number of transport tasks and m the number of log-trucks. The sums over ρ_{ir} count how often attributes (i,r) that are element of $s°$ were part of a solution s.

- The neighbor solution $s°$ that has the lowest value of $g(s°)$ is chosen and called s'.
- After having found the best neighbor solution s' the algorithm continues with the following steps:
  - For each attribute (i,r), which was part of solution s but is not part of s' set $\tau_{ir} := \kappa + \theta$. The tabu duration θ is calculated with Equation (21).

$$\theta = \lceil (\log(n \cdot m))^2 \cdot 4 \rceil \tag{21}$$

We obtained this formula after a number of parameterization approaches for θ. The value of θ is dependent on the size of the problem according to this formula.

- For each attribute (i,r) that is part of the best neighbor solution s' set $\rho_{ir} := \rho_{ir} + 1$.
- If s' is a feasible solution and $c(s') < c(s^*)$ set $s^* := s'$ and $c(s^*) := c(s')$; otherwise, leave the values of $c(s^*)$ and s^* unchanged.
- If s' is a feasible solution do: for each attribute (i,r) which is part of s' set $\sigma_{ir} := \min\{\sigma_{ir}, c(s')\}$.
- Adjustment of α and β:
  - If $d(s') > 0$ set $\alpha := \alpha \cdot (1 + \delta)$, else set $\alpha := \alpha / (1 + \delta)$.
  - If $h(s') > 0$ set $\beta := \beta \cdot (1 + \delta)$, else set $\beta := \beta / (1 + \delta)$.
  - Set $\kappa := \kappa + 1$ and $s := s'$.
- End For

3.1.4   Post-optimization heuristic

A 2-opt based heuristic is applied as a post-optimization procedure after each iteration step of the Tabu Search algorithm. The algorithm attempts to improve single tours by changing the position of two transport tasks. If improvement is attained, the tour is rebuilt accordingly and the same procedure restarts until no further improvement can be found. Per definition, an improvement of a solution is only tolerated if the solution is feasible. The post-optimization procedure does not influence the Tabu Search algorithm; the input data for the next Tabu Search iteration step remains unchanged even if improvement is attained. Only s^* and $c(s^*)$ are updated if the costs $c(s')$ of the post-optimized solution s' are lower than the current best found costs $c(s^*)$.

3.2   New search strategies

The Tabu Search strategy described in Section 3.1.3 implies a search of the entire neighborhood of a solution in each iteration step. We call this strategy hereafter a Standard Tabu Search. This is a very time-consuming procedure since there are no rules to restrict the search space. Therefore, we developed a search strategy that concentrates on the elimination of bad connections between tasks. Toth and Vigo (2003) proposed the Granular Tabu Search in order to restrict the neighborhood of solutions drastically and reduce computing times. They attempt to limit moves that insert "long" arcs in the current solution. Our approach concentrates on a certain fraction of empty truck movements in the current solution s; only these links are to be removed in neighbor solutions. Other links can only be modified if a task from a removed link is inserted between their starting and ending points.

The procedure functions in the following way: The links are first sorted according to their duration in descending order. Then, a predefined number of links is chosen starting from the one with the longest duration. The number of used links is calculated as a fraction of all the existing empty truck movements; the divider D is set as a parameter. If $D = 4$ this means that one fourth of all the links of a solution s is taken away for being removed in neighborhood solutions.

We call this strategy a Tabu Search with a limited neighborhood. This strategy seems to be myopic since "shorter" links are unaffected directly. To overcome this we merge the Standard Tabu Search and Tabu Search with a limited neighborhood in a new algorithm called a Tabu Search with an alternating strategy. After a predefined number of iteration steps with a restricted neighborhood, an iteration step with a full neighborhood search is set. The parameter A is used to define which iteration steps will be computed with a full neighborhood search. For example, a setting of $A = 8$ means that in every eighth iteration step a full neighborhood search is performed. These new strategies lead to drastic reductions of the computing time. As shown in Section 5, there are also no, or only minimal, losses in the solution quality if the Tabu Search with an alternating strategy is used.

4. NUMERICAL EXPERIMENTS

The small introductory example with eight transport tasks and two trucks can be solved with standard solver software within seconds. Unfortunately, real-life problems have far more trucks and trips to consider. We have observed that regional forest enterprises have to perform approximately 30 transport tasks per day and on average, they operate 10 log-trucks. In the course of a year up to 600 pick-up locations are visited to supply five industrial sites. A large wood processing company in the area operates four sites. In order to ensure a smooth wood supply, up to 250 transport tasks and 80 trucks per day are on order. We estimate their overall yearly number of pick-up locations as 2,500. Murphy (2003) presents a case study with an average of 9 trucks and 35 transport tasks per day for a company situated on the Southern Island of New Zealand. Palmgren et al. (2003) present two case studies for Sweden: one with six trucks and 39 transport tasks, and one with 28 trucks and approximately 85 transport tasks.

In order to test the algorithmic approach for real-life sized problem instances we have developed a random problem generator. Two sets of problem instances have been generated. Each set consists of 20 instances with 30 transport tasks and 10 trucks. The first set of instances has weaker constraints than the second one in terms of the average task duration and the traveling times between the tasks. In the first instance set, the same 10 truck starting points are used for all of the instances. There are three different industrial sites and 560 possible wood storage locations. In the second instance set, the same 10 truck starting points are also used for all of the instances; but they are different from those of instance set 1. In instance set 2 there are four different industrial sites and 560 possible wood storage locations. In instance set 1, we chose the three industrial sites with the lowest average distance to the 560 wood storage locations out of a set of nine industrial sites; whereas in instance set 2 we use four industrial sites out of this set, which belong to one company and are situated less centrally. This is the reason why we have longer distances in instance set 2.

The model formulation was implemented with the software Xpress-MP. The heuristic solution approach was programmed with Visual Basic 6. We tested the algorithm in the following variants: Standard Tabu Search, Tabu Search with a limited neighborhood, and Tabu Search with an alternating strategy. The post-optimization strategy is only applied in some test runs.

All of the computers used are equipped with a Pentium IV processor with 2.52 GHz and 512 MB RAM; their operating system is Windows XP.

The values of the following parameters were varied in the test runs:
- weighting factor λ for the penalty function $p(s)$,
- parameter δ to update α and β,
- number of iteration steps η,
- divider D,
- and parameter A.

The variables α and β are initialized with the value 1. We use the best found solutions and lower bounds computed with Xpress-MP after a certain computing time as a benchmark for the heuristic solutions. It is also necessary to compare the different variants of Tabu Search with respect to computing times and solution quality. Section 5 shows the results of the numerical studies.

5.   RESULTS

The optimal solution for the introductory example with two log-trucks and eight transport tasks can be found within a few iteration steps for all of the variants of the heuristic. The numerical studies were started with a Standard Tabu Search variant, which forbids log-trucks to stay at home and uses no post-optimization strategy. The first test case of each instance set was taken to find the best alues for the parameters

$\lambda \in [0.010, 0.025]$ and $\delta \in [0.1, 0.9]$. The resulting values were taken for further computations. Tables 1 and 2 show the deviation from the best found solution for different parameter values. The algorithm is executed for 10,000 iteration steps. In total, we tested 36 parameter variants. In Table 1 and 2, the first row shows the different values for λ and the first column shows the different values for δ. The highest deviation is written in cursive; the shaded cell marks the best found parameterization.

	0.010	0.015	0.020	0.025
0.1	0.1616%	0.1269%	0.0952%	0.0744%
0.2	0.0744%	0.1578%	0.1269%	0.1371%
0.3	0.2492%	0.2175%	0.1688%	0.2492%
0.4	0.1896%	0.1371%	0.1341%	0.2609%
0.5	0.0744%	0.1269%	0.1325%	0.0627%
0.6	0.3338%	0.1269%	0.3965%	**0.0000%**
0.7	0.3701%	0.4139%	0.3761%	0.0310%
0.8	0.5422%	*0.7416%*	0.6638%	0.1325%
0.9	0.5460%	0.6442%	0.6967%	0.5234%

Table 4-1. Deviation from the best found solution for test case 1 of instance set 1

	0.010	0.015	0.020	0.025
0.1	0.0554%	0.2979%	0.2440%	0.2281%
0.2	0.2042%	0.1332%	0.2245%	0.1781%
0.3	**0.0000%**	0.2799%	0.0914%	0.0462%
0.4	0.1411%	0.2220%	0.0554%	0.0500%
0.5	0.2973%	0.2695%	0.2339%	0.2440%
0.6	0.1518%	0.5159%	0.3205%	0.0481%
0.7	0.3606%	0.2822%	0.4143%	0.1424%
0.8	0.5824%	0.1518%	0.1424%	0.0941%
0.9	0.3205%	*0.9748%*	0.1054%	0.6251%

Table 4-2. Deviation from the best found solution for test case 1 of instance set 2

Table 3 shows the deviation from the best found solution for all test cases of instance set 1 depending on the number of iteration steps. The best found solution is obtained after 1,000,000 iteration steps. If there is no deviation from the best found solution, the cell is shaded. We have adopted the best found parameter values for test instance 1 with $\lambda = 0.025$ and $\delta = 0.6$. Table 3 provides insight into the speed of convergence. The first row shows the number of iteration steps and the first column shows the different test instances.

	10	100	1,000	10,000	100,000
T1	no sol.	1.5116%	0.4928%	**0.0000%**	**0.0000%**
T2	no sol.	0.4799%	0.4799%	0.0838%	**0.0000%**
T3	no sol.	0.6693%	0.6693%	0.1570%	0.1570%
T4	no sol.	0.5808%	0.5808%	0.0474%	0.0147%
T5	no sol.	1.1163%	0.7436%	0.2793%	0.1573%
T6	no sol.	0.4273%	0.4273%	**0.0000%**	**0.0000%**
T7	no sol.	3.5075%	0.7752%	0.5607%	0.2204%
T8	no sol.	1.1864%	0.8319%	0.2374%	**0.0000%**
T9	no sol.	0.5149%	0.5149%	0.3448%	0.0127%
T10	no sol.	0.6192%	0.6192%	0.4733%	0.1605%
T11	no sol.	0.4182%	0.4182%	0.4182%	0.0210%
T12	no sol.	0.8338%	0.8338%	0.2614%	0.0546%
T13	no sol.	1.0375%	0.2896%	0.2127%	**0.0000%**
T14	no sol.	0.2859%	0.2859%	0.0868%	0.0868%
T15	no sol.	0.1770%	0.1770%	0.1770%	**0.0000%**
T16	no sol.	0.5266%	0.1911%	0.1911%	0.1499%
T17	no sol.	0.3748%	0.3748%	0.2933%	0.1484%
T18	no sol.	1.6211%	0.9737%	0.2782%	0.1476%
T19	no sol.	0.2550%	0.2550%	0.1807%	0.0464%
T20	no sol.	0.4012%	0.4012%	**0.0000%**	**0.0000%**

Table 4-3. Deviation from the best found solution after 1,000,000 iteration steps depending on the number of performed iteration steps for instance set 1

With 1,000 iteration steps the solution values of all test instances of set 1 are less than 1% worse than the best found solution. It takes approximately 150 seconds to perform 1,000 iteration steps with the Standard Tabu Search. Since we can estimate a linear relationship between computing times and the number of iteration steps we only need about a tenth part of the computing time for 10,000 iteration steps.

Table 4 shows the same data as Table 3 for all of the test instances of instance set 2 depending on the number of iteration steps. We also used the best found parameterization for test instance 1 with $\lambda = 0.010$ and $\delta = 0.3$.

	10	100	1,000	10,000	100,000
T1	no sol.	2.2515%	0.4010%	**0.0000%**	**0.0000%**
T2	no sol.	1.8074%	1.8074%	1.8074%	0.8177%
T3	no sol.	5.9537%	2.5884%	0.1172%	**0.0000%**
T4	no sol.	3.2998%	0.0548%	0.0438%	**0.0000%**
T5	no sol.	2.6279%	0.9880%	0.1037%	**0.0000%**
T6	no sol.	5.4487%	2.1683%	1.2378%	0.5272%
T7	no sol.	5.5435%	**0.0000%**	**0.0000%**	**0.0000%**
T8	no sol.	0.1990%	0.1990%	0.1990%	0.1944%
T9	no sol.	3.4008%	0.9153%	0.8786%	0.0002%
T10	no sol.	1.0430%	1.0430%	0.8509%	**0.0000%**
T11	no sol.	0.8382%	0.8382%	0.4553%	0.1801%
T12	no sol.	1.3593%	1.2256%	0.2803%	0.2803%
T13	no sol.	1.7042%	1.4113%	0.1400%	0.1400%
T14	no sol.	2.4080%	1.3224%	0.8963%	0.0270%
T15	no sol.	3.1242%	0.9137%	0.9137%	0.0113%
T16	no sol.	9.7840%	2.5547%	1.6086%	0.2714%
T17	no sol.	1.3800%	1.3800%	1.1039%	**0.0000%**
T18	no sol.	6.1360%	2.5321%	1.2930%	0.0285%
T19	no sol.	3.1343%	0.7194%	0.1164%	0.0589%
T20	no sol.	1.5204%	0.7800%	0.1743%	0.0223%

Table 4-4. Deviation from the best found solution after 1,000,000 iteration steps depending on the number of performed iteration steps for instance set 2

Table 5 compares the average deviation from the best found solution for all test cases of instance set 1 and 2 in the range of 100 to 100,000 iteration steps. It summarizes the results of Table 3 and Table 4. One can observe that it is possible to find solutions of good quality in less computing time, for instance set 1. We assume that the tighter constraints of instance set 2 make it more difficult to find feasible and good quality solutions.

	100	1,000	10,000	100,000
instance set 1	0.8272%	0.5168%	0.2141%	0.0689%
instance set 2	3.1482%	1.1921%	0.6110%	0.1280%

Table 4-5. Average deviation from the best found solution after 1,000,000 iteration steps depending on the number of performed iteration steps for instance sets 1 and 2

Furthermore, we compared the different variants of Tabu Search with respect to computing times and solution quality. For performing this, we used test case 1 of instance set 1 to compute 10,000 iteration steps with the different Tabu Search variants. In all of the Tabu Search variants, we did not permit unemployed log-trucks. We also applied different parameter values of the divider D and varied the sequence of full neighborhood search iteration steps. The parameter λ was set to 0.025; δ was set to 0.6. Table 6 shows in its first column the used Tabu Search variant and the respective parameterization, in the second, the time deviation from the lowest computing time, and in the third, the deviation of the solution value from the best found solution value is displayed. All Tabu Search variants in Table 6 are computed without a post-optimization strategy. We also applied the post-optimization heuristic to the Tabu Search with an alternating strategy. It turned out that the post-optimization was able to improve the best found solution within the first iteration steps, but after 10,000 iteration steps we obtained the same results as when not using it. The additional computing time for the post-optimization can only be determined empirically for each test instance; roughly spoken, one can expect an increase of approximately 10%.

The following abbreviations are used for the Tabu Search variants in the below-mentioned text: Standard Tabu Search (TS), Tabu Search with a limited neighborhood (TSLN), and Tabu Search with an alternating strategy (TSAS).

One can observe that the TS has the highest computing time of all the variants and offers a solution quality that is close to the best found solution. We tested four parameterizations of the TSLN. The divider D determines which portion of the connections between the transport tasks is removed in neighboring solutions. The results show that the lowest computing time (203 seconds) is reached with a TSLN and a divider $D = 8$. However, this method also offers the worst solution quality, which is in turn unacceptable. When the TSLN is used with $D = 2$, a quite good solution quality is obtained in reasonable computing time. Nevertheless, the TSLN is a myopic strategy. Some parts of the neighborhood are excluded permanently. The TSAS seems to be a good way to overcome this problem. A look at the results shows that it is able to reduce computing times drastically with little or no loss in solution quality. As the results show, it is sufficient to search the full neighborhood in every eighth iteration step.

Method	Time deviation	Solution value deviation
TS	673.40%	0.18%
TSLN D = 2	227.59%	0.27%
TSLN D = 4	90.64%	2.09%
TSLN D = 6	33.00%	7.31%
TSLN D = 8	0.00% (203 s)	7.69%
TSAS D = 2 A = 2	444.33%	0.29%
TSAS D = 4 A = 2	373.89%	0.00%
TSAS D = 6 A = 2	348.77%	0.08%
TSAS D = 8 A = 2	326.11%	0.03%
TSAS D = 2 A = 8	287.68%	0.05%
TSAS D = 4 A = 8	159.11%	0.00%
TSAS D = 6 A = 8	110.84%	0.06%
TSAS D = 8 A = 8	78.82%	0.00%

Table 4-6. Comparison of the different Tabu Search variants for test case 1 of instance set 1 for 10,000 iteration steps

We also compared the results of the different Tabu Search variants after a fixed computing time. Table 7 shows the results for the same Tabu Search variants and parameterizations as Table 6. The running time was chosen as the average of the running times of the different Tabu Search variants for 10,000 iteration steps (696 seconds). The second column of Table 7 shows the number of iteration steps in this time, and the third, the deviation from the best found solution. The best found solution has the same value in both comparisons.

Method	Iteration steps	Solution value deviation
TS	4,430	0.18%
TSLN D = 2	10,459	0.27%
TSLN D = 4	17,973	2.09%
TSLN D = 6	25,762	7.31%
TSLN D = 8	34,263	7.69%
TSAS D = 2 A = 2	6,295	0.29%
TSAS D = 4 A = 2	7,230	0.15%
TSAS D = 6 A = 2	7,635	0.08%
TSAS D = 8 A = 2	8,041	0.03%
TSAS D = 2 A = 8	8,838	0.05%
TSAS D = 4 A = 8	13,223	0.00%
TSAS D = 6 A = 8	16,251	0.06%
TSAS D = 8 A = 8	19,161	0.00%

Table 4-7. Comparison of the different Tabu Search variants for test case 1 of instance set 1 after 696 seconds running time

In the following the mean solution value of three arbitrary test cases namely 1, 10, and 20 of instance set 1 for 10 to 1,000,000 iteration steps is displayed. We forbade unemployed log-trucks and applied no post-optimization. In Figure 3, the abscissa represents the number of iteration steps; the ordinate shows the deviation from the best found solution; if the deviation is equal to 10%, this means that no feasible solution was found for one or more test cases up to this iteration step. The solution quality does not improve significantly for the TSLN with a divider D equal to 6 and 8 if more than 100 iteration steps are computed; the same is true with more than 10,000 iteration steps for a divider D equal to 2 and 4. With the TSAS and the TS solutions of good quality can be obtained. Even with a very fast Tabu Search variant (TSAS with $D = 8$, full neighborhood search in every eighth iteration step) the deviations from the best found solution are far less than 1% after 1,000 iteration steps. The bars in figures 3 and 4 are ordered in the same way as the sequence of the legend.

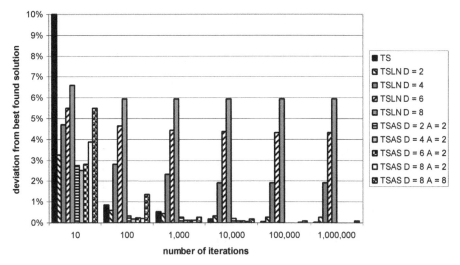

Figure 4-3. Average deviation from the best found solution for test cases 1, 10, and 20 of instance set 1

Figure 4 shows the average deviation from the best found solution for test cases 1, 10, and 20 of instance set 2. Since instance set 2 has tighter constraints, it is more difficult to find feasible solutions. Even after 1,000,000 iteration steps with a TSLN no feasible solution for the dividers $D = 4$, $D = 6$, and $D = 8$ can be obtained. The TSLN achieves a solution of good quality only for a divider $D = 2$. The TS and the TSAS are able to find solutions that are equal to the best found solution with one exception; the TSAS with a divider $D = 8$ and a full neighborhood search in every eighth iteration step seems to be improper for solving problems with tight constraints. However, if the divider is reduced it is also possible to improve the solution quality in this case.

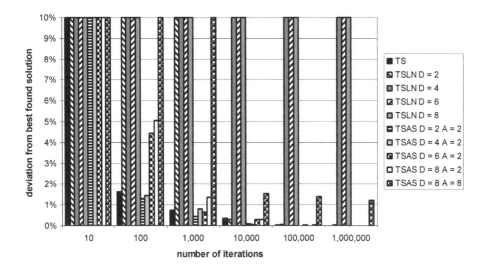

Figure 4- 4. Average deviation from the best found solution for test cases 1, 10, and 20 of instance set 2

Due to the computational complexity of the TTVRP, it is clear that standard solver software is unsuitable for these problems. However, to benchmark the heuristics, we compared in Tables 8 and 9 their best found solutions after 10,000 iteration steps with the best found solution obtained with the solver software Xpress-MP for the same computing time. In this comparison, we permitted log-trucks to stay at home and applied a post-optimization heuristic to the Tabu Search variants. Table 8 shows the results for test case 1 of instance set 1. In Table 8 for example the TSAS with $D = 8$ and a full neighborhood search in every second iteration needs 825 seconds for 10,000 iteration steps. This provides a solution value of 2,603.52. For the same timeframe, the Xpress solver provides a value of 2,640.13. Even after a computing time of 24 hours, the solver software obtains a solution value (2,603.12) that is worse than most heuristic solution values. The lower bound after 24 hours is equal to 2,593.37; but this solution may represent an infeasible solution to the problem.

Method	Time [sec]	sol. val.	sol. val. Xpress Solver	Deviation
TS	1,447	2,597.52	2,616.16	0.7125%
TSAS D = 8 A = 2	825	2,603.52	2,640.13	1.3867%
TSAS D = 4 A = 8	503	2,598.33	2,702.59	3.8578%
TSAS D = 6 A = 8	413	2,599.72	2,767.19	6.0520%
TSAS D = 8 A = 8	328	2,602.79	no solution	no solution

Table 4-8. Comparison of solution values after certain computing times for test case 1 of instance set 1

Table 9 shows the results for test case 1 of instance set 2. The best found parameter values of Tabu Search variants that allow log-trucks to stay at home and variants that do not permit this differ in instance set 2. Therefore the parameter λ was set to 0.015; δ was set to 0.6. Since there are tighter constraints, the solver software could not find feasible solutions within the computing times needed by heuristics. Even after a computing time of 24 hours, the solver software obtains only one feasible solution with a value of 5,617.76 that is much worse than the heuristic solutions. The lower bound after 24 hours is equal to 3,786.93.

Method	Time [sec]	sol. val.	sol. val. Xpress Solver	Deviation
TS	1,440	4,749.97	no solution	no solution
TSAS D = 8 A = 2	820	4,773.07	no solution	no solution
TSAS D = 4 A = 8	497	4,784.72	no solution	no solution
TSAS D = 6 A = 8	438	4,774.76	no solution	no solution
TSAS D = 8 A = 8	344	4,772.40	no solution	no solution

Table 4-9. Comparison of solution values after certain computing times for test case 1 of instance set 2

Additional comparisons were made for other test cases also. It turned out that all of the heuristic solutions were better than the best found solutions obtained with Xpress-MP after the same computing time.

6.   CONCLUSION

In this paper, we presented a formal description of the TTVRP, which was only described verbally in the literature up to now. We have also developed a heuristic solution approach based on a Tabu Search with different neighborhood structures. The numerical studies show that the proposed heuristics are able to solve real-life problem instances in reasonable computing times with good solution quality. The heuristics perform rather well if they are compared to the best found solutions of solver software as a benchmark.

The TSAS is a good method to reduce computing times and keep the solution quality nearly constant. This approach can also be enhanced with a dynamic component; instead of fixing the iteration steps with a full neighborhood search, one can also make the neighborhood structure dependent on the solution quality. If the solution quality does not improve for a certain number of iteration steps with a limited neighborhood (this number could be a function of the total number of iteration steps) one can set an iteration step with a full neighborhood search. The development of further variants of the TSAS could also bring forth benefits for other research areas that use a Tabu Search as a solution method. The TS is recommendable if there are tight constraints, in which feasible solutions are difficult to find; but it is also worth attempting to use the TSAS with frequent iteration steps with a full neighborhood search for such problem instances. Even though the TSLN offers a reduction in computing times compared to the TS, it is not recommendable since it is myopic, and therefore, the search process is locked very often in local optima for a large number of iteration steps.

We can also observe that the improvements in solution quality have not been significantly compared to the additional computing times after 10,000 iteration steps. We can conclude that the heuristic solution approaches quickly converge to solutions with good quality. This fast speed of convergence may be an indication for being locked in local optima; but if we look at the intermediate solutions, we can notice that this is not the case, and the diversification strategy of the Tabu Search heuristics is working well.

Future research will concentrate on an extension of the planning horizon of this scheduling problem. The current method is able to optimize the routing of log-trucks during a given timeframe, which is generally one day. When the planning horizon is extended to one week, an evenly distributed workload among the days of that week cannot be assured. Since this is an important factor for industrial sites in the wood industry, it is necessary to introduce a model formulation that first performs an optimal allocation of the transport tasks to single days of the week. Subsequently, the current method can be reused. As the results show, it also makes sense to put forth additional effort toward the enhancement of the TSAS.

Acknowledgements:
The authors would like to thank the reviewers and the auditorium at the MIC 2005 for their useful comments on this contribution.

REFERENCES

Coja-Oghlan, A., Krumke, S. O., Nierhoff, T. (2004): A Heuristic for the Stacker Crane Problem on Trees which is Almost Surely Exact. Journal of Algorithms, In Press.

Cordeau, J. F., Laporte, G., Mercier, A. (2001): A unified tabu search heuristic for vehicle routing problems with time windows. Journal of the Operational Research Society **52**, 928-936.

Glover, F. and Laguna, M. (1997): Tabu Search. Kluwer Academic Publishers, Boston, USA.

Gronalt, M., Hartl, R., Reimann, M. (2003): New savings based algorithms for time constrained pickup and delivery of full truckloads. European Journal of Operational Research **151(3)**, 520-535.

Karanta, I., Jokinen, O., Mikkola, T., Savola, J., Bounsaythip, C. (2000): Requirements for a Vehicle Routing and Scheduling System in Timber Transport. Proceedings of the IUFRO International Conference: Logistics in the forest sector, 235-250.

Murphy, G. (2003): Reducing Trucks on the Road through Optimal Route Scheduling and Shared Log Transport Services. Southern Journal of Applied Forestry **27(3)**, 198-205.

Palmgren, M., Rönnqvist, M., Värbrand, P. (2003): A solution approach for log truck scheduling based on composite pricing and branch and bound. International Transactions in Operational Research **10**, 433-447.

Righini, G. and Trubian, M. (1999): Data-dependent bounds for the General and the Asymmetric Stacker-Crane problems. Discrete Applied Mathematics **91**, 235-242.

Schwarzbauer, P. (2005): Die österreichischen Holzmärkte: Größenordnungen – Strukturen – Veränderungen. Lignovisionen Band 8, Universität für Bodenkultur Wien.

Toth, P. and Vigo, D. (eds.) (2002): The Vehicle Routing Problem. SIAM, Philadelphia, USA.

Toth, P. and Vigo, D. (2003): The Granular Tabu Search and Its Application to the Vehicle-Routing Problem. INFORMS Journal on Computing **15(4)**, 333-346.

Weintraub, A., Epstein, R., Morales, R., Seron, J., Traverso, P. (1996): A Truck Scheduling System Improves Efficiency in the Forest Industries. Interfaces **26(4)**, 1-12.

Part III

Nature-inspired methods

Chapter 5

SOLVING THE CAPACITATED MULTI-FACILITY WEBER PROBLEM BY SIMULATED ANNEALING, THRESHOLD ACCEPTING AND GENETIC ALGORITHMS

Necati Aras,[†] Sadettin Yumusak,[†] and İ. Kuban Altınel[†]

[†] *Dept. of Industrial Engineering*
Boğaziçi University
34342, İstanbul, Turkey

Abstract   In this paper we focus on the capacitated multi-facility Weber problem with rectilinear, Euclidean, squared Euclidean and ℓ_p distances. This problem deals with locating m capacitated facilities in the Euclidean plane so as to satisfy the demand of n customers at the minimum total transportation cost. The location and the demand of each customer is known a priori and the transportation cost is proportional to the distance and the amount of flow between customers and facilities. We present three new heuristic methods each of which is based on one of the three well-known metaheuristic approaches: simulated annealing, threshold accepting, and genetic algorithms. Computational results on benchmark instances indicate that the heuristics perform well in terms of the quality of solutions they generate. Furthermore, the simulated annealing-based heuristic implemented with the two-variable exchange neighborhood structure outperforms the other heuristics considered in the paper.

Keywords: Location-allocation problems; simulated annealing; threshold accepting; genetic algorithms.

1.   Introduction

Deterministic location-allocation problems are concerned with locating a set of facilities and allocating their capacity to satisfy the demand of a set of customers with known locations so that the total transportation cost is minimized. Supply centers such as plants and warehouses may constitute the facilities while retailers and dealers may be considered as customers. When the facility locations have to be selected from a given set of candidate locations, the corresponding location-allocation problem (LAP) becomes a discrete optimization problem. A continuous LAP is obtained when the facilities can be located anywhere in the Euclidean plane. The latter problem is also known as the multi-facility Weber problem (MFWP) (Wesolowsky, 1993). It is referred to as the single-facility Weber problem if the objective is the determination of an optimal location for a single facility. In some situations, facilities can have capacity constraints, which gives rise to the capacitated multi-facility Weber problem (CMFWP). As can be easily observed, in an optimal solution to the uncapacitated problem each customer is served from the nearest facility, which is not true for the more restricted CMFWP because of the capacity constraints. In the CMFWP formulations, the transportation cost is usually assumed to be proportional to the amount shipped as well as the distance between facilities and customers. The most frequently used distance functions in location theory are the Euclidean, the squared Euclidean, the rectilinear and ℓ_p distances. The ℓ_p distance between two points $\mathbf{u}$ and $\mathbf{v}$ in the two-dimensional Euclidean space is defined as $d_p(\mathbf{u}, \mathbf{v}) = (|u_1 - v_1|^p + |u_2 - v_2|^p)^{1/p}$. In fact, the Euclidean and rectilinear distances are its special cases when $p = 2$ and $p = 1$, respectively. The mathematical programming formulation of the CMFWP can be stated as:

$$\min \sum_{i=1}^{m} \sum_{j=1}^{n} c_{ij} w_{ij} d(\mathbf{x}_i, \mathbf{a}_j) \tag{5.1}$$

s.t.

$$\sum_{i=1}^{m} w_{ij} = q_j \quad j = 1, ..., n \tag{5.2}$$

$$\sum_{j=1}^{n} w_{ij} = s_i \quad i = 1, ..., m \tag{5.3}$$

$$w_{ij} \geq 0 \qquad i = 1, ..., m; j = 1, ..., n \tag{5.4}$$

Here, n is the number of customers and m is the number of facilities to be located. q_j and $\mathbf{a}_j = (a_{j1}, a_{j2})^T$ represent, respectively, the demand and coordinates of customer j. The capacity of facility i is given by s_i and $\mathbf{x}_i = (x_{i1}, x_{i2})^T$ denotes its unknown coordinates. $d(\mathbf{x}_i, \mathbf{a}_j)$ is the distance between facility i and customer j. The allocations w_{ij} are also unknown and represent the amount to be shipped from facility i to customer j with the unit shipment cost per unit distance being c_{ij}. This formulation assumes that the problem is balanced, i.e., the total supply is equal to the total demand. If the total supply is larger than the total demand, the problem can be balanced by adding a dummy customer with zero unit shipment cost. In case the total supply is less than the total demand, there exists no feasible solution. From this formulation it is clear that the demand of a customer can be satisfied from different facilities. In other words, the CMFWP is a multi-source problem. When, due to some additional considerations, each customer has to be served by a single facility, the problem is formulated as a single-source CMFWP by making use of additional binary variables that keep track of which customer is assigned to which facility.

Note that when allocations w_{ij} are known, the CMFWP reduces to a pure location problem that is separable into m single-facility location problems, each of which can be solved by Weiszfeld's algorithm (Weiszfeld, 1937) and its generalizations (Brimberg and Love, 1993). On the other hand, when the locations of the facilities are given, the CM-FWP becomes the classical transportation problem. As a consequence, an optimal solution to the CMFWP always occurs at an extreme point of the transportation polyhedron (5.2)–(5.4), independent of the type of the distance function used. This characteristic of the CMFWP was shown by Cooper (1972). Although pure location and transportation subproblems are easy to solve, the CMFWP belongs to a difficult class of problems. Sherali and Nordai (1988) have shown that the CMFWP with the Euclidean distance is NP-hard even if all the customers are located on a straight-line.

For the last two decades metaheuristics have successfully been applied to the solution of various combinatorial optimization problems. To the best of our knowledge, apart from the strategy of Cooper (1976) that can be seen as a kind of variable neighborhood search, there is no published work on the application of a metaheuristic strategy to the CMFWP. Motivated by this fact, we propose three heuristics which are based on simulated annealing, threshold accepting, and genetic algorithms. In Section 2 we provide a literature review on the existing solution methods of the problem. The new heuristics are described in Section 3. Section 4 contains the computational results that are obtained on a set of benchmark instances. Concluding remarks and directions for future research are given in Section 4.

2.   Literature Review

In his seminal work, Cooper (1972) considered the Euclidean distance CMFWP (ECMFWP) and proposed an exact solution method based on the complete enumeration of all the extreme points of the transportation polyhedron. Since the number of extreme points can be very large, this method is useful only for very small instances. For larger instances, he suggested the alternating transportation-location heuristic that is based on the idea of decomposing the ECMFWP into location and transportation (allocation) subproblems. When the locations of the facilities are fixed, the resulting transportation problem is solved to determine the corresponding optimal capacity allocations. Then, using these allocations new optimal locations are determined for the facilities. The location and transportation problems are alternately solved until no improvement is possible. It is important to note that the solution method of the single-facility location problems depends on the type of the distance function. The median location method (Francis et al., 1992) can be employed to solve the rectilinear distance single-facility location problems in the case of the rectilinear CMFWP (RCMFWP). It has been shown that Weiszfeld's algorithm solves the ℓ_p distance single-facility location problem to optimality for $1 \leq p \leq 2$ (Brimberg and Love, 1993). As a result, Cooper's alternating two-phase idea can also be used to provide approximate solutions to the CMFWP with the ℓ_p distance function (LpCMFWP) for $1 \leq p \leq 2$.

Later, Cooper (1976) proposed a more efficient heuristic for the ECM-FWP, which performs a local search in the space of the set of extreme points of the transportation polyhedron. This is done in the neighborhood of a given basic feasible solution where the neighbor solutions are generated by moving to extreme points which are one, two or three steps away from the current one. In other words, when one of the nonbasic variables is exchanged with a basic variable (i.e., a simplex iteration is carried out), an adjacent extreme point is reached. When two nonbasic variables are inserted into the basis simultaneously, then an extreme point adjacent to the immediate neighbor of the current extreme point is obtained. This heuristic can be regarded as an early implementation of the variable neighborhood search idea (Hansen and Mladenović, 2001).

To the best of our knowledge, apart from Cooper's complete enumeration algorithm (Cooper, 1972), there are two exact methods to solve the ECMFWP. The first one is a biconvex cutting plane procedure and can be found in Selim's unpublished dissertation (Selim, 1979). Although it is more efficient than Cooper's complete enumeration, this procedure can effectively solve only very small instances. The second exact method appears in the recent work by Sherali et al. (2002), in which the authors design a global optimization procedure for the ECMFWP. This is a branch-and-bound algorithm based on partitioning the allocation space, which finitely converges to a global optimum within a specified percentage tolerance. To derive lower bounds on the subproblems obtained at the nodes of the branch-and-bound tree, the authors use two approaches. The first approach involves computing a lower bound via a projected location space subproblem. In the second approach, a specialized variant of the Reformulation Linearization Technique (RLT) (Sherali and Adams, 1999) is applied to transform an equivalent representation of the original nonconvex problem into a higher dimensional linear programming relaxation. They have shown that the latter approach provides much better lower bounds than the former one. Upper bounds are obtained by using Cooper's alternating transportation-location heuristic. The generalization of this exact method to the solution of the LpCM-FWP is also provided.

The application of the RLT in the solution of location and allocation problems is not new. Sherali and Tunçbilek (1992) apply this approach to solve the squared Euclidean distance CMFWP (SECMFWP). As a more recent application, Sherali et al. (1994) consider a mixed-integer nonlinear programming formulation of the RCMFWP and use the RLT to linearize it. This formulation is based on a useful property of the two dimensional rectilinear distance location problem. Namely, optimal locations always occur within the convex hull of the customer locations and at the intersection points of the vertical and horizontal lines drawn through them (Hansen et al., 1980).

The disadvantage of the exact solution methods mentioned above is that they become computationally intensive as the number of variables is increased throughout the procedure. Therefore, efficient heuristic methods are required to solve large-sized instances accurately.

3. New Heuristics for the CMFWP

We mentioned previously that an optimal solution to the CMFWP occurs at an extreme point of the convex feasible region determined by the constraints (2)–(4). Hence, any heuristic designed on this feature of the problem should perform a search in the space of extreme points that correspond to feasible solutions. Note, however, that each extreme point specifies only values for the allocation variables w_{ij}. In order to compute the objective value of an extreme point solution, it is also necessary to determine the values for the location variables $\mathbf{x}_i$. This can be done by solving as many single-facility location problems as the number of facilities once the values of w_{ij} are fixed.

Heuristic based on simulated annealing

The origin of the Simulated Annealing (SA) algorithm is in statistical mechanics. The idea of using the annealing method in optimization problems is due to Kirkpatrick et al. (1983). Since then, SA and its variants are applied to numerous combinatorial and continuous optimization problems.

Given a combinatorial optimization problem with a finite set of solutions and an objective function, the SA algorithm is characterized by a rule to randomly generate a new solution in the neighborhood of the current solution. The new solution is accepted if $\min\left\{1, e^{-\Delta/T}\right\} \geq p$ where p is a uniform random number in the interval $(0, 1)$, and $\Delta = f(\mathbf{s}^n) - f(\mathbf{s}^c)$ is the difference between the objective values of the new solution $\mathbf{s}^n$ and the current solution $\mathbf{s}^c$. Note that if Δ is negative, i.e., then $\min\left\{1, e^{-\Delta/T}\right\} = 1$, which means that the new solution is always accepted. A certain number of iterations L are performed at fixed temperature T and the temperature is reduced according to a cooling schedule. This implies that for a minimization problem, the probability of accepting uphill moves is high at the beginning of the search leading to the exploration of the search space. Then it decreases slowly so that SA becomes a simple iterative improvement algorithm.

As the search must be performed in the space of extreme points, we have to identify a neighborhood structure to move from the current extreme point (current solution) to another one (new solution) in the vicinity of the former. It can easily be seen that the constraints of the CMFWP are the same as those of the well-known transportation problem. An important result in linear programming states that each extreme point can be characterized by $m + n - 1$ basic variables and $mn - (m + n - 1)$ nonbasic variables when the number of constraints is $m + n$. As a matter of fact, each iteration of the transportation simplex method involves moving from the current extreme point to an adjacent one that results in the largest improvement in the objective value. This is achieved by determining an entering variable among the nonbasic variables and a leaving variable among the basic variables. We refer to this operation as the one-variable exchange. The main components of the simulated annealing heuristic is explained below.

To start the SA algorithm we need to have an initial solution. This corresponds to generating an initial extreme point in our problem. We do this by applying the Northwest corner rule (Bazaraa et al., 1990).

The objective value of the initial solution is found by solving a single-facility location problem for each facility. An important consideration is to set the initial value of the temperature T_0. This parameter is instance-dependent and it is not appropriate to always use the same value. Therefore we implement the idea suggested in Ohlmann and Thomas (2006). First we generate n (random) solutions sequentially by starting with the initial solution and applying the one-variable exchange n times. We also calculate the *absolute* differences between the objective values of consecutive solutions. Lastly, we obtain the average of these n differences and assign it to $\bar{\Delta}$, which is ultimately used in determining the value of T_0. The idea is to set a value to T_0 such that the probability of accepting an average bad move early in the algorithm is equal to p_0. Algebraically, $p_0 = e^{-\bar{\Delta}/T_0}$ which implies that $T_0 = \bar{\Delta}/\ln p_0$.

In the SA-based heuristic we use two different neighborhood structures. The first one implements the one-variable exchange while the other one implements the two-variable exchange. In one-variable exchange we randomly select one of the nonbasic variables in the current solution and make it the entering variable. The leaving variable is determined by the stepping stone method (Bazaraa et al., 1990). The neighborhood size is $NS_1 = mn - (m + n - 1)$ because this equation gives the number of nonbasic variables in a basic feasible solution. Figure 5.1 demonstrates the network representation of a one-variable exchange. The edges between nodes i and j mean that the corresponding allocation variable w_{ij} between facility i and customer j is positive.

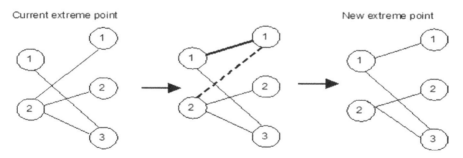

Current extreme point   New extreme point

Figure 5.1. One-variable exchange for SA heuristic.

The two-variable exchange is the application of the one-variable exchange twice. Hence, it corresponds to moving from the current extreme point to a nonadjacent extreme point by performing the one-variable exchange twice. In Figure 5.2 we illustrate the application of the two-variable exchange. First, nonbasic variable w_{11} is added to the basis, which leads to the removal of the basic variable w_{21}. The one-variable exchange is applied once more by selecting nonbasic variable w_{23} as the entering variable and determining basic variable w_{32} as the leaving one. It is important to emphasize that the neighborhood size defined by the two variable exchange is larger than that of the one-variable exchange and is given as $NS_2 = \binom{NS_1}{2}$. In each run of the SA-based heuristic, we use either the one-variable exchange or the two-variable exchange as the neighborhood structure. It should be pointed out, however, that it is also possible to define the neighborhood structure as a mixture of both exchange methods.

As mentioned before, to compute the objective value corresponding to a solution we have to solve m single-facility location problems. In case of the SECMFWP each location subproblem can be solved analytically by computing the flow-weighted centroid of customer locations. For the RCMFWP we use the median location method. The ECMFWP is solved via the Weiszfeld procedure, while we apply a generalization of the Weiszfeld procedure (Brimberg and Love, 1993) for the LpCMFWP. If the new solution is accepted, then it becomes the current solution. Furthermore, we label it as the best solution if its objective value is lower than that of the best solution found so far.

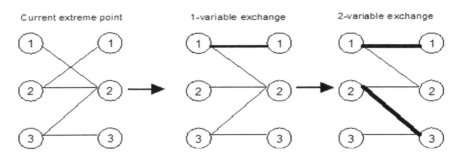

Figure 5.2. Two-variable exchange for SA-based heuristic.

Cooling schedule and the number of iterations to be performed at each temperature (referred to as a cycle) are two important issues in designing a simulated annealing based heuristic. In our implementation the number of iterations L_k performed at temperature T_k is defined as $L_k = r \cdot NS_i$ where r is a parameter that should be tailored to the problem instance and NS_i is the size of the neighborhood with respect to the one-variable exchange ($i = 1$) and two-variable exchange ($i = 2$). After L_k iterations are carried out in each cycle, the value of the temperature is decreased in a geometric manner given as $T_{k+1} = \alpha \cdot T_k$

where $\alpha \in (0,1)$ is called the cooling rate. If the value of α is close to one, cooling occurs slower, which implies that a larger portion of the search space of extreme points is explored.

When the ratio of accepted solutions to all the solutions generated at constant temperature T is less than a threshold value for K consecutive cycles, then the algorithm is terminated. A pseudocode of the SA-based heuristic is given next.

0. Generate an initial solution s^0 by Northwest corner rule and calculate $f(s^0)$. Let $s^{best} = s^0$, $f^{best} = f(s^0)$, $T_0 = \bar{\Delta}/\ln p_0$, $k = 0$, $s^c = s^0$

1. $iter = 0$
   While $iter \leq L_k$ Do
   Generate a solution s^n in the neighborhood of s^c and calculate $f(s^n)$
   Calculate $\Delta = f(s^n) - f(s^c)$.
   If $\min\left\{1, e^{-\Delta/T_k}\right\} \geq U(0,1)$, then $s^c = s^n$
   If $f(s^n) < f^{best}$, then $s^{best} = s^n$ and $f^{best} = f(s^n)$
   $iter \leftarrow iter + 1$

2. $T_{k+1} = \alpha T_k$, $k \leftarrow k + 1$
   If termination criterion is not satisfied go to Step 1

Heuristic based on threshold accepting

Threshold accepting (TA) was first introduced by Dueck and Scheuer (1990) as a deterministic version of the SA algorithm. The difference is in the acceptance rule of inferior solutions. In TA, a new solution is accepted if the difference between the objective values of the new and current solutions is not greater than a threshold term θ. This implies that better solutions are always accepted while worse solutions are also accepted if their objective value is within a certain threshold from the current objective value. In general, θ is gradually decreased to zero as the heuristic proceeds.

In our implementation we adopt the approach used by Yan and Luo (1999) in which the threshold θ is represented as a fraction of the current objective value $f(s^c)$, i.e., $\theta = \mu f(s^c)$ where $\mu \in (0,1)$. Hence the condition for the acceptance of a new solution is that $f(s^n) - f(s^c) = \Delta \leq \mu f(s^c)$. The value of parameter μ is decreased geometrically by multiplying it by coefficient α every L_k iterations, i.e., $\mu_{k+1} = \alpha \mu_k$. The initial value of μ, μ_0, is set by a similar procedure used in the SA-based heuristic for setting the initial value T_0 of the temperature. First, we generate n random solutions sequentially by starting with the initial solution that is obtained by the Northwest corner rule and applying the one-variable exchange n times. For each move we compute the corresponding value of μ using the expression $\mu = |\Delta|/f(s^c)$ that follows from the definition of the acceptance rule. Using the mean and standard deviation of μ values we determine $\mu_0 = \bar{\mu} + 2\sigma_{\bar{\mu}}$.

In the TA-based heuristic we employ the same neighborhood structures that were used in the SA-based heuristic, i.e., the one-variable exchange and two-variable exchange. The steps of the TA-based heuristic are given below.

0.   Generate an initial solution s^0 by Northwest corner rule and calculate $f\left(s^0\right)$. Let $s^{best} = s^0$, $f^{best} = f\left(s^0\right)$, $k = 0, s^c = s^0$

1. $iter = 0$

   While $iter \leq L_k$ Do

      Generate a solution s^n in the neighborhood s^c and calculate $f\left(s^n\right)$

      Calculate $\Delta = f\left(s^n\right) - f\left(s^c\right)$.

      If $\Delta \leq \mu_k f\left(s^c\right)$, then $s^c = s^n$

      If $f\left(s^n\right) < f^{best}$, then $s^{best} = s^n$ and $f^{best} = f\left(s^n\right)$

      $iter \leftarrow iter + 1$

2. $\mu_{k+1} = \alpha\mu_k$, $k \leftarrow k + 1$

   If termination criterion is not satisfied go to Step 1

Heuristic based on genetic algorithms

Genetic algorithms mimic the genetic evolution of a species. The main difference with the SA and TA algorithms is that GAs do not explore the neighborhood of a single solution but perform a search in the neighborhood of a population of solutions. Throughout the algorithm, solutions that are also called individuals or chromosomes take part in a reproductive process in which they interact, mix together and produce offspring that retain the good characteristics of their parents. The reproductive process which involves the creation of new solutions is based on the selection, crossover, and mutation operators. Depending on the optimization problem at hand, there is a need to encode the solutions as strings so that the three genetic operators can be applied.

There are earlier works that propose GA-based heuristics to solve the fixed charge transportation problem (FCTP) which takes into account not only the linear costs as in the case of the classical transportation problem but also fixed costs. From our perspective it is important that the FCTP has the same constraints as the CMFWP. Different representations were applied to the FCTP such as the Prüfer number representation (Li et al., 1998), matrix representation (Gottlieb and Paulmann, 1998; Gottlieb and Eckert, 2000), permutation representation (Gottlieb and Paulmann, 1998) and direct or edge-based representation (Eckert and Gottlieb, 2002). Eckert and Gottlieb (2002) show that the edge-based representation not only outperforms the others in terms of solution quality but also exhibits superior performance with respect to the locality and heritability properties (Gottlieb et al., 2001). On the basis of these results we opt for using this representation for our GA-based heuristic.

The starting point of the edge-based representation is that an extreme point with $m+n-1$ basic variables corresponds to a spanning tree with $m+n-1$ edges in the underlying transportation graph. This means that each extreme point can be represented by a chromosome that consists of the set of edges of the corresponding spanning tree and the associated values of the allocation (flow) variables. For example, the chromosome consisting of edges $\{(1,1),(1,2),(2,2),(2,3)\}$ would encode the extreme point which has w_{11}, w_{12}, w_{22}, w_{23} as the basic variables.

The fitness function is taken the same as the objective function of the CMFWP. Hence, the fitness of each chromosome can be calculated by first solving the location subproblems and computing the objective value, as was the case in the SA-based and TA-based heuristics. The main components of the GA-based heuristic are explained in detail below.

After the population size is fixed, the first generation of solutions is created by applying the one-variable exchange successively to the solution found by the Northwest corner rule. During this phase, if a newly generated chromosome encodes an already existing solution, it is discarded to eliminate duplicate solutions within the population. For each parent, we calculate the fitness function by solving the location subproblems. Parents that will take part in the reproduction process are determined using the binary tournament selection, where two solutions are randomly picked from the population and the better solution is selected.

The crossover operator is based on the idea that the offspring should be formed from the edges of the parents and it has to be roughly at the same distance from both parents. Given parents P_1 and P_2, about half of the edges of parent P_2 that do not exist in P_1 are selected. These edges are then inserted into the chromosome of P_1 one by one. At each step a selected edge of P_2 is inserted into P_1 while one of the edges creating a cycle in the transportation tree of P_1 is deleted. The edge to be removed corresponds to the leaving basic variable that is determined by the stepping stone method that was also applied during the one-variable exchange in SA and TA-based heuristics. The main advantage of this method is that we always get a basic feasible solution after one move. The crossover operator is illustrated in Figure 5.3 where the set of edges from P_2 that do not exist in P_1 is given as $E(P_2) \setminus E(P_1) = \{(1,1),(2,3),(3,1)\}$. Two of them, i.e., edges $(1,1)$ and $(2,3)$ depicted in bold lines in the figure, are selected randomly and inserted sequentially into the first parent. Edges $(2,1)$ and $(3,2)$ shown in dashed lines are removed from the first parent in order to obtain a feasible solution.

The mutation operator applied to the offspring created by the crossover operator is defined as follows. A new edge is selected randomly from

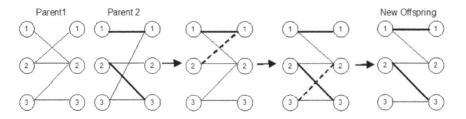

Figure 5.3.   Crossover operator.

the set of edges that do not exist in the offspring and added to it. As this
addition creates a cycle in the transportation tree, we apply again the
stepping stone method to determine the edge that should be deleted to
form a new tree. As soon as an offspring (an extreme point in the space
of allocation variables **w**) is generated using the crossover and mutation
operators, we solve the location subproblems to find the coordinates of
the facilities and consequently the corresponding fitness value. We use a
steady-state replacement scheme where the offspring replaces the worst
individual in the population if the following two conditions are satisfied.
First, the fitness of the offspring (i.e., its cost) should be lower than that
of the worst individual. Second, the offspring should not have duplicates
in the population. Hence the population in the next generation differs
from the current one in at most one solution.

When the number of generations reaches a certain value, we terminate
the algorithm and the individual with the best fitness value is reported
as the solution of the GA-based heuristic. In general, a larger number
of generations results in a better solution at the expense of increased
computation time.

4.   Computational Results

In this section we assess the performance of the new heuristics both in
terms of solution quality and running time efficiency on a number of test
instances. The instances are classified in five different groups and they
are referred to by the same labels as used in the original papers. The first
group consists of RCMFWP instances. Instances R8, R9 and R15 are
from Sherali et al. (2002) while instances R16, R23, R26, R29, R30 are
from Sherali et al. (1994). The second group includes small ECMFWP
instances (instances E2–E8) while the third group contains larger-sized
ECMFWP instances (instances E9–E20). These instances are given in
Al-Loughani (1997) and in Sherali et al. (2002). The fourth group

consists of three SECMFWP instances given in Sherali and Tunçbilek (1992). Finally, three LpCMFWP instances with $p = 1.25$, $p = 1.50$, and $p = 1.75$ obtained from Sherali et al. (2002) form the last group.

Table 5.1 displays all the instances, where prefixes "R", "E", "SE", and "Lp" denote respectively the rectilinear, Euclidean, squared Euclidean, and ℓ_p distances. For each instance we provide the number of facilities to be located (m), the number of customers (n), the best objective value and the CPU time required to find this value. It is important to note that the CPU times are measured in different hardware configurations. R16, R23, R26, R29, R30 and SECMFWP instances were solved on an IBM 3090 (Sherali et al., 1994; Sherali and Tunçbilek, 1992) whereas the experiments for R8, R9 and R15 as well as all ECMFWP and LpCMFWP instances were conducted on a Sun Ultra 1 workstation having 256 Megabytes of RAM (Sherali et al., 2002). We therefore convert the original CPU times into equivalent CPU seconds that would be required if all these experiments were carried out on our hardware configuration. To this end, we use the performance measures given in Mflops/s units provided in Dongarra (2006) and report the converted CPU times in the last column of Table 5.1.

The new heuristics are coded in Microsoft Visual Basic 6.0 and run on a notebook computer with 1.7 GHz Pentium Centrino processor and 256 Megabytes of RAM. Each subsection below is dedicated to the results obtained by running one of the new heuristics 10 times. For each instance we present the results in terms of the best, average, and worst percent deviation from the best known objective value as well as the CPU time needed for all 10 runs.

SA-based heuristic

We mentioned previously that in the implementation of the SA-based heuristic the number of iterations performed at a fixed temperature is determined as a constant times the neighborhood size, i.e., $L_k = r \cdot N S_i$. We set $r = 4$ for the one-variable exchange and $r = 1$ for the two-variable exchange. The value of the cooling rate α is taken equal to 0.9. The algorithm is terminated when the percentage of accepted solutions is less than 5% for five consecutive cycles. The results are presented in Table 5.2.

The results clearly indicate that if the two-variable exchange is used as the neighborhood structure, much better solutions can be obtained in comparison with the one-variable exchange. However, the computational effort is also higher for the two-variable exchange. Clearly there is a compromise between the solution quality and CPU time with respect

Table 5.1. Test instances.

Instance	(m,n)	Best known	CPU time	Converted CPU time
R8	(4,8)	793	107.30 sec	53.65 sec
R9	(5,15)	9619	419.13 sec	209.57 sec
R15	(5,10)	3427	158.00 sec	79.00 sec
R16	(4,10)	259	28.43 sec	7.11 sec
R23	(5,8)	238	26.34 sec	6.59 sec
R26	(5,12)	284	203.08 sec	50.77 sec
R29	(5,15)	729	310.28 sec	77.57 sec
R30	(5,20)	745	35.02 sec	8.76 sec
Avg.				**61.63 sec**
E2	(2,4)	247.28	0.20 sec	0.10 sec
E3	(2,4)	214.34	0.90 sec	0.45 sec
E4	(3,5)	24.00	2.30 sec	1.15 sec
E5	(3,5)	73.96	2.00 sec	1.00 sec
E6	(3,9)	221.40	66.40 sec	33.20 sec
E7	(3,9)	871.62	42.20 sec	21.10 sec
E8	(4,8)	609.23	6 min	3 min
Avg.				**33.86 sec**
E9	(5,15)	8169.79	23 min	11.50 min
E10	(5,20)	12846.87	134 min	67 min
E11	(5,20)	1107.18	73 min	36.50 min
E15	(5,10)	2595.47	8 min	4 min
E16	(6,10)	7797.21	9 min	4.50 min
E17	(7,10)	6967.90	315 min	157.50 min
E18	(8,10)	1564.46	468 min	234 min
E19	(9,10)	3250.68	12 min	6 min
E20	(10,10)	7719.00	462 min	231 min
Avg.				**83.56 min**
SE9	(4,8)	875.34	227.09 sec	56.77 sec
SE16	(4,15)	3591.53	14.75 sec	3.69 sec
SE21	(4,24)	6805.43	98.02 sec	24.51 sec
Avg.				**28.32 sec**
Lp8, $p{=}1.25$	(4,8)	710.20	114.49 sec	57.25 sec
Lp8, $p{=}1.50$	(4,8)	661.90	126.71 sec	63.36 sec
Lp8, $p{=}1.75$	(4,8)	630.72	141.94 sec	70.97 sec
Lp9, $p{=}1.25$	(5,15)	8998.93	998.44 sec	499.22 sec
Lp9, $p{=}1.50$	(5,15)	8609.12	588.50 sec	294.25 sec
Lp9, $p{=}1.75$	(5,15)	8350.95	806.43 sec	403.22 sec
Lp5, $p{=}1.25$	(5,10)	3046.07	254.37 sec	127.19 sec
Lp15, $p{=}1.50$	(5,10)	2827.55	294.39 sec	147.29 sec
Lp15, $p{=}1.75$	(5,10)	2689.12	215.42 sec	107.71 sec
Avg.				**196.72 sec**

Table 5.2. Results obtained by the SA-based heuristic.

Instance	One-variable exchange				Two-variable exchange			
	Best %Dev.	Avg. %Dev.	Worst %Dev.	CPU time (sec)	Best %Dev.	Avg. %Dev.	Worst %Dev.	CPU time (sec)
R8	0.00	0.21	2.14	6	0.00	0.21	2.14	21
R9	0.00	0.05	0.16	31	0.00	0.00	0.02	257
R15	0.00	3.87	4.61	18	0.00	1.74	4.26	114
R16	0.00	8.26	35.52	10	0.00	1.24	9.27	43
R23	0.00	0.00	0.00	10	0.00	0.00	0.00	46
R26	1.41	10.00	26.06	19	0.00	2.54	7.75	135
R29	0.96	6.30	11.52	27	0.00	1.04	3.43	270
R30	0.94	2.64	4.02	48	0.67	2.02	3.75	605
Avg.	**0.41**	**3.92**	**10.50**	**21**	**0.08**	**1.10**	**3.83**	**186**
E2	0.00	0.00	0.00	1	0.00	0.00	0.00	1
E3	0.00	0.00	0.00	2	0.01	0.01	0.00	1
E4	0.00	0.00	0.00	8	0.00	0.01	0.06	10
E5	0.00	0.00	0.00	11	0.00	0.00	0.00	13
E6	0.00	0.01	0.06	16	0.00	0.00	0.00	95
E7	0.00	0.00	0.00	23	0.00	0.00	0.00	81
E8	0.00	1.98	8.15	51	0.00	0.21	1.34	163
Avg.	**0.00**	**0.28**	**1.17**	**16**	**0.00**	**0.03**	**0.20**	**52**
E9	0.00	0.58	3.39	525	0.00	0.00	0.99	3823
E10	0.00	0.05	0.09	413	0.00	0.00	0.60	5479
E11	0.00	20.31	41.19	605	0.00	4.01	25.37	8217
E15	0.00	8.17	24.32	110	0.00	0.24	1.22	647
E16	0.00	5.56	9.41	153	0.00	4.04	11.07	1147
E17	3.61	10.18	15.39	184	0.00	2.47	4.22	1782
E18	2.45	38.83	89.60	408	0.00	3.38	9.24	3287
E19	11.55	20.11	43.59	451	0.00	8.08	15.35	4778
E20	0.01	3.65	14.55	724	0.01	0.01	0.01	10087
Avg.	**1.96**	**11.94**	**26.84**	**397**	**0.00**	**2.47**	**7.56**	**4361**
SE9	0.00	1.75	15.03	8	0.00	0.00	0.00	26
SE16	0.00	0.69	6.34	24	0.00	0.16	0.92	164
SE21	0.00	11.13	21.31	55	4.72	8.14	10.74	590
Avg.	**0.00**	**4.52**	**14.23**	**29**	**1.57**	**2.77**	**3.89**	**260**
Lp8,p=1.25	0.00	3.84	7.41	127	0.00	0.48	2.52	493
Lp8,p=1.50	0.00	2.16	7.36	133	0.00	0.00	0.01	412
Lp8,p=1.75	0.00	1.46	8.85	111	0.00	0.12	1.15	335
Lp9,p=1.25	0.01	0.17	0.73	1933	0.01	0.01	0.03	15771
Lp9,p=1.50	0.00	0.23	1.45	1721	0.00	0.00	0.40	11171
Lp9,p=1.75	0.00	0.09	0.42	1382	0.00	0.00	0.01	11300
Lp15,p=1.25	0.00	4.58	6.69	260	0.00	1.77	11.00	1634
Lp15,p=1.50	0.00	7.05	8.61	288	0.00	1.77	15.30	1768
Lp15,p=1.75	1.21	8.72	10.68	251	0.00	1.22	9.78	1799
Avg.	**0.14**	**3.14**	**5.80**	**690**	**0.00**	**0.60**	**4.47**	**4965**

to the two neighborhood structures. The best and average percent deviations remain respectively within 1.96% and 11.94% for the one-variable exchange and within 1.57% and 2.77% for the two-variable exchange. It is possible to obtain the best solution with the two-variable exchange neighborhood structure for all instances except instance R30 and SE21.

TA-based heuristic

In the TA-based heuristic parameters r and α are assigned the same values as those used in the SA-based heuristic. The termination criterion is also the same. Table 5.3 summarizes the results.

We can observe that the results are in parallel to those obtained in the previous section in the sense that the two-variable exchange neighborhood provides better solutions than the one-variable exchange at the expense of more computation time. When we compare the TA-based heuristic with the SA-based heuristic with respect to the solution quality, we can observe that the former one yields slightly inferior solutions.

In order to have a better understanding of the trade-off between the solution quality and computation time with respect to the neighborhood structures, we carried out additional experiments by running the heuristics with the one-variable exchange for an amount of time that is required by the SA-based heuristic with the two-variable exchange. Although there have been some improvements in the solutions, i.e., smaller best, average, and worst percent deviations are obtained, the results are still inferior to those given by the two-variable neighborhood structure.

GA-based heuristic

In the implementation of the GA-based heuristic, the population size is taken as the minimum of 100 and the lower bound on the number of extreme points given as $n!/(n-m+1)!$ in Cooper (1976). As mentioned before, the parent selection is performed with binary tournament method and crossover as well as mutation operators are applied with probability one. A steady-state replacement scheme is employed where the offspring replaces the worst individual in the population provided that its fitness value is lower and there is no duplicate of the offspring in the population.

An important point here is that in order to have a better comparison between the GA-based heuristic and the SA-based heuristic which appears to be slightly better than the TA-based heuristic, we limit the running time of the GA-based heuristic by the CPU time of the SA-based heuristic with the two-variable exchange. Therefore, a different

Table 5.3. Results obtained by the TA-based heuristic.

	One-variable exchange				Two-variable exchange			
Instance	Best %Dev.	Avg. %Dev.	Worst %Dev.	CPU time (sec)	Best %Dev.	Avg. %Dev.	Worst %Dev.	CPU time (sec)
R8	0.00	3.33	14.00	11	0.00	0.00	0.00	26
R9	0.00	0.08	0.31	79	0.00	0.00	0.00	306
R15	0.00	4.29	9.69	27	0.00	2.45	4.26	93
R16	0.00	11.97	33.98	13	0.00	4.32	17.76	42
R23	0.00	0.25	2.52	35	0.00	0.00	0.00	122
R26	0.00	19.51	28.52	60	0.00	3.17	17.61	285
R29	0.00	4.09	11.52	84	0.00	1.18	4.25	482
R30	0.00	1.91	2.68	145	0.00	2.11	4.56	1240
Avg.	**0.00**	**5.68**	**12.90**	**57**	**0.00**	**1.65**	**6.06**	**325**
E2	0.00	0.00	0.00	1	0.00	0.00	0.00	1
E3	0.00	0.00	0.00	1	0.00	0.02	0.21	1
E4	0.00	0.00	0.00	15	0.00	0.00	0.00	17
E5	0.00	0.00	0.00	10	0.00	0.00	0.00	12
E6	0.00	0.01	0.08	12	0.00	0.00	0.00	94
E7	0.00	4.98	13.75	32	0.00	0.00	0.00	80
E8	0.00	5.44	31.72	49	0.00	0.14	1.4	141
Avg.	**0.00**	**1.49**	**6.51**	**17**	**0.00**	**0.02**	**0.23**	**49**
E9	0.00	3.66	34.31	566	0.00	0.06	0.56	3611
E10	0.00	0.06	0.09	179	0.00	0.00	0.00	2657
E11	0.00	29.41	62.08	423	0.00	7.62	25.37	5310
E15	0.00	14.12	23.53	100	0.00	0.28	1.22	561
E16	0.00	5.98	15.43	119	3.09	4.59	6.33	1001
E17	0.00	1.75	6.80	227	0.00	2.39	10.66	1865
E18	0.00	6.27	30.99	350	0.00	2.91	19.01	2897
E19	11.7	20.62	54.51	511	0.00	3.8	13.3	5415
E20	0.01	5.25	33.30	1372	0.01	0.07	0.61	15121
Avg.	**1.30**	**9.68**	**29.00**	**427**	**0.34**	**2.41**	**8.56**	**4271**
SE9	0.00	9.93	34.61	20	0.00	0.00	0.00	66
SE16	0.00	10.93	64.63	71	0.00	1.27	6.33	390
SE21	3.04	13.98	16.64	126	0.00	7.01	15.55	1170
Avg.	**1.01**	**11.61**	**38.63**	**72**	**0.00**	**2.76**	**7.29**	**542**
Lp8,p=1.25	0.00	0.31	3.03	94	0.00	0.44	2.23	298
Lp8,p=1.50	0.00	3.42	9.69	112	0.00	0.00	0.03	277
Lp8,p=1.75	0.00	4.23	21.48	90	0.00	0.41	4.06	199
Lp9,p=1.25	0.01	0.10	0.91	1580	0.01	0.01	0.03	11821
Lp9,p=1.50	0.00	0.22	0.44	1926	0.00	0.04	0.44	8504
Lp9,p=1.75	0.00	7.54	37.19	1266	0.00	0.00	0.01	8356
Lp15,p=1.25	0.00	8.55	11.05	176	0.00	3.54	11.05	730
Lp15,p=1.50	0.00	6.73	16.26	189	0.00	1.60	15.95	845
Lp15,p=1.75	0.00	13.61	20.33	169	0.00	0.47	2.29	955
Avg	**0.00**	**4.97**	**13.38**	**622**	**0.00**	**0.72**	**4.01**	**3554**

value is reported in Table 5.4 for the number of generations associated with each problem instance.

The best, average, and worst percent deviations provided by the GA-based heuristic are within 1.89%, 9.16%, and 16.90%, respectively. These results are approximately the same as those obtained by the SA-based heuristic with the one-variable exchange when it is run for a time limit identical to SA with two-variable exchange. Hence, we can conclude that all the three heuristics yield solutions of more or less the same quality when they are allowed the same amount of time to run.

When the two-variable exchange is used as the mutation operator, the results are not changed for most of the problem instances while insignificant improvements are obtained for others. Most probably, this is related to the steady-state replacement scheme adopted in this study. As a result, we can say that when implemented with the two-variable exchange as the mutation operator, the GA-based heuristic is outperformed by the SA-based heuristic for which the best, average, and worst percent deviations are 1.57%, 2.77%, and 7.56%, respectively.

5.   Conclusions

In this paper we develop three heuristics based on simulated annealing, threshold accepting, and genetic algorithms for the solution of the NP-hard CMFWP with rectilinear, Euclidean, squared Euclidean and ℓ_p distances. All the heuristics we propose are based on the characteristic of the CMFWP that an optimal solution always occurs at an extreme point of the feasible region defined by constraints of the problem. Since each extreme point corresponds to a set of feasible values for the allocation variables and the CMFWP reduces to single-facility location problems when the values of the allocation variables are fixed, all the heuristics are designed to perform a search in the space of extreme points. The objective value corresponding to an extreme point can be calculated by solving the single-facility location problems by a suitable method depending on the distance function used.

The SA-based, TA-based, and GA-based heuristics are tested in terms of both their solution quality and computation time on benchmark instances available in the literature. The results are very accurate for the SA-based and TA-based heuristics with the two-variable exchange neighborhood structure. We also observe that when the one-variable exchange is adopted as the neighborhood structure in the TA-based and SA-based heuristics and as the mutation operator in the GA-based heuristic, there is no clear-cut difference among the three heuristics. Moreover, the GA-based heuristic does not benefit from the two-variable exchange used

Table 5.4. Results obtained by the GA-based heuristic.

Instance	Best % Dev.	Average % Dev.	Worst % Dev.	No. of Generations
R8	0.00	0.00	0.00	3658
R9	0.00	0.05	0.16	25383
R15	0.00	2.98	4.26	15200
R16	0.00	7.80	18.53	6880
R23	0.00	0.00	0.00	8000
R26	0.00	4.08	22.89	16615
R29	0.00	4.80	11.25	25414
R30	2.01	3.07	11.41	42461
Avg	**0.25**	**2.85**	**8.56**	**17951**
E2	0.00	0.00	0.00	531
E3	0.00	0.00	0.00	263
E4	0.00	19.35	29.68	1709
E5	0.00	0.00	0.00	2152
E6	0.00	0.01	0.07	5315
E7	0.00	0.00	0.00	9971
E8	0.00	1.56	7.11	8936
Avg	**0.00**	**2.99**	**5.27**	**4125**
E9	0.00	0.00	0.00	82217
E10	0.00	0.03	0.08	148591
E11	16.96	28.63	42.79	261901
E15	0.00	8.47	23.53	43850
E16	0.00	4.82	10.70	48809
E17	0.00	4.64	15.39	69881
E18	0.00	19.64	33.93	119525
E19	0.07	14.33	16.34	142629
E20	0.01	1.90	9.30	286132
Avg.	**1.89**	**9.16**	**16.90**	**133726**
SE9	0.00	0.00	0.00	4960
SE16	0.00	1.90	6.34	18480
SE21	2.36	13.21	21.24	42910
Avg.	**0.79**	**5.04**	**9.19**	**22117**
Lp8,p=1.25	0.00	0.00	0.01	14880
Lp8,p=1.50	0.00	0.00	0.01	16913
Lp8,p=1.75	0.00	0.83	8.26	17385
Lp9,p=1.25	0.02	0.09	0.34	1005
Lp9,p=1.50	0.00	0.00	0.00	1517
Lp9,p=1.75	0.00	0.14	1.05	1221
Lp15,p=1.25	0.00	7.62	11.05	2169
Lp15,p=1.50	0.00	11.38	16.26	1891
Lp15,p=1.75	0.00	12.20	20.33	1828
Avg.	**0.00**	**3.58**	**6.37**	**6534**

as the mutation operator. It produces worse results than the SA-based heuristic employed with the two-variable exchange.

In fact, there is another alternative way to design heuristics for the CMFWP. Instead of performing the search in the discrete space of extreme points and solving single-facility location problems, one can opt for making the search in the continuous space of location variables. When the location variables are assigned a set of values, i.e., the locations of the facilities are given, the optimum values of the allocation variables and the corresponding objective value can be found by solving a classical transportation problem. The preliminary results obtained by a genetic algorithm which directly encodes the facility locations by their coordinates are not very satisfactory.

References

Al-Loughani, I., 1997, *Algorithmic approaches for solving the Euclidean distance location-allocation problems*, PhD Dissertation, Industrial and System Engineering, Virginia Polytechnic Institute and State University, Blacksburgh, Virginia.

Bazaraa, M.S., Jarvis, J.J., and Sherali, H.D., 1990, *Linear Programming and Network Flows*. John Wiley and Sons Inc., Singapore.

Brimberg, J. and Love, R.F., 1993, Global convergence of a generalized iterative procedure for the minisum location problem with l_p distances, *Operations Research* **41**:1153–1163.

Cooper, L., 1972, The transportation-location problem, *Operations Research* **20**:94–108.

Cooper, L., 1976, An efficient heuristic algorithm for the transportation-location problem, *Journal of Regional Science* **16**:309–315.

Dongarra, J., 2006, Performance of various computers using standard linear equations software, Technical Report available online at http://www.netlib.org/benchmark/performance.ps

Dueck, G. and Scheuer, T., 1990, Threshold accepting: A general purpose optimization algorithm appearing superior to simulated annealing, *Journal of Computational Physics* **90**:161–175.

Eckert, C. and Gottlieb, J., 2002, Direct Representation and Variation Operators for the Fixed Charge Transportation Problem, in: *Proceedings of 7th International Conference on Parallel Problem Solving from Nature – PPSN VII*, Granada, Spain, pp. 77–87.

Francis R.L., McGinnis, L.F., and White, J.A., 1992, *Facility Layout and Location: An Analytical Approach*, 2nd edition, Prentice Hall, Upper Saddle River, NJ.

Gottlieb J. and Paulmann, L., 1998, Genetic algorithms for the fixed charge transportation problem, in: *Proceedings of the 1998 IEEE International Conference on Evolutionary Computation*, Anchorage, Alaska, pp. 330–335.

Gottlieb, J. and Eckert, C., 2000, A comparison of two representations for the fixed charge transportation problem, in: *Proceedings of 6th International Conference on Parallel Problem Solving from Nature–PPSN VI*, Berlin, Germany, pp. 345–354.

Gottlieb, J., Julstrom, B., Raidl, G., and Rothlauf, F., 2001, Prüfer numbers: A poor representation of spanning trees for evolutionary search, in: *Proceedings of the 2001 Genetic and Evolutionary Computation Conference*, San Francisco, California, pp. 343–350.

Hansen, P. and Mladenović, N., 2001, Variable neighborhood search: Principles and applications. *European Journal of Operational Research* **130**:449–467.

Hansen, P., Perreur, J., and Thisse, F., 1980, Location theory, dominance and convexity: Some further results, *Operations Research* **28**:1241–1250.

Kirkpatrick, S., Gelatt, C.D., and Vecchi, M.P., 1983, Optimization by simulated annealing, *Science*, **4598**:671–680.

Li, Y., Gen, M., and Ida, K., 1998, Fixed charge transportation problem by spanning tree-based genetic algorithm, *Beijing Mathematics* 4:239–249.

Ohlmann, J.W., and Thomas, B.W., 2006, A compressed annealing approach to the traveling salesman problem with time windows, *INFORMS Journal on Computing*, forthcoming.

Selim, S., 1979, *Biconvex programming and deterministic and stochastic location allocation problems*, Ph. D. dissertation, School of Industrial and Systems Engineering, Georgia Institute of Technology, Atlanta, Georgia.

Sherali, H.D. and Nordai, F.L., 1988, NP-hard, capacitated, balanced p-median problems on a chain graph with a continuum of link demands, *Mathematics of Operations Research* **13**:32–49.

Sherali, H.D. and Tunçbilek, C.H., 1992, A squared-Euclidean distance location-allocation problem. *Naval Research Logistics* **39**:447–469.

Sherali, H.D. and Adams, W.P., 1999, *A reformulation-linearization technique for solving discrete and continuous nonconvex problems*, Kluwer Academic Publishers, The Netherlands.

Sherali, H.D., Ramachandran, S., and Kim, S., 1994, A localization and reformulation discrete programming approach for the rectilinear distance location-allocation problem, *Discrete Applied Mathematics* **49**:357–378.

Sherali, H.D., Al-Loughani, I., and Subramanian, S., 2002, Global op-
    timization procedures for the capacitated Euclidean and ℓ_p distance
    multifacility location-allocation problems, *Operations Research*
    50:433–448.

Weiszfeld, E., 1937, Sur le point lequel la somme des distances de n points
    donné est minimum, *Tôhoku Mathematics Journal* **43**:355–386.

Wesolowsky, G., 1993, The Weber problem: history and perspectives,
    Location Science **1**:5–23.

Yan, S. and Luo, S.C., 1999, Probabilistic local search algorithms for
    concave cost transportation network problems, *European Journal of
    Operational Research* **117**:511–521.

Chapter 6

REVIEWER ASSIGNMENT FOR SCIENTIFIC ARTICLES USING MEMETIC ALGORITHMS

Alexander Schirrer

Karl F. Doerner

Richard F. Hartl
Department of Management Science, University of Vienna
Bruenner Strasse 72, A-1210 Vienna, Austria
alexander.schirrer@tuwien.ac.at,{karl.doerner,richard.hartl} @univie.ac.at

Abstract   In this work we modelled and solved the assignment problem appearing in MIC's paper review process using metaheuristic methods. Each given paper has to be reviewed by several different reviewers before being accepted for the conference. We implemented a memetic algorithm to solve that assignment problem and evaluated different model variants against their real world performance, using valuable feedback from many reviewers. While solutions generated by the solver alone already led to remarkable results compared to random solutions, making use of more expert knowledge throughout the solving process further improved solution quality. One way to achieve this was to fixate, prohibit or change solution parts manually and thus to iteratively build up a tuned solution.

Keywords: Assignment Problem, Memetic Algorithm

1.   Introduction

Most scientific conferences, such as the Metaheuristics International Conference 2005 (MIC 2005), apply a refereeing process in order to select the papers for presentation. The papers are submitted in advance and have to be reviewed by a board of experts. Usually, these referees are members of the program committee. Each paper is assigned to usually three referees, and based on their evaluations the decision upon acceptance or rejection is made.

For the assignment of papers to referees, one should observe that each paper should be handled by the referees most competent for that paper. On the other hand, referees should get papers that lie in their area of interest, otherwise they would not be willing or able to review them. Finally, one should have a fair allocation of workload, i.e. all referees should receive approximately the same number of papers. Also, coauthors, the paper authors' close friends or enemies, and maybe even their colleagues from the same institution or country should not be selected as referees. It is not straightforward how to capture all these aspects in a mathematical model.

This assignment is usually done manually, based on the organizer's or the conference chair's tacit knowledge of the areas of interest and competence of the program committee members.

For some conferences, attempts have been reported to do these assignments partially automatically (see the Paperdyne Conference Management System, for example[1]). However, there are no publications available that deal with the assignment problem of papers and referees in detail. Note that strong advantages arise from streamlining and optimizing this process. Each expert work hour saved here can be used valuably in the programme committee's or conference chair's main fields of work - the conference organization and the peer review process.

The purpose of this contribution is to

- present general modeling approaches for this particular assignment problem

- compare the solutions obtained by standard MIP solvers and meta-heuristics

- provide a case study for the MIC 2005 as well as the former conferences MIC 2003 and MIC 2001

- evaluate modeling approaches using feedback of program committee members on various solutions.

The related well-known generalized assignment problem (GAP) seeks a minimum cost or maximum profit assignment of n jobs to m agents subject to a resource constraint for each agent. In a GAP, each job is

---

[1]http://www.paperdyne.com/

assigned to exactly one agent. Existing literature covering the GAP is discussed below. Our problem can be viewed as an extension of the GAP with additional constraints. The task is to assign a defined number (in our case, 3) of best fitting reviewers to each paper while keeping the workload distribution fair, i.e. balanced.

The GAP is a widely known NP-hard problem and has been treated extensively in the literature. Exact approaches can be found in Savelsbergh's branch-and-price algorithm (1997) and, being one of the best performing in the field, Nauss (2003).

Various heuristic and metaheuristic approaches have been proposed. A combination of a greedy method and local search was used by Martello and Toth (1981), and set partitioning heuristic was proposed by Cattrysse, Salomon, and Van Wassenhove (1994). Amini and Racer developed a variable depth search (1995). Further approaches are a tabu search by Laguna, Kelly, González-Velarde, and Glover (1995), a tabu search and simulated annealing by Osman (1995), and a genetic algorithm by Chu and Beasley (1997). Yagiura, Yamaguchi, and Ibaraki tackled the problem with a variable depth search algorithm (1999). More recent publications are a grasp and ant system by Lourenço and Serra (2002), a well-performing tabu search approach with ejection chains by Yagiura, Ibaraki, and Glover (2004), and a recent follow-up publication by the same authors using a path-relinking approach with ejection chains (2006).

Our modeling approach is based on a property matching scheme. Papers shall be assigned to those reviewers who fit the topic and key properties of the paper best. In order to reflect this within the model, a fine-grained utility measure to assess paper-reviewer assignments is defined. The objective is to maximize the resulting total utility value.

Model variants and solution approaches using an exact IP model and a memetic algorithm are developed. The performance of the model is evaluated, based on results of a survey among reviewers.

2. Modeling Paper-Reviewer Matching and Assignment Fairness

The first important issue is the extraction of significant indicators whether and how good a paper matches a reviewer in terms of content. This matching can be done ad-hoc via expert knowledge, but it is impractical to do this for a high number of paper-reviewer combinations. Instead, we model the utility of a given pairing based on a weighted property matching strategy. Each paper and reviewer are characterized by a set of congruent properties making up a content or expertise profile.

The property set is defined according to the conference topic and should depict each relevant subtopic or area of interest. For the MIC, the property set includes 51 properties, categorized into "Methodology" (Guided Local Search, Iterated LS, Large Neighborhood Search, Simulated Annealing, Tabu Search, Variable Neigborhood Search, Distribution Estimation Algorithms, Evolution Strategies, Evolutionary Programming, Genetic Algorithms, Genetic Programming, Memetic Algorithms, Ant Colony Optimization, Cross Entropy Method, GRASP, Artificial Neural Nets, Constraint Satisfaction, Constraint Programming, Corridor Method, Hybridisation with Exact Methods, Hyperheuristics, Local Branching, Path Relinking, Pilot Method, Scatter Search), "Problem Type" (Arc Routing Problems, Telecommunications (Network Routing), TSPs, Vehicle Routing Problems, Activity Scheduling, Machine Scheduling, Project Scheduling, Staff Scheduling, Timetabling, Unit Commitment, Assignment Problems, Location Problems, Knapsack Problems, Portfolio Selection, Bioinformatics, Cutting and Packing, Partitioning Problems, Search based SW Eng., Set Covering), and other problem characteristics (Dynamic Problem, Multi Objective, Parallel Computing, SW Eng., Statistical Testing, Stochastic Problem, Theoretical Foundation).

Note that this choice of properties and their categorization fully depends on the area of research treated by the conference. The MICs put emphasis on the methodological areas, while for other conferences a substantially different property set might be chosen. Deriving utility from property sets significantly reduces overall data acquisition effort and expert judgement effort for real-world problem sizes (hundreds of papers and reviewers).

Discrete Matching Approach

In our initial approach, the property values are discrete and can be assigned the values "no", "neutral" or "yes" (1,2,3, respectively). The property values for papers and reviewers are

$$\pi(p, i) \in \{1, 2, 3\} \qquad \forall p \in P, i \in \Pi \tag{1}$$

$$\vartheta(r, i) \in \{1, 2, 3\} \qquad \forall r \in R, i \in \Pi, \tag{2}$$

where Π is the set of defined properties, and $\pi(p, i)$ and $\vartheta(r, i)$ contain the i-th property values of paper p or reviewer r, respectively.

Below, this approach is referred to as the "discrete approach". The utility value u_D of a paper-reviewer combination with paper p and reviewer

r for the j-th assignment of a reviewer to that paper can be defined as

$$u_D\left(p,r,j\right) = \sum_{i\in\Pi} w(i,j) \cdot n_{\pi(p,i),\vartheta(r,i)} \tag{3}$$

where $w(i,j)$ can be used to weight properties differently for e.g. the first, second and third assigned reviewer. The idea is to guarantee an appropriate reviewer in each category. A paper dealing with stochastic vehicle routing using tabu search should get assigned an expert in TS, one in vehicle routing and one being an expert in stochastic aspects. Hence, in $u_D(p,r,j)$, the matching values for "methodological" properties are multiplied with a higher weight than the other properties for the first assigned reviewer $j = 1$. The same is done for "application area" properties for $j = 2$, and "other aspects" properties for $j = 3$. This way, reviewer 1 generates higher weighted utility if the matching is particularly strong in the methodological properties, likewise for the other reviewers and categories. Thus we drive the solution towards assignments in which each of these property groups have at least one appropriate referee per paper.

The 3×3-matrix $N = (n_{kl})$ reflects how much each property value pair is desired or undesired, respectively. An exemplary choice of N is shown in Eqn. (4), retrieved from interpreting each property value combination and estimating a utility value. This setting was used in our calculations.

$$N = (n_{kl}) = \begin{pmatrix} 0 & 0 & 0 \\ 0 & 1 & 2 \\ -1 & 2 & 3 \end{pmatrix} \tag{4}$$

Typically the highest utility for property i will be obtained, if both reviewer r (indexing columns) and paper p (indexing rows) have the highest property values $\vartheta(r,i) = 3$ and $\pi(p,i) = 3$. Thus, the n_{33} element will be the largest. If paper p does not have property i at all ($\vartheta(p,i) = 1$), it does not matter whether the referee has this property, i.e., $n_{11} = n_{12} = n_{13} = 0$. Finally, the most undesired matching is given when paper p has the highest property value ($\vartheta(p,i) = 3$), whereas reviewer r has the lowest one ($\pi(r,i) = 1$). Hence, element n_{31} is set to a negative value in order to penalize such assignments. The remaining elements can be interpolated or set specifically, according to further interpretations of property combinations.

Continous Matching Approach

In an alternative modeling approach, we use continuous property values. They can now be any positive real numbers which enables the model to resolve properties better and in more detail. A major advantage is the possibility to "boost" certain properties which gives way for human-guided tuning of the properties (in order to facilitate the exploit of tacit knowledge at the property level). This approach is called "continuous approach" in the following. Now,

$$\pi(p,i) \in \mathbb{R}, \ \pi(p,i) \geq 0 \ \forall p \in P, i \in \Pi \tag{5}$$

$$\vartheta(r,i) \in \mathbb{R}, \ \vartheta(r,i) \geq 0 \ \forall r \in R, i \in \Pi \tag{6}$$

The utility function (3) must now be defined for all non-negative real values of the properties. For this, the lookup matrix $N = (n_{kl})$ can be interpolated by any appropriate smooth function. However, using the possibility to boost the utility of an assignment has to be reflected in that function as well, so large property values have to yield large utility values.

A straightforward choice which fulfills the desired behavior is the multiplication operation, i.e.

$$u_C(p,r,j) = \sum_{i \in \Pi} w(i,j) \cdot \pi(p,i) \cdot \vartheta(r,i) \tag{7}$$

Aggregating the matching utility values of all assignments that constitute a solution yields the overall matching utility that shall be maximized.

Defining a powerful matching strategy and seeking high input data quality turned out to be of great importance in order to gain solution quality in terms of applicability and reviewer satisfaction.

Modeling Assignment Fairness by Measuring Imbalance

Seeking only maximal paper-reviewer matching utility can result in an unfair assignment in the sense of imbalanced workload distribution. It lies in the interest of both the programme committee and the conference chair to balance the workload distribution across all reviewers. Even if this is not directly visible for the referees, it can be seen as an act of organizational fairness and might positively affect the conference's reputation in the long run. In order to drive the solution towards a balanced workload, a solution imbalance function acts as fitness penalization. We define the review effort of paper p as e_p, which is assumed the same for each reviewer. However, the effort depends on the paper, since these can be very different in structure and complexity. The paper effort e_p can be estimated out of characteristic quantities of the papers, such as page count, number of sub-problems treated and number of optimization methods in discussion. This data can be entered directly by the paper authors at submission, or extracted out of the papers, with reasonably low effort. In our approach, the paper effort is a linear combination of the mentioned quantitites.

A straightforward (linearizable) imbalance measure is the absolute value deviation of a reviewer's assigned workload from a target, or average, workload per reviewer, aggregated over all reviewers. With l_r as the aggregated workload of referee r of all his/her assigned papers, and l_r^* as his/her target workload, the imbalance penalty term yields

$$\eta = \sum_{r \in R} |l_r - l_r^*| \tag{8}$$

In our approach, as shown in (9), the target workload was set to the average workload per reviewer. An ideally, uniformly balanced solution would have the least possible imbalance penalty value.

$$l_r^* := \bar{l}^* = \frac{k}{|R|} \cdot \sum_{p \in P} e_p, \tag{9}$$

where k denotes the number of referee assignments per paper.

3. IP Formulation

Using the above building blocks, the assignment problem can be modeled as an Integer Programming (IP) program. The binary decision variables x_{prj} are introduced to represent a solution. We set $x_{prj} = 1$ if the assignment of paper p to reviewer r is done as the j-th assignment, and $x_{prj} = 0$ otherwise. The objective function is made up of the utility term $\alpha \cdot U$ and the imbalance penalty term $\beta \cdot \eta$:

$$z_{MIP} = \alpha \cdot U - \beta \cdot \eta \tag{10}$$

$$z_{MIP} = \alpha \cdot \sum_{\substack{r \in R \\ p \in P \\ j=1,\dots,k}} x_{prj} \cdot u\,(p,r,j) - \beta \cdot \sum_{r \in R} \left| \sum_{\substack{p \in P \\ j=1,\dots,k}} x_{prj} \cdot e_p - l_r^* \right| \tag{11}$$

with the precalculated $u(p,r,j) = u_D(p,r,j)$ in the discrete and $u(p,r,j) = u_C(p,r,j)$ in the continous case. The factor $\alpha > 0$ is the utility weight, and $\beta > 0$ stands for the imbalance weight, two free control parameters that enable us to freely shape the solution structure.

The problem formulation could also be tackled as a multi-objective problem when defining the matching utility as well as solution balance as partial objectives. However, viewed from the conference organizer's perspective, matching is the main goal to reach, while the imbalance penalization is seen as a safety net against an unwanted, unfair workload distribution. For that reason, we do not follow a multiobjective approach in the current model, but this might be an option for the future.

The following constraints have to be fulfilled:

$$\sum_{r \in R} x_{prj} = 1 \qquad \forall p \in P, j \in \{1, \ldots, k\} \qquad (12)$$

$$\sum_{j=1}^{k} x_{prj} \leq 1 \qquad \forall p \in P, r \in R \qquad (13)$$

$$x_{prj} = 0 \qquad \forall(p, r, j) \in E \qquad (14)$$

$$x_{prj} = 1 \qquad \forall(p, r, j) \in I \qquad (15)$$

and $\forall p \in P, r \in R, j \in \{1, \ldots, k\}$:

$$x_{prj} \in \{0, 1\} \qquad (16)$$

In equation (12) it is ensured that each paper and each slot have exactly one reviewer assigned. Equation (13) makes sure that the same reviewer is not assigned more than one time to the same paper. Using a freely definable "exclusion list" E, specific assignments can be prohibited (in (14)). The set E contains triples (p, r, j) denoting that paper number p must not be assigned to reviewer r as j-th reviewer. By using k entries in E, a referee can be completely prohibited for a given paper. This allows preprocessing of fine-grained constraints, for example that papers cannot be reviewed by their authors or affiliated people. Likewise, assignments can be prescribed using an "inclusion list" I containing triples (p, r, j) (in (15)) that have to be part of a feasible solution. Prohibited and fixed assignments are important to enable iterative problem solving: well-fitting solution parts can be fixed, and seemingly bad-fitting parts can be prohibited for subsequent iterations. Clearly, the sets E and I have to be disjoint, and only valid assignments are allowed in the inclusion list. Finally, (16) requires the decision variables to be binary.

Comparison to the Minimum Cost Flow Problem

The model presented above can be transformed into the minimum cost flow (MCF) problem, if all paper efforts are the same (i.e., if $e_p \equiv 1 \ \forall p \in P$). This problem is polynomially solvable, Edmonds and Karp (1972) showed an early scaling algorithm, and many later approaches build up on it (e.g. Orlin (1993), and Goldfarb and Jin (1999)). Korte and Vygen (2006) list a number of polynomially performing approaches.

It is possible to transform the entire model with all constraints into an equivalent minimum cost flow problem, but we only sketch the main part of the transformation and use them to accelerate the MIP solving process.

Figure 1 shows the MCF graph which is built up in three main layers. Starting from the source node S with a supply of the total workload, now being $|P| \cdot k$, we introduce nodes P_{pj} for each paper p and slot j, and connect all of them to the source by edges with capacity 1 and cost 0. Then, we introduce nodes Q_{pr}, collecting all respective flows from $P_{p1}, \ldots, P_{pk}$ with unit-capacity edges and the negative utility values $-\alpha \cdot u(p, r, j)$ as costs. In order to impose constraint (13), these nodes Q_{pr} are connected to nodes R_r with unit-capacity, zero-cost edges from all $Q_{1r}, \ldots, Q_{|P|r}$. Finally, each reviewer workload node R_r is connected to the sink T with 3 edges, as seen in Figure 2. Their capacities are $\lfloor l_r^* \rfloor$, 1 and unlimited, respectively, and their costs are set to $-\beta$, $\beta \cdot (\lceil l_r^* \rceil + \lfloor l_r^* \rfloor - 2 \cdot l_r^*)$ and β. These edges efficiently bound the solution space for the MIP solver. Figure 3 illustrates the costs imposed by a certain assigned workload, and valid solutions can either only lie on the first or third line segment, or their intersections with the middle segment. The lines' slopes are the edge costs in Figure 2. The fitness value of the reduced minimum cost flow problem (MCF) is connected to the original objective function value through $z_{MCF} + |R| \cdot \beta \cdot l_r^* = -z_{MIP}$.

We implemented the imbalance function as in the MCF and as shown in Figure 3, whereas the middle line segment is omitted in the general (variable-paper-effort) case.

4.   Memetic Algorithm

We developed a memetic algorithm (MA), based on the ideas of Moscato and Cotta (2003), in order to improve solution quality and optimization speed. It proved to perform better than the IP approach in the general case with variable paper efforts and desired parameter settings. Moreover, the MA is able to quickly construct very good solutions, which the IP approach cannot accomplish in the same time. This is particularly important in our iterative solution approach, where the use of tacit knowledge is facilitated and which requires short solving times between the iterations.

Memetic algorithms make use of all available domain knowledge to extend and improve the plain genetic evolutionary search approach. This includes not only problem-specific construction and search operators, but can also extend to the problem representation itself. Using problem representations that exploit features of the problem domain can aid the

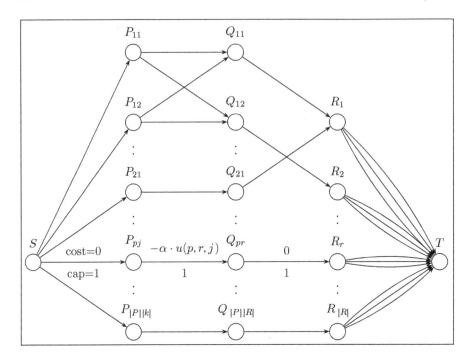

Figure 1. Graph of Minimum Cost Flow Problem

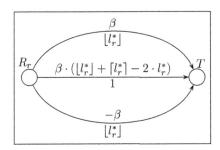

Figure 2. Realization of Imbalance Penalization in MCF formulation

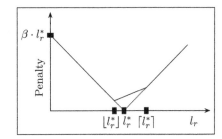

Figure 3. Imbalance Penalty and Clipping for Unit Paper Effort

algorithm in finding or improving solutions faster.

We use an assignment matrix (y_{pj}) to represent a solution. The element y_{pj} holds the index of the j-th reviewer assigned to paper p. An element y_{pj} will be called "paper slot" in the following. Using reviewer indexes cuts down variable count per solution from $|R| \times |P| \times k$ to $|P| \times k$. Constraint (12) is now implicitly fulfilled, so solution calculation and validation requires less computational effort. A pseudo code listing of the MA is given below, in Figure 4.

Figure 4. Pseudo Code - Memetic Algorithm

> **Input:** Problem Data: $u(p, r)$ and solver parameters
> **Output:** Optimized Solution x_{best}
>
> **initialization**();
> Population $Pop \leftarrow$ **createStartSolutions**();
> **sort**(Pop by fitness);
> $x_{best} \leftarrow x_{max}$ from Pop;
> *For* all iterations
> $S \leftarrow$ **select**(Pop);
>   *For* all x in Pop except elite solutions
>   *With* probability $p_{crossover}$
>   *Do* $x \leftarrow$ recombine($s1, s2 \in S$); (randomly chosen)
>   *Else* $x \leftarrow s \in S$; (randomly chosen)
>   *End-With*
> $Pop \leftarrow x$;
>   *End-For*
>   *For* all x in Pop
> $x \leftarrow$ mutate(x);
>   *With* probability p_{new} *Do* $x \leftarrow$ **createStartSolution**();
>   *With* probability $p_{lSearch}$ *Do* $x \leftarrow$ **localSearch**(x);
> $Pop \leftarrow x$;
>   *End-For*
>   **sort**(Pop by fitness);
>   *If* $x_{max} \in Pop > x_{best}$ *Then* $x_{best} \leftarrow x_{max}$;
> *End-For*

Initialization

The memetic algorithm initializes its population with solutions constructed by a greedy-randomized heuristic. First, inclusion list entries have to be considered (prescribed assignments). Then, for each remaining paper slot y_{pj}, one of the $1 \leq k_{init} \leq |R|$ best matching and –in this context– valid reviewers is chosen randomly. The idea stems from "Greedy Randomized Adaptive Search Procedures" (GRASP) that represent a large field of metaheuristics (see Resende and Ribeiro (2003)).

Maintaining solution feasibility is accomplished by adding only valid reviewers to the partial solutions. A reviewer choice is valid if the chosen reviewer r has not yet been chosen for the same paper and it is not prohibited by the exclusion list. Additionally, the robustness of this method can be increased by adapting k_{init}. If no valid reviewers can be found within several tries and if $k_{init} < |R|$, the choice interval size k_{init} is increased.

Finally, the initial solutions are locally optimized. The applied local search operation is described below in detail, since it is identically used in the memetic search step.

Selection Step

The population size is kept constant at all times. We select a fixed number of good solutions from the total population for reproduction. This "selection set" serves as genetic pool for the next generation. A small number of elite solutions is kept, while all other solutions in the population are replaced by offspring solutions - generated with a given recombination probability -, or by surviving solutions, i.e. selection set members.

A fixed number of solutions are probabilistically selected and copied into the selection set S. We apply a fitness-proportional selection method called "stochastic universal sampling" (SUS). The SUS is an $O(n)$ - implementation of the well-known roulette-wheel method, with $n = |S|$ being the number of items to select and under the assumption that the fitness evaluation and comparison can be done in constant time.

A pseudo code of the sampling algorithm is given below in Figure 5. In order to select solutions stochastically based on their fitness, all solutions of the population are evaluated and the fitness values are linearly normalized and scaled, so that the scaled total fitness sum of the population becomes 1. Similar to the roulette wheel method, the solutions' lined up scaled fitness values define intervals in the $[0; 1]$-interval. The SUS method starts at a random position u in the $[0; \frac{1}{|S|})$ interval and takes $|S|$ samples with equal distances of $\frac{1}{|S|}$. Solutions corresponding to the sample values are selected. This method is computationally fast and can be proven to be equivalent to an all-random sampling such as the standard roulette-wheel method (see Baker (1987)).

Recombination and Mutation Steps

We implemented recombination operators that use a "paper-wise" crossover strategy to produce offspring solutions. All $j = 1, \ldots, k$ slots y_{pj} of a paper p of two parent solutions s_1, $s_2 \in S$ are inherited at once,

Figure 5. Pseudo Code - Stochastic Universal Sampling

Input: Population *Pop* of solutions
Output: Selection Set *S* of given size

FitnessScaling(*Pop*)
Set $u := \frac{random(0,1)}{|S|}$
Set $step := \frac{1}{|S|}$
Set $i := 0$ (solution index)
Set $sum := $ **ScaledFitness**(*Pop*(*i*))
For all solutions in *S*
   While sum < *u*
     Set $i := i + 1$
     Set $sum := sum + $ **ScaledFitness**(*Pop*(*i*))
   End-While
   $S \leftarrow S \cup Pop(i)$
   $u := u + step$
End-For

for each paper. The choice of which paper slot set to inherit depends on the matching utility sum of the existing assignments. The advantage of this approach is its speed, since the in-paper constraints (Equation (13)) remain valid and need not be checked during crossover. When the decision of which parent to inherit from is done probabilistically, biased towards the better matching paper slot set, the selection pressure is effectively increased. This leads to a faster improvement of the population as compared to random inheritance. Our observations show that a greedy approach (always taking the better performing paper slot set) introduces even more effective selection pressure and can cut down genetic variety in the solution pool too much. However, leveraged by an increased mutation rate, that operator yields the best results for real data sets.

Following the recombination, each solution of the population is subject to mutation. A random set of assignments is chosen to be replaced by different random, but valid reviewer assignments. The mutation of an assignment is only realized if the new, mutated assignment gives a valid solution (i.e., all constraints are still fulfilled). Due to that, the mutation operation is a problem-aware shaking step rather than random mutation, a concept which complies to the memetic algorithm idea as a whole.

Figure 6. Pseudo Code - Paper-Wise Recombination Operator

> **Input:** 2 randomly chosen solutions $s_1, s_2 \in S$
> **Output:** 1 child solution c, made up of s_1 and s_2
>
> *For* all papers $p \in P$
>   choose parent $s_x \in \{s_1, s_2\}$ to take p from
>   inherit reviewer assignments of p from s_x to c
> *End-For*
> evaluate solution c

Random Restart and Local Search

A number of additional solution improvement algorithms were developed. With little probability p_{new} (as seen in the MA pseudo code in Figure 4), new genetic material is introduced into the population by replacing solutions in the population by randomized start solutions. Finally, a local search in a move-neighborhood with "first-improvement" policy and limited iteration depth is applied to the solutions. A move in that neighborhood is defined by reassigning another, valid reviewer to a paper slot. One local search iteration is completed when all paper slots have been tested for the first improving move. This is accelerated by trying the reviewers in descending order of utility.

5.   Computational Results

In this section, the solver implemenations and their performance are evaluated. Benchmarks of the IP modeling approach and the memetic algorithm are reported and compared, using randomly generated test data as well as test data derived from real data of MIC 2001, MIC 2003 and MIC 2005.

The second part of the result section evaluates the modeling approaches themselves. Based on MIC 2005 real data, a survey among all current reviewers was carried out. Their feedback is used to estimate the effects of different model parameter choices.

Implementation Details

We used $Xpress^{MP}$ optimizer version 16.01.08 as IP solver[2]. The memetic algorithm produced good solutions early in the optimization process, so all real-data results were retrieved using the memetic

---

[2]http://www.dashoptimization.com/

algorithm. The significant advantage of the MA over the IP is its ability to provide good solutions very quickly, a critical issue in an iterative solution approach.

The MA's engine was implemented in C++, combined with a Java^{TM} front-end for data management, pre and post processing[3]. The static problem structure was exploited, using pre-calculated utility values and incremental calculation steps wherever possible. The fitness changes for local search moves can be evaluated in constant time. However, due to accumulating rounding errors, the entire solution has to be reevaluated between the local search steps. This effect is observed especially in the continous matching approach (due to the larger dynamic range of the values) and can be reduced by using high-precision floating-point arithmetics. It it typically of the order of 10^{-6} of the fitness value, or less, and thus does not affect search behavior.

The MA implementation is object-oriented and kept flexible to ease future reuse and maintainbility. The solver core is designed to have a low memory footprint. The problem data is stored efficiently, so very large populations can be maintained. The memory usage is of the order $O(|P| \cdot |Pop|, |P| \cdot |R|)$, for MIC 2005 - sized problems and a typical population size $|Pop| = 100$ about 1 MByte RAM is used. When considering very large instances (e.g. $|P| = 3000$, $|R| = 1000$ and $|Pop| = 100$), which might arise with large conferences such as the INFORMS annual meeting, a desktop workstation with 512 MByte of RAM can still cope with that problem.

Solver Benchmark Results

The performance of the memetic algorithm was benchmarked against the IP model. For one series of benchmarks, randomly generated test data was used, the other benchmark was based on real MIC papers and reviewers (see Table 1 for results of the weighted, i.e. paper specific effort approach). In an earlier approach, we solved unweighted instances (i.e., with unit paper efforts). The MA performed significantly better than the original IP implementation. Using the clipping trick however (see Figure 3), the IP approach could solve these instances quickly to optimality. In the following, instances with paper specific effort are used. For the real data instances the paper weights were determined by counting the papers' pages and the number of treated problems and methods. Solution quality and computation times are reported for both the exact approach using XPress^{TM}, as well as our MA implementation. The

---

[3]http://java.sun.com/

computation times are taken on an Intel® Pentium® 4 HT / 2.4GHz desktop PC, whereas both algorithms were single-threaded and thus used 50% CPU load. The times t_{CPU} show when the best solutions were found, total time given was $45min$ for both approaches. If not stated otherwise, all gaps refer to the best LP-relaxation bounds found for that instance.

Three different random data sets were used. Their sizes ranged from the original MIC 2005 problem size up to 1000 papers and 200 reviewers. It is observed that the memetic algorithm can exploit its constructive strength especially in difficult parameter settings (e.g., with high imbalance weights), see the lines with high values of Beta in Table 1. In the simple case of imbalance weight 0 both approaches find optimal matching instantly. At low imbalance penalization, the IP approach achieves a smaller gap, but at high imbalance weight, the MA outperforms the IP approach. It is also observed that MA solutions are more balanced than their IP counterparts, at the same data and parameter settings. This indicates differently shaped search regions of both solution approaches. Three real data instances, based on MIC 2001, MIC 2003, and MIC 2005 data were created. The performance characteristics of the memetic algorithm is similar to the previous benchmark. Since these problems are significantly smaller, computation times are generally shorter.

Solving the MIC 2005 Assignment Problem Instance

The MIC 2005 problem instance consisted of $|P| = 169$ papers and $|R| = 66$ reviewers. Each paper had to be assigned to $k_p = 3$ different reviewers, and $|\Pi| = 51$ properties were defined for each paper as well as each reviewer. The properties were originally set by the papers' authors and the reviewers themselves and were afterwards fine-tuned using model and expert knowledge. This instance was calculated with unit paper efforts.

For most cases, the solution quality was sufficient after 100 iterations with a population size of 50 solutions. The MA runtime primarily depends on the depth of the local search step (which typically accounts for more than 90% of total computation time).

For the real data runs, the problem was solved in several iterations to introduce as much expert knowledge as possible: After an optimized solution was retrieved, well fitting parts of the solution as well as good assignments found by expert knowledge were made compulsory by adding them to the inclusion list. Likewise, undesirable assignments were added to the exclusion list, and the next iteration was run. For the published

solution), and it can be observed that the continous approach can increase solution quality compared to the discrete approach.

Figure 8 depicts the distribution of the given marks. While the random solution curve has a completely different shape, the other solutions' graphs are of an optimized shape: the mark distribution is maximized in the "good" region of marks ≥ 3.

It turns out that the detailed matching approach depicts the problem well enough to automatically generate solutions that are almost as good as manually tuned solutions, at significantly lower human post-processing effort. Moreover, these solutions assign a uniform workload to all reviewers.

6. Conclusion

In this work, a modeling strategy is proposed to find the optimal assignments of papers to review and referees, in terms of content matching. Therefore, property matching strategies are outlined and the mathematical optimization problem formulation is given. The problem is then solved with an IP and a memetic algorithm approach, where the latter proves to be better in difficult parameter settings. The results for random data as well as real data are given, backed by survey results for the MIC 2005. They validate the modeling approaches and show that automatic optimization leads to high-quality results, which can further be improved by introducing expert knowledge in an iterative solution process.

Table 2. Solution Performance

Solution	Avg. Mark	Exclusion	Matching	Generation
D-m14	3,48	full	discrete	14× manually tuned
C-m3	3,41	relaxed	continuous	3× manually tuned
C-auto	3,32	relaxed	continuous	automatically
D-auto	3,19	full	discrete	automatically
Random	2,56	full	–	randomly

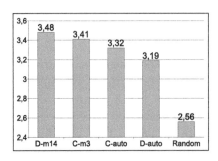

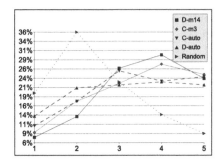

Figure 7. Average Mark of Selected Solutions

Figure 8. Solution Mark Distribution

Acknowledgments

We would like to thank all members of the board of reviewers for giving valuable feedback and providing vital information on their areas of interest. We thank Monika Treipl for designing and implementing the review management system. We are grateful to Günther Raidl who drew our attention to the idea of using a memetic algorithm. We also thank our three anonymous referees for pointing out many details and the minimum cost flow transformation. Their feedback deeply influenced this contribution and increased its quality. We are also grateful for valuable comments of Pablo Moscato, especially during the MIC 2005 real data and survey phase.

References

Amini, M. M.; Racer, M. (1995): "A hybrid heuristic for the generalized assignment problem". In: European Journal of Operations Research **87** Heuristics: Theory & Applications. Kluwer Academic Publishers, Boston, USA, 343-348.

Baker, J.E.(1987): "Reducing Bias and Inefficiency in the Selection Algorithm". In: Grefenstette, J.J.: Proceedings of the Second International Conference on Genetic Algorithms and their Application, Hillsdale, New Jersey, USA: Lawrence Erlbaum Associates, 1987, 14-21.

Cattrysse, D.G.; Salomon, M.; Van Wassenhove, L.N. (1994): "A set partitioning heuristic for the generalized assignment problem". In: European Journal of Operations Research **72**, 167-174.

Chu, P.C.; Beasley, J.E. (1997): "A genetic algorithm for the generalized assignment problem". Computational Operations Research **24**, 17-23.

Edmonds, Jack; Karp, Richard M. (1972): "Theoretical Improvements in Algorithmic Efficiency for Network Flow Problems". In: Journal of the Association for Computing Machinery, Vol.19, No.2, 248–264.

Goldfarb, Donald; Jin, Zhiying (1999): "A new scaling algorithm for the minimum cost network flow problem". In: Operations Research Letters **25**, Elsevier Science B.V., 205–211.

Korte, B.; Vygen, J. (2006): "Combinatorial Optimization: Theory and Algorithms". 3rd Edition, Springer, 191–214

Laguna, M.; Kelly, J.P.; González-Velarde, J.L.; Glover, F. (1995): "Tabu search for the multilevel generalized assignment problem". European Journal of Operations Research **82**, 176-189.

Lourenco, H.R.; Serra, D. (2002): "Adaptive search heuristics for the generalized assignment problem". Mathware Soft Computing **9**, 209-234.

Martello, S.; Toth, P. (1981): "An algorithm for the generalized assignment problem". Proceedings of the Ninth IFORS Internat. Conf. on Operational Research. Hamburg, Germany. North-Holland, Amsterdam, The Netherlands, 589-603.

Moscato, P.; Cotta, C. (2003): "A Gentle Introduction To Memetic Algorithms". In: Glover, Fred, and Kochenberger, Gary A. (eds.): *Handbook of Metaheuristics*. (International Series in Operations Research & Management Science, vol. 57). Kluwer Academic Publishers, Boston, USA, 105–144.

Nauss, R.M. (2003): "'Solving the generalized assignment problem: An optimizing and heuristic approach"'. INFORMS Journal on Computing **15**, 249-266.

Orlin, J.B. (1993): "A faster strongly polynomial minimum cost flow algorithm". In: Operations Research, Vol.41,Issue 2, 338–350.

Osman, I.H. (1995): "Heuristics for the generalized assignment problem: Simulated annealing and tabu search approaches". OR Spektrum **17**, 211-225.

Resende, M.G.C.; Ribeiro, C.C. (2003): "Greedy Randomized Adaptive Search Procedures". In: Glover, Fred, and Kochenberger, Gary A. (eds.): *Handbook of Metaheuristics*. (International Series in Operations Research & Management Science, vol. 57). Kluwer Academic Publishers, Boston, USA, 219–249.

Savelsbergh, M.W.P. (1997): "A branch-and-price algorithm for the generalized assignment problem". Operations Research **45**, 831–841

Yagiura, M.; Ibaraki, T.; Glover, F. (2004): "An Ejection Chain Approach for the Generalized Assignment Problem". In: *INFORMS* (INFORMS Journal of Computing , vol. 16, no. 2), 133–151.

Yagiura, M.; Ibaraki, T.; Glover, F. (2006): "A Path Relinking Approach with Ejection Chains for the Generalized Assignment Problem". European Journal of Operational Research, **169**, 548–569.

Yagiura, M.; Yamaguchi, T.; Ibaraki, T. (1999): "A variable depth search algorithm for the generalized assignment problem". In: Meta-Heuristics: *Advances and Trends in Local Search Paradigms for Optimization*, Kluwer Academic Publishers, Boston, USA, 459–471.

Part IV

GRASP and Iterative Methods

Chapter 7

GRASP WITH PATH-RELINKING FOR THE TSP

Elizabeth F. Gouvêa Goldbarg, Marco C. Goldbarg and João P. F. Farias
Department of Informatics and Applied Mathematics, Federal University of Rio Grande do Norte, Campus Universitário Lagoa Nova, Natal, Brazil

Abstract: This paper presents a new GRASP for the TSP. Two versions utilizing two distinct implementations of the LK neighborhood are introduced. Those heuristics are hybridized with a path-relinking procedure. A distance metric, previously proposed to fitness analysis landscape is utilized to decide whether to apply path-relinking between a pair of solutions. A computational experiment is reported to support conclusions about efficiency of the proposed approach.

Key words: GRASP, path-relinking, traveling salesman problem, Lin-Kernighan neighborhood

1. INTRODUCTION

The Traveling Salesman is a classical NP-hard combinatorial problem that has been an important test ground for most algorithms. Given a graph $G = (N,E)$, where $N = \{1,...,n\}$ is the set of nodes and $E = \{1,...,m\}$ is the set of edges of G, and costs, c_{ij}, associated with each edge linking vertices i and j, the problem consists on finding the minimal total length Hamiltonian cycle. The length is calculated by the summation of the costs of the edges in a cycle. If for all pairs of nodes $\{i, j\}$, the costs c_{ij} and c_{ji} are equal then the problem is said to be symmetric, otherwise it is said to be asymmetric.

The main importance of TSP regarding applicability is due to its variations, nevertheless some applications of the basic problem in real world problems are reported for different areas such as VLSI chip fabrication (Korte, 1989), X-ray crystallography (Bland and Shalcross, 1989), genome map (Guyon et al., 2003), DNA sequence (Gonnet et al., 2000) and broadcast schedule (Yajima et al., 2001), among others.

The most effective exact algorithm for the TSP is based on a branch and cut strategy presented by Applegate et al. (1999). Currently, the largest TSP instance exactly solved has 24978 cities and represents the shortest tour among Swedish cities.

Once to exactly solve the TSP is a hard and time consuming task, a number of heuristic algorithms were presented in the last decades to find good sub-optimal solutions for the problem. Burkard (2002) divides the approaches for constructing heuristic algorithms for the TSP in three classes: construction methods, improvement methods and metaheuristics. Construction methods build a tour iteratively and are based on some greedy criterion. A well known constructive method for the TSP is the nearest neighbor algorithm presented by Bellmore and Nemhauser (1968). As pointed out by Burkard (2002), although it is very easy to generate instances where the nearest neighbor algorithm performs arbitrarily bad, this is a well suited procedure to generate solutions to be processed by an improvement method. A randomized adaptive version of the nearest neighbor is utilized in this work in the constructive phase of the GRASP algorithms.

Local search algorithms constitute the class of improvement methods. Given a neighborhood structure defined over a search space, a local search procedure begins with a solution and search the neighborhood of the current solution for an improvement. Some well known neighborhood structures for the TSP are: 2-opt (Flood, 1956), 3-opt and Lin-Kernighan (Lin and Kernighan, 1973).

Finally, the metaheuristics are general frameworks to design heuristic algorithms. A number of techniques based on several metaheuristic approaches were proposed to solve the TSP, such as: Simulated Annealing, Neural Networks, Tabu Search, Genetic and Memetic Algorithms, Ant System, Scatter Search and Variable Search Neighborhood. A survey of the TSP and the solution methods utilized to solve it is presented by Gutin and Punnen (2002).

Although metaheuristic approaches are widely used to solve the TSP, only recently a GRASP was proposed for this problem (Marinakis et al., 2005).

GRASP is a multi-start search procedure. First introduced by Feo and Resende (1989), it consists on a method that repeatedly applies local search from different starting solutions of a search space. Initial solutions are built with a greedy randomized adaptive procedure, that is, a construction method adapted to incorporate randomness and to be adaptive. In this first phase, a solution is constructed iteratively by the addition of elements that are randomly chosen from a restricted candidate list. The restricted candidate list is built according to a greedy criterion that evaluates the attractiveness of each element for the solution. One element of this list is randomly chosen, in general, with a uniform probability distribution. The addition of a new element to the solution modifies the attractiveness of the remaining elements out of the solution (this incorporates the adaptive character to the algorithm). Once a solution is built, the algorithm proceeds to an improvement phase. In this second phase, a local search method is utilized to improve the solution built on the first phase. In general, the stop criterion is a given number of iterations (construction/local search). The fundamentals of GRASP, enhancements, hybridization with other methods and several applications to combinatorial problems are surveyed in the work of Resende and Ribeiro(2003a).

Path-relinking is an intensification technique which ideas were originally proposed by Glover (1963) in the context of scheduling methods to obtain improved local decision rules for job shop scheduling problems (Glover, Laguna and Marti, 2000). The strategy consists on generating a path between two solutions creating new solutions. Given an origin, x_s, and a target solution, x_t, a path from x_s to x_t leads to a sequence x_s, $x_s(1)$, $x_s(2)$, …, $x_s(r) = x_t$, where $x_s(i+1)$ is obtained from $x_s(i)$ by a move that introduces in $x_s(i+1)$ an attribute that reduces the distance between attributes of the origin and target solutions. The roles of origin and target can be interchangeable. Resende and Ribeiro (2003b) identify some strategies for considering such roles:

- forward: the worst among x_s and x_t is the origin and the other is the target solution;
- backward: the best among x_s and x_t is the origin and the other is the target solution;
- back and forward: two different trajectories are explored, the first using the best among x_s and x_t as the initial solution and the second using the other in this role;
- mixed: two paths are simultaneously explored, the first starting at the best and the second starting at the worst among x_s and x_t, until they meet at an intermediary solution equidistant from x_s and x_t.

The hybridization of GRASP and path relinking is very promising and has been investigated in a number of papers with applications to several problems such as the 2-layer straight line crossing minimization (Laguna and Marti, 1999), the three index assignment (Aiex et al., 2005), the prize collector Steiner tree (Canuto et al., 2001), the channel assignment in mobile phone networks (Gomes et al., 2001), the job shop scheduling (Aiex et al., 2003), the capacitated minimum spanning tree (Souza et al 2003), the railway planning (Delorme et al., 2004), the matrix bandwith minimization (Piñana et al., 2004), the rural road network development (Scaparra and

Church, 2005), the weighted maximum satisfiability (Festa et al., 2005), the single machine total tardiness scheduling (Gupta and Smith, 2005) and the uncapacitated facility location (Resende and Werneck, 2005).

As far as the authors' knowledge is concerned, no work reports an application of GRASP with path-relinking for the TSP. In this paper, a hybrid heuristic with those approaches is proposed. The GRASP algorithm utilizes a LK neighborhood (Lin and Kernighan, 1973). Two distinct implementations of the LK define two algorithm's versions. In the hybrid approach, path-relinking operations are done with the best solutions found by GRASP in a given number of iterations.

Computational experiments compare the algorithms proposed in this paper with other efficient heuristics proposed for the same problem. A comparison of the results found by the GRASP algorithms with the recent GRASP proposed by Marinakis et al. (2005) is also presented. The experiments show that the proposed algorithms find high quality tours.

The paper is organized as follows. The proposed algorithm is described in Section 2. The results of the computational experiments are reported in Section 3. Finally, some concluding remarks are drawn in Section 4.

2.   THE ALGORITHMS

The initial solutions of the GRASP procedure are constructed with an adaptive randomized version of the nearest neighbor algorithm. The main difference between the original algorithm and the one proposed in this work for the constructive phase of GRASP is that elements are randomly chosen from a restricted candidate list to build a solution. The restricted candidate list is built with the cities closest to the last element that entered in a solution. The size of the restricted candidate list was set to $0.05n$ for all the instances of the experiment.

The algorithm of Lin-Kernighan (1973), LK, is a recognized efficient improvement method for the TSP. The basic LK algorithm has a number of decisions to be made and depending on the strategies adopted by programmers distinct implementations of this algorithm may result on different performances. The literature contains reports of many LK implementations with widely varying behavior (Johnson and McGeoh, 2002). In this work two distinct implementations of LK are utilized as local search procedures to generate two versions of the proposed GRASP algorithm: LK-ABCC (Applegate et al., 1999) and LK-H (Helsgaun, 2000). The former is the implementation utilized in the Concorde solver (http://www.tsp.gatech.edu/concorde.html). It produces good tour qualities and low runtimes. The latter is a very effective LK implementation where 2-

opt moves are extended to sequential 5-opt moves. A briefly description of each technique and the main differences among them are presented by Johnson and McGeoh (2002). The DIMACS TSP Challenge web page (http://www.research.att.com/~dsj/chtsp/)) presents a comparison of those heuristics regarding tour quality and processing times. Figures 1 and 2, extracted from that site, summarize graphically the relative behavior of LK-ABCC and LK-H for TSPLIB instances with $n \geq 1000$.

Figure 1 shows the percent difference among the tour lengths found by those heuristics. The LK-H heuristic is referenced as "Helsgaun" on the top horizontal legend. The points show the percent difference regarding tour lengths between the solutions found by the two investigated heuristics. The points over line "0.0" show the percent differences for instances where the LK-H presents the best tours. Similarly, the point under line "0.0" shows the percent difference of the unique instance where the LK-ABCC presents a better solution than LK-H.

The chart of figure 2 shows the comparison of processing times. Points under line "1.0" indicate the instances where LK-ABCC presented better processing times than LK-H, and points over that line should indicate the contrary, but there are none. LK-ABCC presents lower runtimes than LK-H for all the investigated instances, however the differences are very small.

In this work the algorithms utilized the LK implementations made available by their authors in the internet.

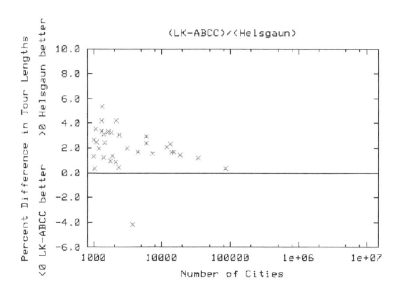

Figure 7-1. Solution quality comparison between LK-ABCC and LK-H

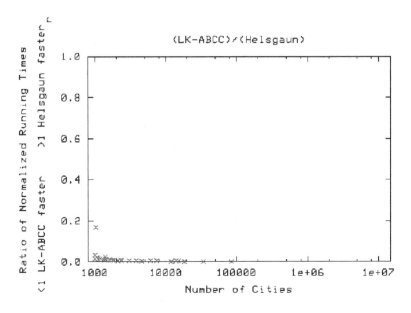

Figure 7-2. Runtime comparison for LK-ABCC and LK-H

The path-relinking phase utilizes a pool with *p* GRASP solutions. The pool is formed with the *p* best solutions among *p* iterations and the best current solution. Preliminary experiments showed that the best strategy for the case investigated in this work was the back and forward relinking. A high level pseudo-code of the heuristic approach proposed in this paper is given in Table 1. During the exploration of a given path, if a new solution improves the origin and target solutions, then the LK procedure is applied to the new generated solution.

Table 7-1. High level pseudo-code of the proposed algorithm

Algorithm: GRASP with path-relinking
$j \leftarrow 0$
best_solution $\leftarrow$ Grasp( )
Repeat
For $i \leftarrow 1$ to *p* do
$S[i] \leftarrow$ Grasp( )
$j \leftarrow j + 1$
$x \leftarrow$ Path_relink(*S*,*best_solution*)
if $(f(x) < f(best\_solution)$ then
best_solution $\leftarrow x$
until ($j =$ #grasp_iterations)

In function *Path_relink(S,best_solution)*, path-relinking operations are done between pairs of solutions of a pool defined on set *S* and the best current solution, *best_solution*. This function returns the best solution found in the path-relinking operations. In this function, distances are calculated between all pairs of solutions, and the closest pair is chosen for the path-relinking operation. Given a sequence x_s, $x_s(1)$, $x_s(2)$, ..., $x_s(r) = x_t$, the best solution of the sequence replaces the origin and the target solutions. If a pool has p solutions, this method leads to $O(p)$ path-relinking operations.

Two metrics were investigated for calculating the distance between a pair of solutions. The first is presented in Aiex and Resende (2005) and defines the difference between two permutations s and s' as the set $\delta(s,s') = \{i | s(i) \neq s'(i)\}$. The distance is given by $d(x_s, x_t) = |\delta(x_s, x_t)|$. The second metric was introduced by Boese (1995) who investigated the fitness landscape of TSP instances. Given two solutions x_s and x_t, the distance $d(x_s,x_t)$ is n minus the number of edges contained both in x_s and x_t. Experimental analysis showed that the latter metric yielded, in average, better results than the former. Thus, the results presented in the computational experiments were obtained with the implementations that used Boese's metric.

The relinking operator utilized in this work is illustrated in figure 3. The operator swaps one element with its right (left) neighbor. The steps of a path-relinking procedure that explores solutions in the path between the origin solution (1 2 3 4 5) and the target solution (3 5 1 2 4) are shown. At first, the element 3 is moved to the first position by swapping it with elements 2 (Figure 3(a)) and 1 (Figure 3(b)). At this point, element 5 has to be moved to the second position. It is swapped with element 4 (Figure 3(c)), element 2 (Figure 3(d)) and, finally, it is swapped with element 1, when the target solution is reached. The swap operators lead to $O(n^2)$ procedures.

The stop criterion adopted in the experiments was a given number of GRASP iterations.

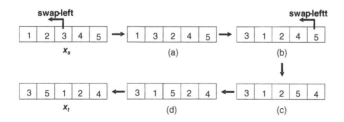

Figure 7-3. Swap-left operator for path-relinking

Table 2 summarizes the versions and abbreviations of the proposed algorithms that are utilized in the computational experiment reported in the next section.

Table 7-2. Versions of the proposed algorithm

Algorithm	Description
GRASP-ABCC	GRASP with local search LK-ABCC
GRASP-H	GRASP with local search LK-H
GRASP_PR-ABCC	GRASP-ABCC with path-relinking
GRASP_PR-H	GRASP-H with path-relinking

3.   COMPUTATIONAL EXPERIMENTS

The algorithms were implemented in C++ on a Pentium IV (3.0 GHz and 512 Mb of RAM) running Linux. The algorithm was applied to symmetric instances of the benchmark TSPLIB (TSPLIB; http://www.iwr.uni-heidelberg.de/iwr/comopt/software/TSPLIB95/) with sizes ranging from 51 to 7397. The stop criterion was 80 GRASP iterations for instances with $n < 1000$ and 20 GRASP iterations for the remaining instances. The length of the RCL was $0.05n$.

Table 7-3. Improvement of GRASP in LK procedures

Instance	n	LK-ABCC	GRASP-ABCC	LK-H	GRASP-H
dsj1000	1000	0.2973	0.0002	0.0290	0
pr1002	1002	0.1318	0	0.0001	0
u1060	1060	0.1786	0.0038	0.0002	0
vm1084	1084	0.0669	0.0005	0.0103	0
pcb1173	1173	0.1814	0	0.0022	0
d1291	1291	0.4333	0.0006	0.0251	0
rl1304	1304	0.3984	0.0001	0.0512	0.0069
rl1323	1323	0.2300	0.0008	0.0169	0.0010
nrw1379	1379	0.1354	0.0116	0.0026	0.0001
fl1400	1400	0.1215	0	0.1950	0.1838
fl1577	1577	2.2974	0.0424	0.0146	0
vm1748	1748	0.1311	0.0002	0.0139	0.0012
u1817	1817	0.5938	0.0680	0.0900	0.0254
rl1889	1889	0.3844	0.0059	0.0149	0
d2103	2103	0.3085	0.0213	0.0259	0.0088
u2152	2152	0.5548	0.0549	0.0403	0
pr2392	2392	0.3904	0.0411	0	0
pcb3038	3038	0.2568	0.0526	0.0056	0
fl3795	3795	1.0920	0.0065	0.1577	0.0134
fnl4461	4461	0.1717	0.0757	0.0019	0
rl5915	5915	0.5343	0.0993	0.0265	0.0084
rl5934	5934	0.4761	0.0820	0.0768	0.0307
pla7397	7397	0.2912	0.0025	0.0052	0

Table 7-4. Comparison of GRASP algorithms for the TSP

Instance	n	GRASP_M	GRASP-ABCC	GRASP-H
eil101	101	0	0	0
lin105	105	0	0	0
pr107	107	0	0	0
pr124	124	0	0	0
bier127	127	0.03	0	0
ch130	130	0	0	0
pr136	136	0	0	0
pr144	144	0	0	0
ch150	150	0	0	0
kroA150	150	0	0	0
pr152	152	0	0	0
rat195	195	0.34	0	0
d198	198	0.05	0	0
kroA200	200	0.04	0	0
kroB200	200	0.15	0	0
ts225	225	0	0	0
pr226	226	0.05	0	0
gil262	262	0.29	0	0
pr264	264	0	0	0
a280	280	0.38	0	0
pr299	299	0.09	0	0
rd400	400	0.68	0	0
fl417	417	0.28	0	0
pr439	439	0.17	0	0
pcb442	442	0.33	0	0
d493	493	0.71	0	0
rat575	575	1.32	0	0
p654	654	0.18	0	0
d657	657	1.26	0.002	0
rat783	783	1.03	0	0
pr1002	1002	1.16	0	0
pcb1173	1173	1.37	0	0
d1291	1291	1.60	0	0
rl1304	1304	0.88	0	0
rl1323	1323	1.07	0	0
fl1400	1400	0.90	0	0.1838
fl1577	1577	0.80	0.0315	0
rl1889	1889	0.85	0.0022	0
d2103	2103	1.07	0	0
pr2392	2392	2.11	0.0090	0

A first experiment investigated the improvement of the GRASP (without path-relinking) approach for the LK procedures. Table 3 shows the results for 23 symmetric TSPLIB instances with n from 1000 to 7397. The first two columns of table 3 show the names and the actual sizes of the TSPLIB instances, the remaining four columns show the average percent deviation from the optimal solution given by $(v_{heur} - v_{opt}) \times 100/ v_{opt}$, where v_{opt} is the value of the optimal solution and v_{heur} is the average of the best solutions obtained on 20 independent runs of each algorithm. The stop criterion for the two versions of GRASP was 20 iterations.

Table 3 shows that all solutions are improved when the LK-ABCC is utilized within the GRASP and, except for instance pr2392, the same occurs for the LK-H.

Marinakis et al (2005) applied their GRASP to 51 symmetric TSP instantances. Table 4 shows the best deviations from the optimal solutions reported in their paper for instances with $n > 100$ in column GRASP_M. The other two columns show the best solutions (percent deviation from the optimum) found by the two GRASP versions presented in this work.

Among the 40 instances summarized in table 4, the three GRASP algorithms find the same tour quality for 16 instances. The versions GRASP-ABCC and GRASP-H find the optimal solution of 36 and 39 instances, respectively. The averages of the columns corresponding to each algorithm are: 0.4798 (GRASP_M), 0.0011 (GRASP-ABCC) and 0.0046 (GRASP-H).

Table 7-5. Comparison of the results of GRASP-ABCC with and without path-relinking

Instance	n	GRASP-ABCC		GRASP_PR-ABCC	
		Min	Average	Min	Average
dsj1000	1000	0	0.0002	0	0
pr1002	1002	0	0	0	0
u1060	1060	0	0.0038	0	0
vm1084	1084	0	0.0005	0	0
pcb1173	1173	0	0	0	0
d1291	1291	0	0.0006	0	0
rl1304	1304	0	0.0001	0	0
rl1323	1323	0	0.0008	0	0
nrw1379	1379	0	0.0116	0	0
fl1400	1400	0	0	0	0
fl1577	1577	0.0315	0.0424	0	0
vm1748	1748	0	0.0002	0	0
u1817	1817	0	0.0680	0	0.0130
rl1889	1889	0.0022	0.0059	0	0.0012
d2103	2103	0	0.0213	0	0.0011
u2152	2152	0.0187	0.0549	0	0
pr2392	2392	0.0090	0.0411	0	0.0096
pcb3038	3038	0.0334	0.0526	0.0029	0.0129
fl3795	3795	0	0.0065	0	0
fnl4461	4461	0.0668	0.0757	0.0197	0.0303
rl5915	5915	0.0752	0.0993	0.0074	0.0235
rl5934	5934	0.0611	0.0820	0.0050	0.0117
pla7397	7397	0	0.0025	0	0.0011

Another experiment was conducted in order to conclude about the amount of improvement obtained with the inclusion of the path-relinking in the GRASP algorithms. Tables 5 and 6 show the results of the computational experiment that compared the proposed algorithms with and without the path-relinking procedures for 23 symmetric instances with n ranging from 1000 to 7397. The columns of both tables show the minimum and average percent difference from the optimum of the proposed algorithms without and with the path-relinking procedure.

Table 7-6. Comparison of the results of GRASP-H with and without path-relinking

Instance	n	GRASP-H		GRASP_PR-H	
		Min	Average	Min	Average
dsj1000	1000	0	0	0	0
pr1002	1002	0	0	0	0
u1060	1060	0	0	0	0
vm1084	1084	0	0	0	0
pcb1173	1173	0	0	0	0
d1291	1291	0	0	0	0
rl1304	1304	0	0.0069	0	0
rl1323	1323	0	0.0010	0	0.0005
nrw1379	1379	0	0.0001	0	0
fl1400	1400	0.1838	0.1838	0.1838	0.1838
fl1577	1577	0	0	0	0
vm1748	1748	0	0.0020	0	0
u1817	1817	0	0.0254	0	0.0179
rl1889	1889	0	0	0	0
d2103	2103	0	0.0088	0	0.0031
u2152	2152	0	0	0	0
pr2392	2392	0	0	0	0
pcb3038	3038	0	0	0	0
fl3795	3795	0	0.01338	0	0
fnl4461	4461	0	0	0	0
rl5915	5915	0.0041	0.0084	0	0.0054
rl5934	5934	0	0.0307	0	0.0003
pla7397	7397	0	0	0	0

Among the 23 instances of table 5 GRASP-ABCC finds the optimal solution of 15 instances. An increasing of 4 new optimal solutions is obtained when the path-relinking is added to GRASP-ABCC. The best solutions found by the algorithm for the remaining 4 instances are also improved. The improvement is, in average, 86%. Except for three instances that already had average deviation zero, the version of the algorithm with path-relinking improves all the average solutions. The mean improvement of the average solutions is 82%.

Among the 23 instances of table 6, GRASP-H finds the optimal solution of 21 instances. The inclusion of path-relinking does not modify the results found for instance fl1400, but improves the best solution of instance rl5915. GRASP-H also finds percent deviation zero for 13 of the 23 instances. All the remaining average results are improved with the inclusion of path-relinking.

Table 7-7. Comparison of heuristics for the TSP

Instance	n	GRASP_PR-ABCC		GRASP_PR-H		Tourmerge		ILKJM Nb10
		Min	Av	Min	Av	Min	Av	Min
dsj1000	1000	0	0	0	0	0.0027	0.0478	0.0063
pr1002	1002	0	0	0	0	0	0.0197	0.1482
u1060	1060	0	0	0	0	0	0.0049	0.0210
vm1084	1084	0	0	0	0	0	0.0013	0.0217
pcb1173	1173	0	0	0	0	0	0.0018	0.0088
d1291	1291	0	0	0	0	0	0.0492	0
rl1304	1304	0	0	0	0	0	0.1150	0
rl1323	1323	0	0	0	0.0005	0.0100	0.0411	0
nrw1379	1379	0	0	0	0	0	0.0071	0.0018
fl1400	1400	0	0	0.1838	0.1838	0	0	0
fl1577	1577	0	0	0	0	0	0.0225	0
vm1748	1748	0	0	0	0	0	0	0
u1817	1817	0	0.0130	0	0.0179	0.0332	0.0804	0.2657
rl1889	1889	0	0.0012	0	0	0.0082	0.0682	0.0041
d2103	2103	0	0.0011	0	0.0031	0.0199	0.3170	0
u2152	2152	0	0	0	0	0	0.0794	0.1743
pr2392	2392	0	0.0096	0	0	0	0.0019	0.1495
pcb3038	3038	0.0029	0.0129	0	0	0.0036	0.0327	0.1213
fl3795	3795	0	0	0	0	0	0.0556	0.0104
fnl4461	4461	0.0197	0.0303	0	0	---	---	0.1358
rl5915	5915	0.0074	0.0235	0	0.0054	0.0057	0.0237	0.0168
rl5934	5934	0.0050	0.0117	0	0.0003	0.0023	0.0104	0.1723
pla7397	7397	0	0.0011	0	0	---	---	0.0497

Table 7 shows a comparison of the results obtained by the proposed algorithms and two effective heuristics: Tourmerge (Cook and Seymour, 2003) and JM iterated Lin-Kernighan variant (Johnson and McGeoh, 2002). The results of those heuristics were obtained in the DIMACS Challenge page (http://www.research.att.com/~dsj/chtsp/ results.html). The columns related to GRASP with path-relinking show the best and average tours found in 20 independent runs. The columns related to the Tourmerge algorithm show the best and average tours obtained in five independent runs. Results are not reported for instances fnl446 and pla7397. The column related to JM iterated Lin-Kernighan variant (ILKJM Nb10) shows the best tours obtained in ten n iterations runs.

From the 21 instances for which Tourmerge presented results, GRASP_PR-ABCC finds 6 best minimal results, Tourmerge finds 2 best minimal results and both algorithms find the same quality tours for 13 instances. Comparing Tourmerge and GRASP_PR-H, the latter algorithm finds the best tours of 8 instances, there are 7 ties and the former algorithm finds the best solution of 1 instance.

The JM iterated Lin-Kernighan variant finds only one better result than GRASP_PR-H and no better result than the GRASP_PR-ABCC.

Regarding average results, GRASP_PR-ABCC and GRASP_PR-H find, respectively, 17 and 19 better results than Tourmerge on 21 instances.

The averages of the "Min" and "Average" columns for the twenty-one instances where the two proposed algorithm's versions and the Tourmerge were applied are 0.0007 and 0.0035 for GRASP_PR-ABCC, 0.0088 and 0.0100 for GRASP_PR-H, and 0.0041 and 0.0467 for Tourmerge. These results show that the proposed algorithms present a better performance than the Tourmerge, although the latter presents a better average regarding the "Min" column than GRASP_PR-H. The higher value presented by GRASP_PR-H is due to instance fl1400 which results indicate that there is a strong attractor for the LK-H neighborhood not overcome by the path-relinking procedure.

Table 7-8. Comparison of processing times between GRASP_PR-H and LK-H

Instance	n	GRASP_PR-H		LK-H	
		Min	Average	Min	Average
dsj1000	1000	23.2	32.4	26.9	38.6
pr1002	1002	2.2	2.5	2.7	3.3
u1060	1060	25.2	35.2	33.3	47.2
vm1084	1084	9.1	27.9	11.2	20.4
pcb1173	1173	7.4	10.2	8.8	12.7
d1291	1291	24.2	39.8	28.1	45.1
rl1304	1304	7.9	20.1	8.0	12.1
rl1323	1323	14.4	99.2	8.7	16.4
nrw1379	1379	12.9	38.4	17.1	24.0
fl1400	1400	2941.4	3783.9	152.3	230.3
fl1577	1577	265.3	575.5	309.4	446.6
vm1748	1748	30.9	75.1	29.1	45.8
u1817	1817	78.6	1099.0	60.8	74.0
rl1889	1889	27.8	55.3	33.1	52.1
d2103	2103	1721.9	9671.2	242.0	436.0
u2152	2152	95.0	278.1	105.3	133.9
pr2392	2392	88.0	96.3	109.3	122.6
pcb3038	3038	237.4	427.1	272.5	349.0
fl3795	3795	1929.2	9360.8	2103.7	3128.3
fnl4461	4461	438.4	754.7	367.2	409.2
rl5915	5915	16284.0	35156.4	569.5	827.2
rl5934	5934	1171.4	19301.3	738.7	1233.8
pla7397	7397	4913.3	10561.2	7954.4	10001.2

Table 8 summarizes the minimal and average runtimes in seconds of GRASP_PR-H and LK-H for the instances with $n \geq 1000$, where it is showed that the processing times are comparable. The relative results of GRASP_PR-ABCC and LK-ABCC are similar.

An analysis of the time spent on different parts of the search showed that the processing time spent with the path-relinking procedure of GRASP_PR-ABCC, increased with n. The instances were divided into 4 classes: with $n < 000$, with n between 1000 and 2000, with n between 2000 and 4000, with $n > 4000$. In average, the percentages of processing time spent with path-relinking for the instances of those classes are 20%, 58%, 75% and 90%, respectively. The same behavior was not observed for GRASP_PR-H. Among the instances were the path-relinking procedure improved the results obtained by the version of GRASP that utilized the LK-H, in average, 1.7% of the processing time was spent with the path-relinking procedure.

4.   CONCLUSION

This paper presented a hybrid approach with GRASP and path-relinking to solve the TSP. Two versions of a GRASP algorithm were introduced, utilizing, each of them, a distinct implementation of the LK neighborhood as the local search procedure.

Experiments were conducted to conclude about the improvement the GRASP approach brought to the LK neighborhood. The results showed that improvements of 76% and 65% were obtained with GRASP-ABCC and GRASP-H regarding the algorithms LK-ABCC and LK-H, respectively.

A comparison with the results obtained in 40 instances with $n > 100$ by Marinakis et al. (2005) was also showed, where the two versions of the proposed GRASP algorithm reached, in average, solutions more than 90% better than the former algorithm.

Another experiment investigated whether the inclusion of a path-relinking procedure was able or not to improve the GRASP results. The results showed that both versions of GRASP were improved, with significant better average values.

Finally, a comparison with two effective heuristics reported in the literature for the TSP was done, where it was showed that the proposed algorithms are able to obtain best tour qualities on the majority of the instances of the experiment.

REFERENCES

Aiex, RM., Binato, S., Resende, M.C.G., 2003, Parallel GRASP with path-relinking for job shop scheduling, *Parallel Computing* **29**:393-430.

Aiex, RM., Resende, M.C.G., 2005, Parallel strategies for GRASP with path-relinking, in: *Metaheuristics: Progress as Real Problem Solvers*, T. Ibaraki, K. Nonobe and M. Yagiura, Eds., Springer, pp. 301-331.

Aiex, R.M., Resende, M.G.C., Pardalos, P.M. and Toraldo, G., 2005, GRASP with path relinking for the three-index assignment problem, *INFORMS Journal on Computing* **17**(2): 224-247.

Applegate, D., Bixby, R., Chvatal, V., and Cook, W., 1999, *Finding Tours in the TSP*, Tech. Rep. TR99-05, Dept. Comput. Appl. Math., Rice University.

Applegate, D., Bixby, R. E., Chvatal, V., and Cook, W., 2001, TSP cuts which do not conform to the template paradigm, in: *Computational Combinatorial Optimization*, Junger, M., and Naddef, D., Eds., Springer, pp. 261-304.

Bellmore, M., and Nemhauser, G. L., 1968, The traveling salesman problem: A survey, *Operations Research* **16**: 538–582.

Bland, R.G., and Shalcross, D.F., 1989, Large traveling salesman problems arising from experiments in x-ray crystallography: A preliminary report on computation, *Operations Research Letters* **8**: 125-128.

Boese, K. D., 1995, *Cost Versus Distance in the Traveling Salesman Problem*, Technical Report TR 950018, UCLA CS Department, California.

Burkard, R. E., 2002, The traveling salesman problem, in: *Handbook of Applied Optimization* P. M. Pardalos and M. G. C. Resende, Eds, Oxford University Press, pp. 616-624.

Canuto, S.A., Resende, M.C.G., Ribeiro, C.C., 2001, Local search with perturbation for the prize collector Steiner tree problems in graphs, *Networks* **38**:50-58.

Cook, W.J., Seymour, P., 2003, Tour merging via branch-decomposition, *INFORMS Journal on Computing* **15**: 233-248.

Delorme, X., Gandibleux, X., and Rodriguez J., 2004, GRASP for set packing problems, *European Journal of Operational Research* **153**(3): 564-580.

Feo, T.A., and Resende, M.G.C., 1989, A probabilistic heuristic for a computationally difficult set covering problem, *Operations Research Letters* **8**: 67–71.

Festa, P., Pardalos, P. M., Pitsoulis, L. S., and Resende, M. G. C., 2005, GRASP with path-relinking for the weighted maximum satisfiability problem, *Lecture Notes in Computer Science* **3503**: 367-379.

Flood, M.M., 1956, The traveling-salesman problem, *Operations Research* **4**:61-75.

Glover, F., 1963, Parametric combinations of local job shop rules, Chapter IV, ONR Research Memorandum no. 117, GSIA, Carnegie Mellon University, Pittsburgh, PA.

Glover, F., Laguna, M., and Martí, R., 2000, Fundamentals of scatter search and path relinking, *Control and Cybernetics* **29**(3): 653-684.

Gomes, F. C., Pardalos, P. M., Oliveira, C. S., and Resende, M. G. C., 2001, Reactive GRASP with path relinking for channel assignment in mobile phone networks, *Proceedings of the 5th International Workshop on Discrete Algorithms and Methods for Mobile Computing and Communications*, ACM, 60-67.

Gonnet, G., Korotensky, C., and Benner, S., 2000, Evaluation measures of multiple sequence alignments, *Journal of Computational Biology* **7**(1-2), 261-276.

Gupta, S. R., and Smith, J. S., 2005, Algorithms for single machine total tardiness scheduling with sequence dependent setups, *European Journal of Operational Research*, article in press.

Gutin, G., and Punnen, A.P., 2002, *Traveling Salesman Problem and Its Variations*, Kluwer Academic Publishers, Dordrecht.

Guyon, R., Lorentzen, T. D., Hitte, C., Kim, L., Cadieu, E., Parker, H. G., Quignon, P., et al., 2003, A 1-Mb resolution radiation hybrid map of the canine genome, *Proceedings of the National Academy of Sciences of the United States of America* **100**(9): 5296-5301.

Helsgaun, K., 2000, An effective implementation of the Lin-Kernighan traveling salesman heuristic, *European Journal of Operational Research* **126**: 106–130.

Johnson, D. S., and McGeoh, L. A., 2002, Experimental analysis of heuristics for the STSP, in: *The Traveling Salesman Problem and Its Variations*, G. Guttin and A. Punnen, Eds, Kluwer Academic Publishers, Dordrecht, pp. 369-443.

Korte, B. H., 1989, Applications of combinatorial cptimization, in: *Mathematical Programming: Recent Developments and Applications*, M. Iri and K. Tanabe, Eds., Kluwer, Dordrecht, pp. 1-55.

Laguna, M., and Marti, R., 1999, GRASP and path relinking for 2-layer straight line crossing minimization, *INFORMS Journal on Computing* **11**(1): 44-52.

Lin, S., Kernighan, B., 1973, An effective heuristic algorithm for the traveling-salesman problem, *Operations Research* **21**: 498-516.

Marinakis, Y., Migdalas, A. ,and Pardalos, P. M., 2005, Expanding neighborhood GRASP for the traveling salesman problem, *Computational Optimization and Applications* **32**: 231-257.

Piñana, E., Plana, I., Campos, V., and Marti, R., 2004, GRASP and path relinking for the matrix bandwidth minimization, *European Journal of Operational Research* **153**(1): 200-210.

Resende, M.G.C., and Ribeiro, C.C., 2003XX, GRASP and path-relinking: Recent advances and applications, in: *Proceedings of the Fifth Metaheuristics International Conference (MIC2003)*, T. Ibaraki and Y. Yoshitomi, Eds., T6–1 – T6–6.

Resende, M.G.C., and Ribeiro, C.C., 2003XX, Greedy randomized adaptive search procedures, in: *Handbook of Metaheuristics*, Glover, F. and Kochenberger, G., Eds., Kluwer Academic Publishers, pp. 219-249.

Resende M.G.C. and Werneck, R.F., 2005, A hybrid multistart heuristic for the uncapacitated facility location problem, *European Journal of Operational Research*, article in press.

Scaparra, M. P., and Church, R. L., 2005, A GRASP and path relinking heuristic for rural road network development, *Journal of Heuristics* **11**(1): 89-108.

Souza, M.C., Duhamel, C., and Ribeiro, C.C., 2003, A GRASP with path-relinking heuristic for the capacitated minimum spanning tree problem, in: *Metaheuristics: Computer Decision Making*, Resende, M.G.C., and Souza, J., Eds., Kluwer Academic Publishers, pp. 627–658.

Yajima, E., Hara, T., Tsukamoto, M., and Nishio, S., 2001, Scheduling and caching Strategies for correlated data in push-based information systems, *ACM SIGAPP Applied Computing Review* **9**(1): 22-28.

Chapter 8

USING A RANDOMISED ITERATIVE IMPROVEMENT ALGORITHM WITH COMPOSITE NEIGHBOURHOOD STRUCTURES FOR THE UNIVERSITY COURSE TIMETABLING PROBLEM

Salwani Abdullah,* Edmund K. Burke,* Barry McCollum†

*Automated Scheduling, Optimisation and Planning Research Group, School of Computer Science & Information Technology, University of Nottingham, Jubilee Campus, Wollaton Road, Nottingham NG8 1BB, United Kingdom{sqa,ekb}@cs.nott.ac.uk; †Department of Computer Science, Queen's University Belfast, Belfast BT7 1NN United Kingdomb.mccollum@qub.ac.uk

Abstract: The course timetabling problem deals with the assignment of a set of courses to specific timeslots and rooms within a working week subject to a variety of hard and soft constraints. Solutions which satisfy the hard constraints are called feasible. The goal is to satisfy as many of the soft constraints as possible whilst constructing a feasible schedule. In this paper, we present a composite neighbourhood structure with a randomised iterative improvement algorithm. This algorithm always accepts an improved solution and a worse solution is accepted with a certain probability. The algorithm is tested over eleven benchmark datasets (representing one large, five medium and five small problems). The results demonstrate that our approach is able to produce solutions that have lower penalty on all the small problems and two of the medium problems when compared against other techniques from the literature. However, in the case of the medium problems, this is at the expense of significantly increased computational time.

1.   INTRODUCTION

In this paper, a randomised iterative improvement algorithm with composite neighbourhood structures for university course timetabling is presented. The approach is tested over eleven benchmark datasets that were introduced by Socha et al. (2002). The results demonstrate that our approach is capable of producing high quality solutions against others that appear in the literature. An extended abstract that describes this work was published in Abdullah et al. (2005b). The paper is organised as follows. The next section describes the university course timetabling problem in general and very briefly discusses the relevant timetabling literature. Section 3 presents a discussion of the literature on composite neighbourhood structures with a particular emphasis upon the employment of such structures in a variety of applications. Section 4 describes, in some detail, our randomised iterative improvement algorithm. The pseudo code of the implemented algorithm is also presented in this section. Experiments and results to evaluate the performance of the heuristic are discussed in Section 5. Section 6 presents a brief summary of the paper.

2.   THE UNIVERSITY COURSE TIMETABLING PROBLEM

Carter and Laporte (1998) defined course timetabling as:

> *"a multi-dimensional assignment problem in*
> *which students, teachers (or faculty members)*
> *are assigned to courses, course sections or*
> *classes; events (individual meetings between*
> *students and teachers) are assigned to*
> *classrooms and times"*

In university course timetabling, a set of courses is scheduled into a given number of rooms and timeslots within a week and, at the same time, students and teachers are assigned to courses so that the meetings can take place.

The course timetabling problem is subject to a variety of hard and soft constraints. Hard constraints need to be satisfied in order to produce a *feasible* solution. In this paper, we test our approach on the problem instances introduced by Socha et al. (2002) who present the following hard constraints:

- *No student can be assigned to more than one course at the same time.*
- *The room should satisfy the features required by the course.*
- *The number of students attending the course should be less than or equal to the capacity of the room.*
- *No more than one course is allowed to be assigned to a timeslot in each room.*

Socha et al. also present the following soft constraints that are equally penalised:

- *A student has a course scheduled in the last timeslot of the day.*
- *A student has more than 2 consecutive courses.*
- *A student has a single course on a day.*

The problem has

- A set of N courses, $e = \{e_1,\ldots,e_N\}$.
- 45 timeslots.
- A set of R rooms.
- A set of F room features.
- A set of M students.

The objective of this problem is to satisfy the hard constraints and to minimise the violation of the soft constraints.

In the last few years, several university course timetabling papers have appeared in the literature. Socha et al. (2002) presented a local search technique and an ant based methodology. They tested their approach on eleven test problems. These eleven problems were produced by Paechter's[1] course timetabling test instance generator and are the instances used to evaluate the method described in this paper. Since then, several papers have appeared which have tested their results on the same instances. Burke *et al.* (2003a) introduced a tabu-search hyperheuristic where a set of low level heuristics compete with each other. The goal was to raise the level of generality of search systems and the method was tested on a nurse rostering problem in addition to course timetabling. A graph hyper-heuristic was presented by Burke *et al.* (2006) where, within a generic hyper-heuristic framework, a tabu search approach is employed to search for permutations of constructive heuristics (graph colouring heuristics). Abdullah *et al.* (2005a) employed a variable neighbourhood search with a fixed tabu list which is used to penalise the unperformed neighbourhood structures. Other papers which test against these instances can be seen in Socha et al. (2003) who discuss ant algorithm methodologies at length and Rossi-Doria et al. (2003) who compare several metaheuristic methods.

In addition to the problem instances introduced by Socha et al (2002), Paechter's generator was also used to produce the problem sets for a timetabling competition held in 2002 (see http://www.idsia.ch/Files/ttcomp2002). They generated twenty instances for the competition itself and another three unseen instances to further check the performance of the algorithms. Some papers have recently appeared which test their methodologies on these competition problems. Kostuch (2005) presented a three phase approach which employs Simulated Annealing. This approach won the competition mentioned above and had 13 best results of the 20

[1] http://www.dcs.napier.ac.uk/~benp/

instances in the competition. Burke et al. (2003b) employed a Great Deluge method which generated 7 best results out of the 20 competition problems mentioned above. This method also produced some poor results on some problems which is why it came 3rd in the competition (because the competition used an average measure). The hybrid local search methodology which came 4th in the competition is described in Di Gaspero and Schaerf (2006). Arntzen and Løkketangen (2004) developed a tabu search method which came 5th in the competition. Lewis and Paechter (2004) designed several crossover operators and tested them against the competition datasets. They concluded that their results were not "state of the art". A hybrid metaheurstic approach has recently appeared in the literature which is tested on these competition problems and which produces improved results to those generated by the competition (Chiarandini et al. 2006). Also, Kostuch and Socha (2004) investigated the possibility of using a statistical model to predict the difficulty of timetabling problems and they employed the competition instances.

In 2005, Lewis and Paechter used the same instance generator to create another sixty "hard" test instances (Lewis and Paechter 2005). They tested their grouping genetic algorithm on these sixty instances but were concerned only with feasibility.

In addition to the university course timetabling papers which have used problems produced by Paechter's generator, several other articles have recently appeared which represent case studies on real university timetabling instances. Examples include Avella and Vasil'Ev (2005), Daskalaki et al. (2004), Dimopoulou and Miliotis (2004) and Santiago-Mozos et al. (2005).

Other aspects of university course timetabling have been widely discussed in the literature over the last thirty years or so. A survey of practical approaches to the problem, up to 1998, can be seen in Carter and Laporte (1998). The following papers represent a comprehensive list of surveys and overviews of educational timetabling (which include issues related to University course timetabling) i.e. Bardadym (1996), Burke et al. (1997), Burke and Petrovic (2002), Burke et al. (2004), Carter (2001), Petrovic and Burke (2004), Schaerf (1999), de Werra (1985) and Wren (1996).

3.   COMPOSITE NEIGHBOURHOOD STRUCTURES: RESEARCH AND DEVELOPMENTS

A composite neighbourhood structure subsumes two or more neighbourhood structures. The advantage of combining several neighbourhood structures is that it helps to compensate against the ineffectiveness of using each type of structure in isolation (Grabowski and Pempera, 2000 and Liaw 2003). For example, a solution space that is easily accessible by insertion moves may be difficult to reach using swap moves. Some examples of composite neighbourhood structures that are available in the literature are discussed here.

Grabowski and Pempera (2000) applied a composite neighbourhood structure for sequencing jobs in a production system that consists of exchanges and the insertion of elements. Gopalakrishnan et al. (2001) used three moves (swap, add and drop) in a tabu search heuristic for preventive maintenance scheduling. The decision on which move to use depends on the current state of the search. The interaction of the moves makes it possible to carry out a strategic search. The computational results show that the approach can improve the solution quality when compared to the local heuristics employed by Gopalakrishnan et al. (1997).

Liaw (2003) also used a composite neighbourhood structure in the tabu search approach for the two-machine preemptive open shop scheduling problem. The tabu search switches to the other neighbourhood structures (between an insertion move that shifts one job from its current position to a new position and a swap move that exchanges the position of two jobs) after a number of iterations without any improvements. Computational experiments have shown that this scheme significantly improves the performance of tabu search in terms of solution quality. The neighbourhood used in Ouelhadj (2003) has a composite structure where the tabu search approach, applied to the dynamic scheduling of a hot strip mill agent, employed three neighbourhood schemes (swap, shift and inversion) alternately. Computational experiments showed that the composite structure improves the solution quality compared with tabu search using a single neighbourhood. Another example of a composite neighbourhood structure was presented by Landa Silva (2003). He employed several neighbourhood structures (relocate, swap and interchange) in different metaheuristics (iterative improvement, simulated annealing and tabu search) and applied this to a space allocation problem in an academic institution.

Bilge et al. (2004) used a "hybrid" neighbourhood structure in a tabu search algorithm for the parallel machine total tardiness problem. The "hybrid" structure consists of the complete "insert neighbourhood" with the addition of a partial "swap neighbourhood". In an insert move operation, two jobs are identified and the first job is placed in the location that precedes the location of the second job. Then, a swap move places each job in the location that was previously occupied by the other job.

4.   THE RANDOMISED ITERATIVE IMPROVEMENT ALGORITHM

This algorithm presented here always accepts an improved solution and a worse solution is accepted with a certain probability.

4.1   The Neighbourhood Structures

The different neighbourhood structures and their explanation can be outlined as follows:

N1:   Select two courses at random and swap timeslots.

N2:   Choose a single course at random and move to a new random feasible timeslot.

N3:   Select two timeslots at random and simply swap all the courses in one timeslot with all the courses in the other timeslot.

N4:   Take 2 timeslots (selected at random), say t_i and t_j (where $j>i$) where the timeslots are ordered t_1, t_2, …, t_{45}. Take all the exams in t_i and allocate them to t_j. Now take the exams that were in t_j and allocate them to t_{j-1}. Then allocate those that were in t_{j-1} to t_{j-2} and so on until we allocate those that were in t_{i+1} to t_i and terminate the process.

N5:   Move the highest penalty course from a random 10% selection of the courses to a random feasible timeslot.

N6:   Carry out the same process as in N5 but with 20% of the courses.

N7:   Move the highest penalty course from a random 10% selection of the courses to a new feasible timeslot which can generate the lowest penalty cost.

N8:   Carry out the same process as in N7 but with 20% of the courses.

N9:   Select one course at random, select a timeslot at random (distinct from the one that was assigned to the selected course) and then apply the kempe chain from Thompson and Dowsland (1996).

N10:   This is the same as N9 except the highest penalty course from 5% selection of the courses is selected at random.

N11:   Carry out the same process as in N9 but with 20% of the courses.

4.2   The Algorithm

In the approach presented in this paper, a set of the neighbourhood structures outlined in subsection 4.1 is applied. The hard constraints are never violated during the timetabling process.

The pseudo code for the algorithm implemented in this paper is given in Figure 1. The algorithm starts with a feasible initial solution which is generated by a constructive heuristic as discussed in Abdullah et al. (2005a). Let K be the total number of neighbourhood structures to be used in the search (K is set to be 11 in this implementation) and $f(Sol)$ is the quality measure of the solution *Sol*. At the start, the best solution, Sol_{best} is set to be *Sol*. In a *do-while* loop, each neighbourhood i where $i \in \{1,...,K\}$ is applied to *Sol* to obtain $TempSol_i$. The best solution among $TempSol_i$ is identified, and is set to be the new solution *Sol**. If *Sol** is better than the best solution in hand, Sol_{best}, then *Sol** is accepted. Otherwise, the exponential Monte Carlo acceptance criterion is applied. This accepts a worse solution with a certain probability. The criterion is discussed in Ayob and Kendall (2003). The new solution *Sol** is accepted if the generated random number in [0,1], *RandNum*, is less than the probability which is computed by $e^{-\delta}$ where δ is the difference between the cost of the old and new solutions (i.e. $\delta = f(Sol^*)$ $- f(Sol)$). The Monte Carlo method will exponentially increase the acceptance probability if δ is small. The process is repeated and stops when the termination criteria is met (in this work the termination criteria is set as the number of evaluations i.e. 200000 evaluations or when the penalty cost is zero).

```
Set the initial solution Sol by employing a constructive
heuristic;
Calculate initial cost function f(Sol);
Set best solution Sol_best ← Sol;
do while (not termination criteria)
   for i = 1 to K where K is the total number of neighbourhood
   structures
      Apply neighbourhood structure i on Sol, TempSol_i;
      Calculate cost function f(TempSol_i);
   end for
   Find the best solution among TempSol_i where i ∈ {1,…,K}
   call new solution Sol*;
   if (f(Sol*) < f(Sol_best))
      Sol ← Sol*;
      Sol_best ← Sol*;
   else
      Apply an exponential Monte Carlo where:
      δ = f(Sol*) -  f(Sol));
      Generate RandNum, a random number in [0,1];
      if (RandNum < e^-δ )
         Sol ← Sol*;
   end if
end do
```

Figure 1. The pseudo code for the randomised iterative improvement algorithm

5. EXPERIMENTS AND RESULTS

The approaches are coded in Microsoft Visual C++ version 6 under Windows. All experiments were run on an Athlon machine with a 1.2GHz processor and 256 MB RAM running under Microsoft Windows 2000 version 5. We evaluate our results on the instances taken from Socha et al (2002) and which are available at http://iridia.ulb.ac.be/~msampels/tt.data/. We employed the same initial solutions as in Abdullah et al. (2005a). The experiments were run for 200000 iterations which takes approximately eight hours for each of the medium datasets and at most 50 seconds for the small datasets. Note that course timetabling is a problem that is usually tackled several months before the schedule is required. An eight hours run for course timetabling is perfectly acceptable in a real world environment. This is a scheduling problem where the time taken to solve the problem is not critical. The emphasis in this paper is on generating good quality solutions and the price to pay for this can be taken as being a large amount of computational time.

The experiments for the course timetabling problem discussed in this paper were tested on the benchmark course timetabling problems proposed by the Metaheuristics Network that need to schedule 100-400 courses into a timetable with 45 timeslots corresponding to 5 days of 9 hours each, whilst satisfying room features and room capacity constraints. They are divided into three categories: *small*, *medium* and *large*. We deal with 11 instances: 5 *small*, 5 *medium* and 1 *large*. The parameter values defining the categories are given in Table 1.

Table 1. The parameter values for the course timetabling problem categories

Category	*Small*	*Medium*	*Large*
Number of courses	100	400	400
Number of rooms	5	10	10
Number of features	5	5	10
Number of students	80	200	400
Maximum courses per student	20	20	20
Maximum student per courses	20	50	100
Approximation features per room	3	3	5
Percentage feature use	70	80	90

The best results out of 5 runs obtained are presented. Table 2 shows the comparison of the approach in this paper with other available approaches in the literature on the five small problems. Table 3 illustrates our comparison on the medium/large problems. The term "x%Inf." in Table 3 indicates a percentage of runs that failed to obtain feasible solutions.

The best results are presented in bold in both tables. Note that the only methods that were able to obtain feasible solutions for the large problem were the ant method (Socha et al, 2002) and the graph based hyper-heuristic (Burke et al, 2006) with the ant method being better.

It can be seen that the randomised iterative improvement algorithm has better results than Abdullah et al. (2005a) on all five medium datasets with the same (best result) penalty cost for the *small* instances. Our approach has better results than the local search method (Socha et al, 2002) on three of the medium instances and on all five of the small datasets. Our method has higher quality results when compared against the ant approach (Socha et al, 2002) on four of the small problems, with both approaches being able to obtain zero penalty on the other. Our algorithm gets better results than the ant technique on two of the medium instances. The iterative improvement approach is has better penalty values than the tabu search hyper-heuristic (Burke et al. 2003a) on three of the small datasets and both methods get zero penalty on the other two. It was better values on just two of the medium sets. The iterative approach obtained better results than the graph based hyper-heuristic (Burke et al. 2006) on all datasets except the large one.

Note that our approach has the very best results across seven of the eleven datsets (although it does perform very poorly on the large one). It is particularly effective on the *small* problems, taking approximately 50 seconds to obtain zero penalties as opposed to, for example, the algorithms of (Socha et al) which take 90 seconds. It is quite effective on the *medium* problems but at the expense of a high level of computational time. It takes our algorithm about 8 hours to produce these solutions for the medium problems whereas, for example, it takes the (Socha et al, 2002) methods 900 seconds (15 minutes). The need for the long run time is probably due to some neighbourhood structures in our method being less effective on this type of problem.

Table 2. Comparison of results on the small datasets

Data Set	Initial Sollution	Randomised Iterative Improvement Algorithm		VNS with tabu (Abdullah et al. 2005a)	Local search (Socha et al. 2002)	Ant Algorithm (Socha et al. 2002)	Tabu-based hyper-heuristic (Burke et al. 2003a)	Graph hyper-heuristic (Burke et al. 2006)
		Best	Median	(Best)	(Median)	(Median)	(Best)	(Best)
s1	261	0	0	0	8	1	1	6
s2	245	0	0	0	11	3	2	7
s3	232	0	0	0	8	1	0	3
s4	158	0	0	0	7	1	1	3
s5	421	0	0	0	5	0	0	4

Table 3. Comparison of results on the medium/large datasets

Data Set	Initial Sollution	Randomised Iterative Improvement Algorithm		VNS with tabu (Abdull ah et al. 2005a) (Best)	Local search (Socha et al. 2002) (Median)	Ant Algorithm (Socha et al. 2002) (Median)	Tabu-based hyper-heuristic (Burke et al. 2003a) (Best)	Graph hyper-heuristic (Burke et al. 2006) (Best)
		Best	Median					
s1	261	0	**0**	**0**	8	1	1	6
m1	914	2	24	31	199	195	**146**	372
m2	878	1	**16**	31	202.5	184	173	419
m3	941	2	26	35	77.5	**248**	267	359
m4	865	1	18	24	177.5	**164.5**	169	348
m5	780	1	**15**	29	100%	219.5	303	171
l	100%	-	-	10	100%	**851.5**	80%	1068

Data Set Key: l = large, m1 = medium1, m2 = medium 2 and so on.

Figures 2 and 3 show the behaviour of the randomised iterative improvement algorithm applied to the *small1* and *medium5* datasets, respectively. In all the figures, the x-axis represents the number of evaluations whilst the y-axis represents the penalty cost. The graphs illustrate the exploration of the search space. The curves move up and down because worse solutions are accepted with a certain probability in order to escape from local optima. The penalty cost can be quickly reduced at the beginning of the search where there is (possibly) a lot of room for improvement. It is believed that better solutions can be obtained in these experiments (particularly on the smaller problems) because the composite neighbourhood structures offer flexibility for the search algorithm to explore different regions of the solution space. The graphs for the *small* datasets show that our algorithm is able to obtain zero penalties in less than 1500 evaluations which is an improvement upon Burke et al. (2003a) which set the number of evaluations at 12000 for *small* datasets.

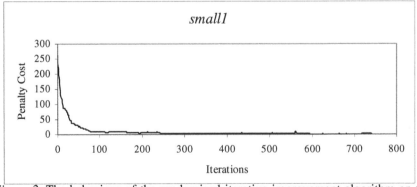

Figure 2. The behaviour of the randomised iterative improvement algorithm on the *small1* dataset

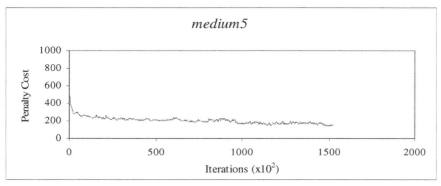

Figure 3. The behaviour of the randomised iterative improvement algorithm on the *medium5* dataset

Figures 4 and 5 show the frequency charts of the neighbourhood structures that have been selected to be used by the randomised iterative improvement algorithm for the *small* and *medium* datasets, respectively. The x-axis represents the datasets while the y-axis represents the frequency of the neighbourhood structures being employed throughout the search.

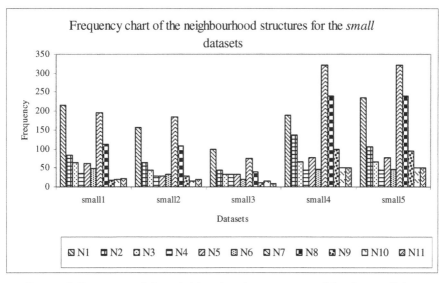

Figure 4. Frequency of the neighbourhood structures used for the *small* datasets

It can be seen, from Figure 4, that the neighbourhood structures "N1", "N2", "N7" and "N8" are the most popular structures used in the algorithm for *small* datasets. The popular structures for the *medium* datasets are "N1", "N2", "N5", "N6", "N7" and "N8" as shown in Figure 5.

This illustrates that the most popular neighbourhood structures that are being supplied to the randomised iterative improvement algorithm are almost the same between the *small* and *medium* datasets (i.e. "N1", "N2", "N7" and

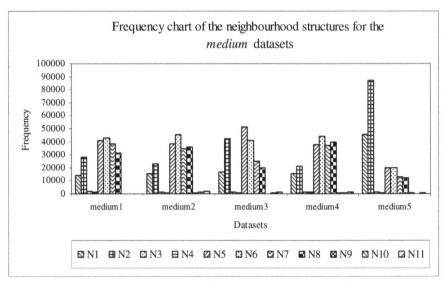

Figure 5. Frequency of the neighbourhood structures used for the *medium*
datasets

"N8"). However, as the problem gets larger, there may be fewer and more sparsely distributed solution points (feasible solutions) in the solution space since too many courses are conflicting with each other. Thus, the approach may need extra neighbourhood structures (i.e. "N5" and "N6" in this case) to force the search algorithm to diversify its exploration of the solution space by moving from one neighbourhood structure to another. Further investigation was carried out to support the claim that the composite neighbourhood structure performs better than the single neighbourhood structure by employing selected neighbourhood structures separately i.e. "N1", "N2", "N5", "N6", "N7" and "N8" (which are the most popular neighbourhood structures used for the *small* and *medium* datasets). The *small* datasets are able to obtain zero penalty in less than 1500 evaluations. Thus, for the experiments carried out here, the number of evaluations for the *small* datasets is set as equal to the number of evaluations where the best solutions are obtained (i.e. 873, 707, 413, 1012 and 1329 evaluations for *small1*, *small2*, *small3*, *small4* and *small5*, respectively). The number of evaluations for the *medium* datasets remains the same. Table 4 gives the comparison of the performance of variants of the randomised iterative improvement algorithm in terms of penalty cost (objective function value). The results demonstrate that the algorithm with composite neighbourhood structures is uniformly the best in terms of penalty cost compared to other randomised iterative improvement algorithm variants.

Table 4. Comparison of the performance of the randomised iterative
improvement algorithm on single and composite neighbourhood structures

Dataset	Initial solution	Randomised iterative improvement algorithm neighbourhoods						Composite
		N1	N2	N5	N6	N7	N8	
small1	261	7	2	2	5	5	8	**0**
small2	245	6	2	4	5	9	6	**0**
small3	232	6	4	6	3	6	1	**0**
small4	158	6	3	4	1	5	9	**0**
small5	421	1	3	4	6	7	1	**0**
medium	914	3	3	5	7	5	7	**242**
medium	878	3	3	5	6	5	6	**161**
medium	941	4	4	7	7	7	7	**265**
medium	865	3	3	5	6	5	6	**181**
medium	780	4	3	6	6	7	6	**151**
large	100%Inf	-	-	-	-	-	-	-

Figures 6 and 7 illustrate the behaviour of the randomised iterative
improvement algorithm using a single neighbourhood structure compared to
the composite neighbourhood structure applied on the *small1* and *medium5*
datasets, respectively.

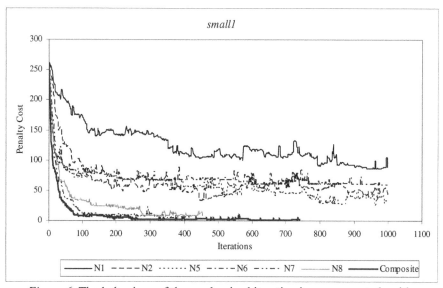

Figure 6. The behaviour of the randomised iterative improvement algorithm
using single and composite neighbourhood structures applied on the *small1* dataset

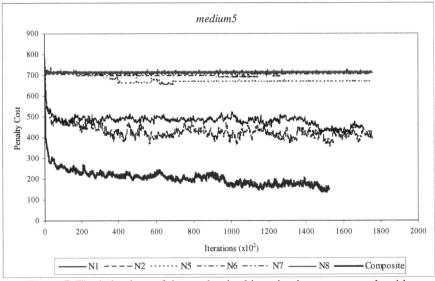

Figure 7. The behaviour of the randomised iterative improvement algorithm using single and composite neighbourhood structures applied on the *medium5* dataset

The diagrams show the convergence of the penalty cost of the algorithm for *small1 and medium5* for a number of evaluations for which the best solution is found. It can be seen that the randomised iterative improvement algorithm with the composite neighbourhood is significantly better than other variants with single neighbourhood in terms of solution quality given the same number of evaluations. All the other problems of the family have the same behaviour as in Figures 6 and 7.

6. CONCLUSION AND FUTURE WORK

This paper has focused on investigating a composite neighbourhood structure with a randomised iterative improvement algorithm for the university course timetabling problem. Preliminary comparisons indicate that this algorithm is competitive with other approaches in the literature. Indeed, it produced seven solutions that were better than or equal to the published penalty values on these eleven instances although it did require significant computational time for the medium/large problems. It is an approach that is particularly effective on smaller problems. Further experiments were carried out to demonstrate that it is more effective to employ composite neighbourhood structures rather than a single neighbourhood structure because of the different ways of search that are represented by various neighbourhood structures.

Future research will be aimed at exploring how the algorithm could intelligently select the most suitable neighbourhood structures according to the characteristics of the problems. Another direction of future research will investigate the integration of a population-based approach with a local search method.

Acknowledgement

This work has been supported by the Public Services Department of Malaysia (JPA) and the University Kebangsaan Malaysia (UKM).

References

[1]   Abdullah S, Burke EK and McCollum B (2005a) An investigation of variable neighbourhood search for university course timetabling. In: Proceedings of *The 2nd Multidisciplinary International Conference on Scheduling: Theory and Applications (MISTA 2005)*, New York, USA, July 18th-21st, pp 413-427.

[2]   Abdullah S, Burke EK and McCollum B (2005b). Using a Randomised Iterative Improvement Algorithm with Composite Neighbourhood Structures for University Course Timetabling. In: Proceedings of the 6th Metaheuristics International Conference (MIC 05), Vienna, Austria, August 22nd-26th, in CD-ROM, 2005.

[3]   Avella P and Vasil'Ev I (2005) A Computational Study of a Cutting Plane Algorithm for University Course Timetabling. *Journal of Scheduling* 8(6), pp 497-514.

[4]   Ayob M and Kendall G (2003) A monte carlo hyper-heuristic to optimise component placement sequencing for multi head placement machine. *Proceedings of the International Conference on Intelligent Technologies, InTech'03*, Thailand, Dec 17th-19th, pp 132-141.

[5]   Bardadym VA (1996) Computer-aided school and university timetabling: A new wave. Practice and Theory of Automated Timetabling V (eds. Burke and Ross), Springer Lecture Notes in Computer Science Volume 1153, pp 22-45.

[6]   Bilge Ü, Kiraç F, Kurtulan M and Pekgün P (2004) A tabu search algorithm for the parallel machine total tardiness problem. *Computers & Operations Research* 31(3), pp 397-414.

[7]   Burke EK, Jackson KS, Kingston JH and Weare RF (1997) Automated timetabling: The state of the art, *The Computer Journal* 40(9), pp 565-571.

[8]   Burke EK and Petrovic S (2002) Recent research direction in automated timetabling. *European Journal of Operational Research* 140, pp 266-280.

[9]   Burke EK, Kendall G and Soubeiga E (2003a) A tabu search hyperheuristic for timetabling and rostering. *Journal of Heuristics* 9(6), pp 451-470.

[10]   Burke EK, Bykov Y, Newall J and Petrovic S (2003b) A Time-Predefined Approach to Course Timetabling, *Yugoslav Journal of Operational Research* (YUJOR), Vol 13, No. 2, pp 139-151.

[11]   Burke, EK, Kingston J and De Werra D (2004) *Applications to Timetabling*,

in the Handbook of Graph Theory, (eds. Gross J and Yellen J), Chapman Hall/CRC Press, pp 445-474,

[12]   Burke, E.K., McCollum, B., Meisels, A., Petrovic, S. and Qu, R., A Graph-Based Hyper Heuristic for Educational Timetabling Problems, *European Journal of Operational Research* 176(1), 1 January 2007, pp 177-192.

[13]   Carter MW (2001) Timetabling, encyclopedia of operations research and management science (eds Gass and Harris), Kluwer, pp 833-836.

[14]   Carter MW and Laporte G (1998) Recent developments in practical course timetabling. Practice and Theory of Automated Timetabling V (eds. Burke and Carter), Springer Lecture Notes in Computer Science Volume 1408, pp 3-19.

[15]   Chiarandini M, Birattari M, Socha K and Rossi-Doria O (2006) An effective hybrid algorithm for university course timetabling. *Journal of Scheduling* 9(5), pp 403-432.

[16]   Daskalaki S, Birbas T and Housos H (2004) An integer programming formulation for a case study in university timetabling. *European Journal of Operational Research* 153(1), pp 117-135.

[17]   Dimopoulou M and Miliotis P (2004) An automated university course timetabling system developed in a distributed environment: A case study. *European Journal of Operational Research* 153(1), pp 136-147.

[18]   de Werra D (1985) An introduction to timetabling. *European Journal of Operational Research* 19, pp 151-162.

[19] Di Gaspero L and Schaerf A (2006) Neighborhood portfolio approach for local search applied to timetabling problems. *Journal of Mathematical Modeling and Algorithms*, 5(1), pp 65-89.

[20]   Gopalakrishnan M, Ahire SL and Miller DM (1997) Maximising the effectiveness of a preventive maintenance system: An adaptive modeling approach. *Management Science*, 43(6), pp 827-840.

[21]   Gopalakrishnan M, Mohan S and He Z. (2001). A tabu search heuristic for preventive maintenance scheduling. *Computers & Industrial Engineering*, 40, pp 149-160.

[22]   Grabowski J and Pempera J (2000) Sequencing of jobs in some production system: Theory and methodology. *European Journal of Operational Research*, 125, pp 535-550.

[23]   Kostuch P (2005) The university course timetabling problem with a three-phase approach. Practice and Theory of Automated Timetabling V (eds. Burke and Trick), Springer Lecture Notes in Computer Science Volume 3616, pp 109-125.

[24]   Kostuch P and Socha K (2004), Hardness Prediction for the University Course Timetabling Problem, Proceedings of the Evolutionary Computation in Combinatorial Optimization (EvoCOP 2004), Coimbra, Portugal, April 5-7, 2004, Springer Lecture Notes in Computer Science Volume 3004, pp 135-144.

[25]   Landa Silva JD (2003) Metaheuristic and Multiobjective Approaches for Space Allocation. *PhD Thesis*, Department of Computer Science, University of Nottingham, United Kingdom.

[26]   Lewis R and Paechter B (2004) New crossover operators for timetabling with evolutionary algorithms. *Proceedings of the 5th International Conference on*

Recent Advances in Soft Computing (ed. Lotfi), UK, December 16[th-]18th, pp 189-194.

[27] Lewis R and Paechter B (2005) Application of the groping genetic algorithm to university course timetabling. Evolutionary Computation in Combinatorial Optimisation (eds. Raidl and Gottlieb), Springer Lecture Notes in Computer Science Volume 3448, pp 144-153.

[28] Liaw CF (2003) An efficient tabu search approach for the two-machine preemptive open shop scheduling problem. *Computers & Operations Research* 30(14), pp 2081-2095.

[29] Ouelhadj D (2003) A multi-agent system for the integrated dynamic scheduling of steel production. PhD Thesis, Department of Computer Science, University of Nottingham, United Kingdom.

[30] Petrovic S and Burke EK (2004) University timetabling, Ch. 45 in the *Handbook of Scheduling: Algorithms, Models, and Performance Analysis* (eds. J. Leung), Chapman Hall/CRC Press.

[31] Rossi-Doria O, Samples M, Birattari M, Chiarandini M, Dorigo M, Gambardella LM, Knowles J, Manfrin M, Mastrolilli M, Paechter B, Paquete L and Stützle T (2003). A comparison of the performance of different meta-heuristics on the timetabling problem. Practice and Theory of Automated Timetabling V (eds. Burke and De Causmaecker), Springer Lecture Notes in Computer Science Volume 2740, pp 329-354.

[32] Santiago-Mozos R, Salcedo-Sanz S, DePrado-Cumplido M, Carlos Bousoalz C. A two-phase heuristic evolutionary algorithm for personalising course timetables: A case study in a Spanish university. *Computers and Operations Research* 32, pp 1761-1776.

[33] Schaerf A (1999) A survey of automated timetabling. *Artificial Intelligence Review* 13(2), pp 87-127.

[34] Socha K, Knowles J and Samples M (2002) A max-min ant system for the university course timetabling problem. *Proceedings of the 3rd International Workshop on Ant Algorithms (ANTS 2002)*, Springer Lecture Notes in Computer Science Volume 2463, pp 1-13.

[35] Socha K, Sampels M and Manfrin M (2003) Ant algorithms for the university course timetabling problem with regard to the state-of-the-art. *Proceedings of 3rd European Workshop on Evolutionary Computation in Combinatorial Optimization (EvoCOP'2003)*, UK, April 14th-16th, Springer Lecture Notes in Computer Science Volume 2611, pp 335-345.

[36] Thompson J and Dowsland K (1996) Various of simulated annealing for the examination timetabling problem. *Annals of Operational Research* 63, pp 105-128.

[37] Wren A (1996) Scheduling, timetabling and rostering – A special relationship? Practice and Theory of Automated Timetabling V (eds. Burke and Ross), Springer Lecture Notes in Computer Science Volume 1153, pp 46-75.

Part V

Dynamic and Stochastic Problems

Chapter 9

VARIABLE NEIGHBORHOOD SEARCH FOR THE PROBABILISTIC SATISFIABILITY PROBLEM

Dejan Jovanović,[1] Nenad Mladenović, [2] and Zoran Ognjanović[1]

[1] *Mathematical Institute*
Kneza Mihaila 35, 11000 Belgrade
Serbia
dejan@mi.sanu.ac.yu, zorano@mi.sanu.ac.yu

[2] *Brunel University*
West London, United Kingdom
Nenad.Mladenovic@brunel.ac.uk

Abstract   This paper presents a new method for the probabilistic logic satisfiability problem, based on the variable neighborhood search metaheuristic. The solution space consists of 0-1 variables, while the associated probabilities are found by our fast approximate variable neighborhood descent procedure combined with the Nelder-Mead nonlinear optimization method. Computational experience shows that, with our approach, problem instances with up to 200 propositional letters can be solved successfully. They are, to the best of our knowledge, the largest solved instances of the PSAT problem that appeared in the literature.

Keywords: Probabilistic Logic, Probabilistic Satisfiability, Variable Neighborhood Search

Introduction

In the field of artificial intelligence researchers have studied uncertain reasoning using different tools. Some of the formalisms for representing and reasoning with uncertain knowledge are based on probabilistic logic [Fagin et al., 1990, Nilsson, 1986, Ognjanović and Rašković,

2000, Rašković, 1993] (See also [Ognjanović et al., 2005] for a more complete list of references). This logic extends the classical propositional language with expressions that speak about probability, while truth values of formulas remain true or false. Probabilistic logic allows inferences to be made in a general framework, without any special assumptions about the underlying probability distributions.

Given a set of linear constraints on probabilities of classical propositional formulas, the probabilistic satisfiability problem (PSAT for short, also called the problem of satisfying conditions of possible experience in [Boole, 1854], the problem of coherence assessment in [de Finetti, 1974], the probabilistic entailment in [Nilsson, 1986], and the decision form of probabilistic satisfiability in [Hansen and Jaumard, 2001]) corresponds to checking consistency of these constraints. Possible applications of PSAT concern knowledge-based systems, reliability checking, game theory and economics, where it is important to take care about players' beliefs, etc. For example, one can use procedures for solving PSAT to check the consistency of rules with associated uncertainty factors, and the corresponding techniques designed to handle uncertain knowledge in expert systems. PSAT is NP-complete [Fagin et al., 1990]. PSAT can be reduced to a linear programming problem. However, solving it by any standard linear system solving procedure is unsuitable in practice. This is due to the exponential growth of the number of variables in the linear system obtained by such a reduction (as expected from NP-completeness). Nevertheless, it is still possible to use more efficient numerical methods, e.g., the column generation procedure of linear programming [Jaumard et al., 1991]. Nilsson [Nilsson, 1986] proposed a heuristic approach for solving large PSAT problem instances. In more recent articles [Ognjanović et al., 2001, Ognjanović et al., 2004] we presented Genetic algorithms (GA) for PSAT.

Variable neighborhood search (VNS) [Mladenović and Hansen, 1997, Hansen and Mladenović, 2001] is a simple, yet very effective metaheuristic that has shown to be very robust on a variety of practical NP-hard problems (for the recent survey on VNS see also [Hansen and Mladenović, 2003]). Among other applications, VNS has already been used for solving the weighted satisfiability problem (WSAT) problem [Hansen et al., 2000]. Also, VNS has been used as a subproblem solver within the column generation framework in [Hansen and Perron, 2004], but for a slightly different version of the PSAT problem. Here we present a new heuristic for PSAT based on VNS metaheuristic rules. Our method alternates solutions in Boolean variables with solutions in continuous variables (probabilities), both obtained by VNS.

The rest of the paper is organized as follows. In Section 1 a brief description of probabilistic logic and the PSAT problem we are dealing with is outlined. In Section 2 our implementation of VNS rules for solving PSAT is described, followed by some experimental results in Section 3. Concluding remarks and directions for further research are given in Section 4.

1. Probabilistic Logic and PSAT

In this section we introduce the PSAT problem formally (for a more detailed description see [Ognjanović and Rašković, 2000]). Let $\mathcal{L} = \{x, y, z, \ldots\}$ be the set of propositional letters (primitive propositions, logical variables). From this set we construct classical propositional formulas in the usual way, using the standard Boolean connectives $\neg$, $\wedge$, $\vee$ and $\rightarrow$.

Let α be a classical propositional formula and $\{x_1, x_2, \ldots, x_k\}$ be the set of all propositional letters that appear in α. An *atom* of α (also called possible world in [Jaumard et al., 1991, Nilsson, 1986]) is defined as formula $at = \pm x_1 \wedge \ldots \wedge \pm x_k$, where $\pm x_i \in \{x_i, \neg x_i\}$. There are 2^k different atoms of a formula containing k primitive propositions. Let $\text{At}(\alpha)$ denote the set $\{at_1, \ldots, at_{2^k}\}$ containing all different atoms of α. We say that atom $at \in \text{At}(\alpha)$ satisfies α (denoted by $at \vDash \alpha$) iff any propositional interpretation I that satisfies the atom at also satisfies α, i.e. $I(at) = \top \Rightarrow I(\alpha) = \top$ for all I. Further, we define $[\alpha] = \{at \in \text{At}(\alpha) : at \vDash \alpha\}$, that is, $[\alpha]$ denotes the set of all atoms that satisfy α.

We extend the basic propositional language with expressions of the form $w(\alpha)$. The intended meaning of $w(\alpha)$ is the probability of α being true. A *weight term* is an expression of the form $a_1 w(\alpha_1) + \ldots + a_n w(\alpha_n)$, where a_i's are rational numbers, and α_i's are classical propositional formulas with propositional letters from $\mathcal{L}$.

A *basic weight formula* has the form $t \geq b$, where t is a weight term, and b is a rational number. We use $t < b$ to denote $\neg(t \geq b)$. A *weight literal* is an expression of the form $t \geq b$ or $t < b$. The set of all *weight formulas* is the minimal set that contains all basic weight formulas, and it is closed under Boolean operations. Similarly as above, we use $\text{At}(f)$ to denote the set of all atoms which contain propositional letters from the weight formula f.

A weight formula is in the *weight conjunctive form* (WCF) if it is a conjunction of weight literals. As every weight formula f is a Boolean combination of basic weight formulas, it can be transformed to a disjunctive normal form (each $\rho_{i,j}$ is either $<$ or $\geq$)

$$\text{DNF}(f) = \bigvee_{j=1}^{m} \bigwedge_{i=1}^{L_j} \left(a_1^{i,j} w(\alpha_1^{i,j}) + \ldots + a_{n_{i,j}}^{i,j} w(\alpha_{n_{i,j}}^{i,j}) \; \rho_{i,j} \; b_{i,j} \right) ,$$

where each disjuncts of $\text{DNF}(f)$ is a formula in WCF. Since a disjunction is satisfiable iff at least one disjunct is satisfiable, we will consider WCF formulas only.

Semantics of a probabilistic formula f is defined with respect to a probability function μ defined on the set $\text{At}(f)$ (probabilities on the possible worlds). Therefore, $\mu : \text{At}(f) \to [0,1]$, and additionally the probabilities $\mu(at)$ over all atoms from $\text{At}(f)$ sum up to 1.

The truth value of f with respect to μ is defined inductively, with the base case being the basic weight formulas of f.

Let $a_1 w(\alpha_1) + \ldots + a_n w(\alpha_n) \geq b$ be a basic weight formula of f. For each classical propositional formula α_i, $\mu(\alpha_i)$ is defined to be the sum of probabilities of the atoms from f that satisfy α_i, i.e.,

$$\mu(\alpha_i) = \sum_{at \in [\alpha_i]} \mu(at) .$$

Having this, μ satisfies the basic weight formula above iff $a_1 \mu(\alpha_1) + \ldots + a_n \mu(\alpha_n) \geq b$. Further, μ satisfies a weight literal $\neg(t \geq b)$ iff it does not satisfy $t \geq b$. Finally, μ satisfies WCF formula f iff it satisfies every weight literal from f.

DEFINITION 9.1 (PSAT PROBLEM) *Given a WCF formula f, is there a probability function μ defined on* $\text{At}(f)$ *which satisfies f?*

In other words, is there a probability function μ defined on the set $\text{At}(f)$ such that for every weight literal $a_1^i w(\alpha_1^i) + \ldots + a_{n_i}^i w(\alpha_{n_i}^i) \; \rho_i \; b_i$ from f

$$a_1^i \sum_{at \in [\alpha_1^i]} \mu(at) + \ldots + a_{n_i}^i \sum_{at \in [\alpha_{n_i}^i]} \mu(at) \; \rho_i \; b_i, \qquad (9.1)$$

with $\mu(at) \geq 0$, for every atom $at \in \text{At}(f)$, and $\sum_{at \in \text{At}(f)} \mu(at) = 1$.

The system (9.1) contains a row for each weight literal of f, and columns that correspond to atoms $at \in \text{At}(f)$ that belong to $[\alpha]$ for at least one propositional formula α from f.

EXAMPLE 9.2 Consider the formula $f = w(p \to q) + w(p) \geq 1.7 \wedge w(q) \geq 0.6$. The set of atoms of f is

$$\text{At}(f) = \{p \wedge q, p \wedge \neg q, \neg p \wedge q, \neg p \wedge \neg q\} .$$

The classical formulas from f are $p \to q$, p, and q, while the sets of atoms satisfying them are:

α	$p \to q$	p	q
$[\alpha]$	$p \wedge q, \neg p \wedge q, \neg p \wedge \neg q$	$p \wedge q, p \wedge \neg q$	$p \wedge q, \neg p \wedge q$

Having the corresponding satisfying sets, the satisfiability of the formula reduces to finding a probability assignment μ over $At(f)$ that satisfies the system

$$\mu(p \wedge \neg q) + \mu(\neg p \wedge q) + \mu(\neg p \wedge \neg q) + 2\mu(p \wedge q) \geq 1.7,$$
$$\mu(p \wedge q) + \mu(\neg p \wedge q) \geq 0.6,$$

and at the same time satisfies the probability constraints:

$$\mu(p \wedge q) + \mu(p \wedge \neg q) + \mu(\neg p \wedge q) + \mu(\neg p \wedge \neg q) = 1,$$
$$\mu(p \wedge q) \geq 0, \ \mu(p \wedge \neg q) \geq 0, \ \mu(\neg p \wedge q) \geq 0, \ \mu(\neg p \wedge \neg q) \geq 0 \ .$$

We can achieve this using the following assignment μ

at	$p \wedge q$	$p \wedge \neg q$	$\neg p \wedge q$	$\neg p \wedge \neg q$
$\mu(at)$	0.8	0.2	0	0

hence the formula is satisfiable.   ■

NP-completeness of PSAT follows from the statement that a system of L linear (in)equalities has a nonnegative solution iff it has a nonnegative solution with at most L entries positive such that the sizes of entries are bounded by a polynomial function of the size of the longest coefficient from the system [Fagin et al., 1990, Georgakopoulos et al., 1988].

2.   VNS for the PSAT

VNS is a recent metaheuristic for solving combinatorial and global optimization problems (e.g, see [Mladenović and Hansen, 1997, Hansen and Mladenović, 2001, Hansen and Mladenović, 2003]). It is a simple, yet very effective metaheuristic that has shown to be very robust on a variety of practical NP-hard problems. The basic idea behind VNS is change of neighborhood structures in the search for a better solution.

In the initialization phase, a set of k_{max} (a parameter) neighborhoods is preselected $(\mathcal{N}_i, i = 1, \ldots, k_{max})$, a stopping condition determined and an initial local solution found. Then the main loop of the method has the steps described in Figure 9.1. To construct different neighborhood structures one needs to supply a metric (or quasi-metric) to the solution space and then induce neighborhoods from it. In the next sections we answer this problem-specific question for the PSAT problem.

Repeat the following steps until the stopping condition is met:

1 Set $k \leftarrow 1$;
2 Until $k = k_{\max}$, repeat the following steps:
   (a) *Shaking.* Generate a point x' at random from the kth neighborhood of x ($x' \in \mathcal{N}_k(x)$);
   (b) *Local Search.* Apply some local search method with x' as initial solution; denote with x'' the so obtained local optimum;
   (c) *Move or not.* If this optimum is better than the the incumbent, move there ($x \leftarrow x''$), and continue the search with $\mathcal{N}_1(k \leftarrow 1)$; otherwise, set $k \leftarrow k + 1$;

Figure 9.1. Main steps of the basic VNS metaheuristic

Contrary to other metaheuristics, VNS does not follow a trajectory but explores increasingly distant neighborhoods of the current incumbent solution, and jumps from this solution to a new one if and only if an improvement has been made. In this way often favorable characteristics of the incumbent solution, e.g., that many variables are already at their optimal value, will be kept and used to obtain promising neighboring solutions. Moreover, a local search routine is applied repeatedly to get from these neighboring solutions to the local optima.

2.1   The Solution Space

As the PSAT problem reduces to a linear program over probabilities, the number of atoms with nonzero probabilities necessary to guarantee a solution, if one exists, is equal to $L+1$. Here L denotes the number of weight literals in the WCF formula. The solution is therefore an array of $L+1$ atoms

$$x = [A_1, A_2, \ldots, A_{L+1}],$$

where A_i, $i = 1, \ldots, L+1$ are atoms from $At(f)$, with assigned probabilities

$$p = [p_1, p_2, \ldots, p_{L+1}] .$$

Probabilities of atoms not in x are taken to be 0.

Atoms are represented as bit strings, with the ith bit of the string equal to 1 iff the ith variable is positive in the atom. A solution (an array of atoms) is the bit string obtained by concatenation of its atom bit strings. Observe that if the solution variable $x = [A_1, A_2, \ldots, A_{L+1}]$ is known, the probabilities of atoms that are not in x are 0, and system

(9.1) can be rewritten compactly as

$$\sum_{j=1}^{L+1} c_{ij}p_j \quad \rho_i \quad b_i, \quad i = 1, \ldots, L \tag{9.2}$$

The coefficients c_{ij} above are the coefficients from (9.1) grouped by atom probabilities, i.e.

$$c_{ij} = \sum_{k:A_j \in [\alpha_k^i]} a_k^i$$

Solving (9.2) gives us a vector of probabilities p.

2.2   Initial Solution

Initial solution is obtained as follows. First, $10 \times (L+1)$ atoms are randomly generated. They are all assigned equal probabilities, i.e. $1/(L+1)$, and assigned grades. From these atoms the $L + 1$ atoms with the best grades are selected to form the initial solution.

The *grade of an atom* is computed as the sum of this atom's contribution to satisfiability of the conjuncts in the WCF formula (i.e. rows of (9.2)). An atom A (with assigned probability p) corresponds to a column $c = [c_1, c_2, \ldots, c_n]^T$ of the linear system (9.2). If a coefficient c_i from the column is positive, and located in a row with $\geq$ as the relational symbol, it can be used to push toward the satisfiability of this row. In such a case we add its value, multiplied with p, to the grade. The $<$ case is not in favor of satisfying the row, so we subtract the coefficient from the grade. Similar reasoning is applied when the coefficient is negative. Thus, we compute the grade of an atom A as

$$grade(A) = p \sum_{i=1}^{L} c_i \cdot sgn(i),$$

with $sgn(i)$ being the sign of the ith conjunct from the formula

$$sgn(i) = \begin{cases} 1 & \text{if } \rho_i \text{ is } \geq, \\ -1 & \text{if } \rho_i \text{ is } <. \end{cases}$$

2.3   Neighborhood Structures (Shaking)

The neighborhood structures are those induced by the Hamming distance on the solution bit strings. The distance between two solutions is the number of corresponding bits that differ in their bit string representations.

With this selection of neighborhood structures a *shake* in the kth neighborhood of a solution is obtained by inverting k bits in the solution's bit string. The grading procedure for atoms, described above, is also used to direct the shake on a solution so it would modify the current solution in the most favorable way. According to the grades of the atoms the bits to be inverted in a shake are selected in such a way that the bits in a atom with a lower grade have greater probability of being inverted.

2.4   Local Search

The local search part of the algorithm scans through the first neighborhood of a solution in pursue of a solution better than the current best. Note that, since the $\mathcal{N}_1$ neighborhood is huge, it would be very inefficient to recompute the probabilities at every point. Therefore, throughout the local search, the probabilities are fixed. Solutions are compared only by the value of the objective function z (given in (9.3) below). This value is kept for the solutions that are in use. As for the newly generated solutions, the objective value z is computed using the current probabilities assignment, but since only one bit of one atom is changed, the updating of probabilities computed is in $O(L)$.

2.5   Finding Probabilities by VND

As local search is performed only on atoms, when the local optimum is obtained, there might exist a better probabilities assignment corresponding to the new atom set in the solution, i.e., one that is closer to satisfying the PSAT problem. At this point, we suggest three procedures within Variable neighborhood descent (VND) framework: two fast heuristics and Nelder-Mead nonlinear programming method.

2.5.1   Nonlinear Optimization Approach.   To find a possibly better probabilities assignment a nonlinear program is defined with the objective function being the distance of the left hand side from the right hand side of the linear system (9.2). Let x be the current solution, we define an unconstrained objective function z to be

$$z(p) = \sum_{i=1}^{L} d_i(p) + g(p), \tag{9.3}$$

where d_i is the distance of the left and the corresponding right hand side of the ith row defined as

$$d_i(p) = \begin{cases} (\sum_{j=1}^{L+1} c_{ij} p_j - b_i)^2 & \text{if the } i^{\text{th}} \text{ row is not satisfied,} \\ 0 & \text{otherwise,} \end{cases}$$

and $g(p)$ is the penalty function

$$g(p) = M \left[(\sum_{p_i < 0} p_i^2) + (1 - \sum_{i=1}^{L+1} p_i)^2 \right]$$

with penalty parameter M. The penalty function is used to transform the constrained nonlinear problem to the unconstrained one.

The value of the objective function is nonnegative and our goal is to make it zero, or as close to zero as possible, under the probability constraints (p_i's are nonnegative and they sum up to 1). If zero is found, the solution has been found. Otherwise, the value of z is used as a measure of quality when comparing solutions in the local search procedure.

The function $z(p)$ is not only non-convex (the difference between two convex function is non convex) but also non differentiable (the functions $d_i(p)$ are "cut off at zero"). Therefore, a minimization method that does not use derivatives is needed. At first we applied the Powell's method [Powell, 1964] but, as it performed poor (in terms of computing time), we switched to the downhill simplex method of Nelder and Mead [Nelder and Mead, 1965]. It performed considerably better. However, in order to speed-up the search, this optimization procedure is used only when the VNS algorithm reaches k_{max}.

In order to improve stability and achieve the required precision (as it is usual in exterior penalty nonlinear programming methods), Nelder-Mead method is run four times. For the parameter M we use values $M_k = 10^{2+k}$, $k = 1, 2, 3, 4$, where k is the iteration number.

2.5.2 Heuristic Approach.

Nonlinear optimization has shown to be too time demanding, so we resorted to a heuristic approach that is used for the majority of the probability optimizations. The heuristic optimization consists of the following two independent heuristics. They try to solve the huge system of linear inequalities heuristically.

H-1: Worst Unsatisfied Projection.

With this heuristic we concentrate on the rows of the system (9.2) that are the most unsatisfied. Five of the rows with the largest values $d_i(p)$ are selected (the most unsatisfied). The equations of these rows define the corresponding hyper-planes that border the solution space. In an attempt to reach the solution space, consecutive projections of the probabilities vector on these hyper-planes are performed. After each projections the probabilities are normalized. Note that the projection of a point $p' = [p'_1, p'_2, \ldots, p'_{L+1}]$

onto a hyperplane that defines the ith row

$$\sum_{j=1}^{L+1} c_{ij} p_j = b_i$$

is obtained by the formula

$$p_j'' = p_j' - \frac{\sum_{k=1}^{L+1} c_{ik} p_k'}{\sum_{k=1}^{L+1} (c_{ik})^2} \ .$$

Using projection formula above, the probabilities are changed towards satisfiability of each worst row in the fastest possible manner, i.e. in the direction normal to that hyper-plane. This procedure is repeated at most 10 times, until no improvement has been achieved, every time selecting the current 5 worst rows.

H-2: Greedy Giveaway. With some statistical analysis of the systems that the method works with during the computation, it can be noticed that the systems are very sparse, i.e. very high percentage (more than 80%) of the system coefficients are zeroes. This led to the idea that we can try to improve the system's value by solving it "by hand".

Again, the worst unsatisfied row is selected (the ith), but also the best satisfied row is selected (the jth). Nonzero coefficients are then found in these two rows, but only those that have zero coefficients at the same position in the other row. For example if k_1 and k_2 is one pair of such coefficient positions then the selected rows look like

$$\ldots + c_{i,k_1} \cdot p_{k_1} + \ldots + 0 \quad\cdot p_{k_2} + \ldots \rho_i \quad b_i$$
$$\ldots + 0 \quad\cdot p_{k_1} + \ldots + c_{j,k_2} \cdot p_{k_2} + \ldots \rho_j \quad b_j \ .$$

Now the probability p_{k_1} can be changed with the ith (the most unsatisfied) row moving towards satisfiability, and p_{k_2} can be changed the opposite way (maintaining the probability constraints), reducing the satisfiability of the most satisfied row. This probabilities giveaway is repeated for all pairs of the coefficient positions from the two selected rows. Since the system is very sparse, the change of probabilities doesn't affect much the satisfiability of other rows.

Ordering within VND. These two heuristics combined together (first one, then the other) yield a great improvement of the objective function value (9.3). Improvements obtained by heuristics are comparable to those obtained by nonlinear optimization, but the computation is much faster. If the nonlinear optimization is performed first, these heuristics do not achieve a notable improvement. Conversely, if heuristic optimization is applied first, and nonlinear optimization afterwards, nonlinear optimization only gives a slight improvement in objective function value. This is why we decided to use the heuristics for the majority of the optimizations, and use the computationally demanding nonlinear optimization only after the $\mathcal{N}_{k_{\max}}$ neighborhood is unsuccessfully explored (step 12 in the Algorithm 1 below).

Heuristic optimization procedures H-1 and H-2 are also used to improve the probabilities of the initial solution. For small size problem instances this has shown to be very efficient: solutions are found even without entering the main VNS loop.

2.6   VNS pseudo-code

Our heuristic may be seen as an alternate procedure:

- for fixed probabilities, we use VNS to find better 0-1 variables, and

- for a given 0-1 variables, we propose a VND heuristic, as well as non-smooth optimization to find probabilities.

This approach allows us to solve large problem instances more effectively and more efficiently than our recent GA based heuristic. The pseudo-code of the VNS method we used is stated in Algorithm 1.

Algorithm 1 VNS for PSAT

1: $x \leftarrow$ initialSolution(); improve(x, heuristic)
2: **while** (**not** done()) **do**
3:      $k \leftarrow 1$
4:      **while** ($k \leq k_{\max}$) **do**
5:          $x' \leftarrow$ shake(x, k); improve(x', heuristic); $x'' \leftarrow$ localSearch(x')
6:          **if** (x'' better than x) **then**
7:              $x \leftarrow x''$; improve(x, heuristic); $k \leftarrow 1$
8:          **else**
9:              $k \leftarrow k + 1$
10:         **end if**
11:     **end while**
12:     improve(x, nelder-mead)
13: **end while**

The meaning of subroutines above are as follows:

- **initialSolution()** finds an initial solution of PSAT, as explained in subsection 2.2;

- **improve(x, heuristic)** improves the probabilities by using the two fast heuristics as given in 2.5.2;

- **shake(x, k)** perturbs the incumbent solution (see 2.3);

- **localSearch(x)** performs local search as explained in 2.4;

- **improve(x, nelder-mead)** improves the probabilities using the Nelder-Mead non-linear programming method described in 2.5.1.

3.   Computational Results

For testing purposes a set of 24 random satisfiable WCF-formulas has been generated. Maximal number of summands in weight terms (S), and the maximal number of disjuncts (D) in the DNFs of propositional formulas has been set to 5. We use N and L to denote the number of propositional letters, and the number of weight literals in a WCF-formula, respectively. Three problem instances were generated for each of the following (N, L) pairs: $(50, 100)$, $(50, 250)$, $(100, 100)$, $(100, 200)$, $(100, 500)$, $(200, 200)$, $(200, 400)$, $(200, 1000)$.

We are not aware of any larger PSAT problem instances reported in the literature. For example, $N = 50$, $L = 70$ in [Kavvadias and Papadimitriou, 1990], $N = 140$, $L = 300$ in [Jaumard et al., 1991], L is up to 500 in [Hansen and Jaumard, 2001], and $N = 200$, $L = 800$ in [Hansen and Perron, 2004]. The instances considered in the mentioned papers have a simpler form than the ones used here, with S - the maximal number of summands in weight terms, and D - the maximal number of disjuncts in DNFs of classical formulas, set to 1 and 4 (or 3) respectively (we used $S = D = 5$). Also, classical propositional formulas in their tests are clauses (disjunctions of propositional letters and their negations - propositional literals). In other words, they use weight terms that contain the probability of only one clause with up to 4 propositional literals.

For comparative purposes we include the results of a previous approach using GAs as well. In the GA-approach [Ognjanović et al., 2004], each individual (chromosome) from the population consists of $L+1$ pairs of the form (atom, probability). Similarly as in the subsection 2.1, the atom is represented as a bit string of length N. The probability of that atom is given as a floating point number. Then, GA applies the genetic operators to population, in order to find the global optima. Crossing-over between 0-1 and continuous variables, taken from the two solutions (chromosomes) in population, is not allowed.

The VNS algorithm was run with $k_{\max}$ set to 30, and we exit if a feasible solution is found or the method doesn't advance in 10 consecu-

Table 9.1. Computational results of the VNS method compared to the previous GA approach. The time column is the average time of the successful runs, N is the number of propositional variables, L is the number of conjuncts in the WCF-formula.

N, L, instance	VNS		GA	
	solved	cpu time	solved	cpu time
50, 100, 1	26/30	29.46	23/30	188.07
50, 100, 2	30/30	0.23	27/30	21.98
50, 100, 3	29/30	17.72	13/30	194.04
50, 250, 1	30/30	32.60	25/30	504.58
50, 250, 2	15/30	14.60	25/30	2515.88
50, 250, 3	30/30	11.37	25/30	664.7
100, 100, 1	30/30	0.03	30/30	6.19
100, 100, 2	30/30	0.13	30/30	13.28
100, 100, 3	30/30	0.03	30/30	66.85
100, 200, 1	30/30	4.23	27/30	858.85
100, 200, 2	27/30	79.07	10/30	1395.62
100, 200, 3	25/30	77.04	10/30	1235.31
100, 500, 1	30/30	230.97	21/30	11369.29
100, 500, 2	23/30	1930.52	19/30	18932.76
100, 500, 3	2/30	30460.00	12/30	17907.2
200, 200, 1	0/30	–	0/30	–
200, 200, 2	30/30	2.30	27/30	253.82
200, 200, 3	30/30	0.27	28/30	48.36
200, 400, 1	30/30	9.73	28/30	524.63
200, 400, 2	12/30	6082.67	16/30	12167.64
200, 400, 3	2/30	5531.50	15/30	11938.51
200, 1000, 1	30/30	966.47	22/30	68437.78
200, 1000, 2	0/30	–	0/30	–
200, 1000, 3	30/30	1942.93	21/30	56081.48

tive iterations. The solver program was run 30 times on each problem instance. The results obtained by both VNS and GA are summarized in Table 9.1. Columns 2 and 4 in the Table 9.1 (i.e., *VNS-solved* and *GA-solved*) report the number of solved instances (out of 30) obtained by VNS and GA respectively.

It appears that VNS outperforms the GA solver in most of the test instances, with increase in the solving success rate and decrease of the execution time; in four instances (i.e., instances $(50, 250, 2)$, $(100, 500, 3)$, $(200, 400, 2)$ and $(200, 400, 3)$) GA had a better success rate. The possible explanation of the different behavior of VNS and GA could be the fact that VNS uses a reduced solution search space: most of the time such reduction pays-off. Moreover, heuristics within VND for finding probabilities appear to be very efficient.

4.   Conclusion

In this paper we suggest a VNS based heuristic for solving the PSAT problem. Although it has a linear programming formulation, the exponential growth of variables with the number of propositional letters suggests the use of a heuristic approach. VNS has already been applied for solving similar problems, i.e., for WSAT and a slightly different version of PSAT, but not in the way we do in this paper. As usual, the neighborhood structures are induced from the Hamming metric in all VNS applications. But, for solving the PSAT, beside logical or 0-1 variables, one has to find probabilities, and then check if the formula is satisfied. Our heuristic may be seen as an alternate procedure: (i) for fixed probabilities, we use VNS to find better 0-1 variables; (ii) given 0-1 variables, we propose a VND heuristic, as well as non-smooth optimization to find probabilities. This approach allows us to solve large problem instances more effectively and more efficiently than our recent GA based heuristic.

There are many directions for further research:

(i) improve efficiency of our VNS based heuristic by reducing the large neighborhoods, or by introducing new ones to be used within local search;

(ii) apply our approach to the so-called interval PSAT [Hansen and Jaumard, 2001] in which weight terms belong to intervals of probabilities (the basic weight formulas are of the form $c_1 \leq t \leq c_2$);

(iii) consider how our approach to PSAT can be extended to fit a more expressible version of probabilistic logic that allows iteration of probabilistic operators [Fagin and Halpern, 1994, Ognjanović and Rašković, 2000] in which PSAT is PSPACE-complete, or for the framework of conditional probabilities;

(iv) similar to the phase transition phenomenon for SAT [Cheeseman et al., 1991], we are still not able to conjecture any relation between the parameters and the hardness of the PSAT problem instance. Thus, an exhaustive empirical study should give better insight into this phenomenon.

References

[Boole, 1854] Boole, George (1854). *An investigation of the laws of thought, on which are founded mathematical theories of logic and probabilities*. Walton and Maberley, London.

[Cheeseman et al., 1991] Cheeseman, Peter, Kanefsky, Bob, and Taylor, William M. (1991). Where the really hard problems are. In *IJCAI*, pages 331–340.

[de Finetti, 1974] de Finetti, Bruno (1974). *Theory of Probability*, volume 1. Wiley, New York.

[Fagin and Halpern, 1994] Fagin, R. and Halpern, J. (1994). Reasoning about knowledge and probability. *Jornal of the ACM*, 41(2):340–365.

[Fagin et al., 1990] Fagin, R., Halpern, J., and Megiddo, N. (1990). A logic for reasoning about probabilities. *Information and Computation*, 87(1/2):78–128.

[Georgakopoulos et al., 1988] Georgakopoulos, George, Kavvadias, Dimitris, and Papadimitriou, Christos H. (1988). Probabilistic satisfiability. *Journal of Complexity*, 4(1):1–11.

[Hansen and Jaumard, 2001] Hansen, P. and Jaumard, B. (2001). Probabilistic satisfiability. In Gabbay, Dov M. and Smets, Philippe, editors, *Algorithms for uncertainty and defeasible reasoning*, volume 5 of *Handbook of Defeasible Reasoning and Uncertainty Management Systems*, pages 321–367. Kluwer Academic Publishers.

[Hansen et al., 2000] Hansen, P., Jaumard, J., Mladenović, N., and Parreira, A.D. (2000). Variable neighbourhood search for maximum weight satisfiability problem. Technical report, GERAD.

[Hansen and Mladenović, 2001] Hansen, P. and Mladenović, N. (2001). Variable neighborhood search: Principles and applications. *European Journal of Operational Research*, 130(3):449–467.

[Hansen and Mladenović, 2003] Hansen, P. and Mladenović, N. (2003). Variable neighborhood search. In Glover, Fred W. and Kochenberger, Gary A., editors, *Handbook of Metaheuristics*, volume 57 of *International Series in Operations Research and Management Science*, pages 145–184. Springer.

[Hansen and Perron, 2004] Hansen, P. and Perron, S. (2004). Merging the local and global approaches to probabilistic satisfiability. Technical report, GERAD.

[Jaumard et al., 1991] Jaumard, Brigitte, Hansen, Pierre, and de Aragão, Marcus Poggi (1991). Column generation methods for probabilistic logic. *ORSA Journal on Computing*, 3:135–147.

[Kavvadias and Papadimitriou, 1990] Kavvadias, Dimitris J. and Papadimitriou, Christos H. (1990). A linear programming approach to reasoning about probabilities. *Annals of Mathematics and Artificial Intelligence*, 1(1-4):189–205.

[Mladenović and Hansen, 1997] Mladenović, N. and Hansen, P. (1997). Variable neighborhood search. *Computers and Operations Research*, 24(11):1097–1100.

[Nelder and Mead, 1965] Nelder, J. A. and Mead, R. (1965). A simplex method for function minimization. *The Computer Journal*, 7(4):308–313.

[Nilsson, 1986] Nilsson, Nils J. (1986). Probabilistic logic. *Artificial Intelligence*, 28(1):71–87.

[Ognjanović et al., 2001] Ognjanović, Z., Kratica, J., and Milovanović, M. (2001). A genetic algorithm for satisfiability problem in a probabilistic logic: A first report. In Benferhat, S. and Besnard, P., editors, *Symbolic and Quantitative Approaches to Reasoning with Uncertainty*, volume 2143 of *Lecture Notes in Computer Science*, pages 805–816. Springer.

[Ognjanović et al., 2004] Ognjanović, Z., Midić, U., and Kratica, J. (2004). A genetic algorithm for probabilistic SAT problem. In Rutkowski, Leszek, Siekmann, J org, Tadeusiewicz, Ryszard, and et al., editors, *Artificial Intelligence and Soft Computing*, volume 3070 of *Lecture Notes in Computer Science*, pages 462–467. Springer.

[Ognjanović and Rašković, 2000] Ognjanović, Z. and Rašković, M. (2000). Some first-order probability logics. *Theoretical Computer Science*, 247(1-2):191–212.

[Ognjanović et al., 2005] Ognjanović, Zoran, Timotijević, Tatjana, and Stanojević, Adam (2005). Database of papers about probability logics. *Mathematical institute Belgrade*, page http://problog.mi.sanu.ac.yu/.

[Powell, 1964] Powell, Michael J. D. (1964). An efficient method for finding the minimum of a function of several variables without calculating derivatives. *The Computer Journal*, 7(2):155–162.

[Rašković, 1993] Rašković, Miodrag (1993). Classical logic with some probability operators. *Publications de l'Institut Mathématique*, 53(67):1–3.

Chapter 10

THE ACO/F-RACE ALGORITHM
FOR COMBINATORIAL OPTIMIZATION
UNDER UNCERTAINTY

Mauro Birattari, Prasanna Balaprakash, and Marco Dorigo
IRIDIA, CoDE, Université Libre de Bruxelles, Brussels, Belgium
{mbiro,pbalapra,mdorigo}@ulb.ac.be

Abstract   The paper introduces *ACO/F-Race*, an algorithm for tackling combinatorial optimization problems under uncertainty. The algorithm is based on *ant colony optimization* and on *F-Race*. The latter is a general method for the comparison of a number of candidates and for the selection of the best one according to a given criterion. Some experimental results on the PROBABILISTIC TRAVELING SALESMAN PROBLEM are presented.

Keywords: Ant colony optimization, combinatorial optimization under uncertainty

1.   Introduction

In a large number of real-world combinatorial optimization problems, the objective function is affected by uncertainty. In order to tackle these problems, it is customary to resort to a probabilistic model of the value of each feasible solution. In other words, a setting is considered in which the cost of each solution is a *random variable*, and the goal is to find the solution that minimizes some *statistics* of the latter. For a number of practical and theoretical reasons, it is customary to optimize with respect to the *expectation*. This reflects a risk-neutral attitude of the decision maker. Theoretically, for a given probabilistic model, the expectation can always be computed but this typically involves particularly complex analytical manipulations and computationally expensive procedures. Two alternatives have been discussed in the literature: *analytical approximation* and *empirical estimation*. While the former explicitly relies on the underlying probabilistic model for approximating the expectation, the latter estimates the expectation through *sampling* or *simulation*.

In this paper we introduce *ACO/F-Race*, an *ant colony optimization* algorithm [8] for tackling combinatorial optimization problems under uncertainty with the *empirical estimation* approach. *F-Race* [6, 5] is an algorithm for the comparison of a number of candidates and for the selection of the best one. It has been specially developed for tuning metaheuristics.[1] In the present paper, *F-Race* is used in an original way as a component of an *ant colony optimization* algorithm. More precisely, it is adopted for selecting the *best-so-far* ant, that is, the ant that is appointed for updating the pheromone matrix.

The main advantage of the *estimation* approach over the one based on *approximation* is generality: Indeed, a sample estimate of the expected cost of a given solution can be simply obtained by averaging a number of realizations of the cost itself. Conversely, computing a profitable approximation is a problem-specific issue and requires a deep understanding of the underlying probabilistic model. Since *ACO/F-Race* is based on the *empirical estimation* approach, it is straightforward to apply it to a large class of combinatorial optimization problems under uncertainty. For definiteness, in this paper we consider an application of *ACO/F-Race* to the PROBABILISTIC TRAVELING SALESMAN PROBLEM, more precisely to its *homogeneous* variant [11]. An instance of the PROBABILISTIC TRAVELING SALESMAN PROBLEM (PTSP) is defined as an instance of the well known TRAVELING SALESMAN PROBLEM (TSP), with the difference that in PTSP each city has a given probability of requiring being visited. In this paper we consider the *homogeneous* variant, in which the probability that a city must be visited is the same for all cities. PTSP is here tackled in the *a priori* optimization sense [1]: The goal is to find an *a priori* tour visiting all the cities, which minimizes the expected length of the associated *a posteriori* tour. The *a priori* tour must be found prior to knowing which cities indeed require being visited. The associated *a posteriori* tour is computed *after* knowing which cities need being visited, and is obtained by visiting them in the order in which they appear in the *a priori* tour. The cities that do not require being visited are simply skipped. This problem was selected as the first problem for testing the *ACO/F-Race* algorithm for two main reasons: First, PTSP is particularly simple to describe and to handle. In particular, the *homogeneous* variant is rather convenient since a single parameter,

---

[1]A public domain implementation of *F-Race* for R is available for download [4]. R is a language and environment for statistical computing that is freely available under the GNU GPL license.

that is, the probability that each city requires being visited, defines the "stochastic character" of an instance: When the probability is one, we fall into the deterministic case; as it decreases, the normalized standard deviation of the cost of a given solution increases steadily. We can informally conclude that an instance of the *homogeneous* PTSP becomes *more and more stochastic* as the probability that cities require being visited decreases. This feature is particularly convenient in the analysis and visualization of experimental results. Second, some variants of *ant colony optimization* have been already applied to PTSP: Bianchi *et al.* [3, 2] proposed *pACS*, a variant of *ant colony system* in which an *approximation* of the expected length of the *a posteriori* tour is optimized; Gutjahr [9, 10] proposed *S-ACO*, in which an *estimation* of the expected length of the *a posteriori* tour is optimized. *ACO/F-Race* is similar to *S-ACO*. The main difference lies in the way solutions are compared and selected.

The rest of the paper is organized as follows: Section 2 discusses the problem of estimating, on the basis of a sample, the cost of a solution in a combinatorial optimization problem under uncertainty. Section 3 introduces the *ACO/F-Race* algorithm. Section 4 reports some results obtained by *ACO/F-Race* on PTSP. Section 5 concludes the paper and highlights future research directions.

2.   The empirical estimation of stochastic costs

For a formal definition of the class of problems that can be tackled by *ACO/F-Race*, we follow [10]:

$$\text{Minimize} \quad F(x) = E\big[f(x, \omega)\big], \qquad \text{subject to} \quad x \in S, \qquad (10.1)$$

where x is a solution, S is the set of feasible solutions, the operator E denotes the mathematical expectation, and f is the cost function which depends on x and also on a random (possibly multivariate) variable ω. The presence of the latter makes the cost $f(x, \omega)$ of a given solution x a random variable.

In the *empirical estimation* approach to stochastic combinatorial optimization, the expectation $F(x)$ of the cost $f(x, \omega)$ for a given solution x is estimated on the basis of a sample $f(x, \omega_1), f(x, \omega_2), \ldots, f(x, \omega_M)$, obtained from M independently-extracted realizations of the random variable ω:

$$\hat{F}_M(x) = \frac{1}{M} \sum_{i=1}^{M} f(x, \omega_i). \qquad (10.2)$$

Clearly, $\hat{F}_M(x)$ is an *unbiased* estimator of $F(x)$.

In the case of PTSP, the elements of the general definition of a stochastic combinatorial optimization problem given above take the following meaning: A feasible solution x is an *a priori* tour visiting once and only once all cities. If cities are numbered from 1 to N, x is a permutation of $1, 2, \ldots, N$. The random variable ω is extracted from an N-variate Bernoulli distribution and prescribes which cities need being visited. In the *homogeneous* variant of PTSP, each element in ω is independently extracted from a same univariate Bernoulli distribution with probability p, where p is a parameter defining the instance. The cost $f(x, \omega)$ is the length of an *a posteriori* tour visiting the cities indicated in ω, in the order in which they appear in x.

3.   The **ACO/F-Race** algorithm

It is straightforward to extend an *ant colony optimization* algorithm for the solution, in the *empirical estimation* sense, of a combinatorial optimization problem under uncertainty. Indeed, it is sufficient to consider one single realization of the random influence ω, say ω', and optimize the function $\hat{F}_1(x) = f(x, \omega')$. Indeed, $\hat{F}_1(x)$ is an *unbiased* estimator of $F(x)$. The risk we run by following this strategy is that we might sample a particularly *atypical* ω' which provides a misleading estimation of $F(x)$. A safer choice consists in considering a different realization of ω at each iteration of the *ant colony optimization* algorithm. The rationale behind this choice is that unfortunate modifications to the pheromone matrix that can be caused by sampling an *atypical* value of ω at a given iteration, will not have a large impact on the overall result and will be corrected in following iterations. In this paper we call *ACO-1* an *ant colony optimization* algorithm for stochastic problems in which the objective function is estimated on the basis of *one single* realization of ω which is sampled anew at each iteration of the algorithm.

A more refined approach has been proposed by Gutjahr [9, 10] and consists in using a large number of realizations for estimating the value of $F(x)$. In Gutjahr's *S-ACO* [9], the solutions produced at a given iteration are compared on the basis of a single realization. The *iteration-best* is then compared with the *best-so-far* solution on the basis of a large number of realizations. The size N_m of the sample is defined by the following equation:

$$N_m = 50 + (0.0001 \cdot n^2) \cdot k \qquad (10.3)$$

where n and k denote the size of the instance and the iteration index, respectively.

A variant of *S-ACO* called *S-ACOa* has been introduced by Gutjahr in [10] in which the size of the sample is determined dynamically on the basis of a parametric statistical test: Further realizations are considered till when either a maximum amount of computation is performed, or when the difference between the sample means for the two solutions being compared is larger than 3 times their estimated standard deviation. The selected solution is stored as the new *best-so-far* for future comparisons and is used for updating the pheromone matrix.

The *ACO/F-Race* algorithm we propose in this paper is inspired by *S-ACOa* and similarly to the latter it considers, at each iteration, a number of realizations for comparing candidate solutions and for selecting the best one which is eventually used for updating the pheromone matrix. The significant difference lies in the algorithm used at each iteration for selecting the best candidate solution. *ACO/F-Race* adopts *F-Race*, an algorithm originally developed for tuning metaheuristics [6, 5]. *F-Race* is itself inspired by a class of *racing* algorithms proposed in the machine learning literature for tackling the model selection problem [13, 14].

A detailed description of the algorithm and its empirical analysis are given in Birattari [5].

Solution methodology

The *ACO/F-Race* algorithm presents many similarities with *S-ACO* and even more with *S-ACOa* [10]. Similarly to *S-ACOa*, at each iteration it considers a number of realizations for comparing candidates solutions and for selecting the best one, which is used for updating the pheromone matrix. The sole difference between the two algorithms lies in the specific technique used to select the best candidate solution at each iteration.

In *S-ACOa*, the solutions produced at a given iteration are compared on the basis of a single realization ω to select the *iteration-best* solution. On the basis of a large sample of realizations, the size of which is computed dynamically, the *iteration-best* solution is then compared with the *best-so-far* solution. For PTSP, the solution with shorter expected *a posteriori* tour length between the two solutions is selected and stored as the new *best-so-far* solution for the subsequent iterations. This solution is used to update the pheromone matrix. In a nutshell, *S-ACOa* exploits sampling techniques and a parametric test.

ACO/F-Race employs F-Race, an algorithm based on a nonparametric test that was originally developed for tuning metaheuristics. In the context of *ACO/F-Race*, the racing procedure consists in a series of steps at each of which a new realization of ω is considered and is used for evaluating the solutions that are still in the race. At each step, a Friedman test is performed and solutions that are statistically dominated by at least another one are discarded from the race. The solution that wins the

---

Algorithm 1 *ACO/F-Race* Algorithm

---

input: an instance C of a PTSP problem

$\tau_{ij} \leftarrow 1, \; \forall (i, j)$

for iteration $k = 1, 2, \ldots$ **do**

    for ant $z = 1, 2, \ldots, m$ **do**

        $s_z \leftarrow$ *a priori* tour of ant z

    end for

    if $(k = 1)$ **then**

        $s_{best} \leftarrow$ F-Race$(s_1, s_2, \ldots, s_m)$

    else

        $s_{best} \leftarrow$ F-Race$(s_1, s_2, \ldots, s_m, s_{best})$

    end if

    $\tau_{ij} \leftarrow (1 - \rho)\tau_{ij}, \; \forall (i, j)$ # evaporation

    $\tau_{ij} \leftarrow \tau_{ij} + c, \; \forall (i, j) \in s_{best}$ # *best-so-far* pheromone update

end for

---

race is used for updating the pheromone and is stored as the *best-so-far*. The race terminates when either one single candidate remains, or when a maximum amount of computation time is reached.

The pseudo-code of *ACO/F-Race* is presented in Algorithm 1. The algorithm starts by initializing to 1 the pheromone on each arc (i, j) of the PTSP. At each iteration of *ACO/F-Race*, m ants, where m is a parameter, construct a solution as it is customary in ant colony optimization. In particular, we have adopted here the *random proportional rule* [8] as shown in Equation 10.4: Ant z, when in city i, moves to city j with a probability given by Equation 10.4, where N_i^z is the set of all cities yet to be visited by ant z.

$$p_{ij}^z = \frac{\tau_{ij}^\alpha \cdot \eta_{ij}^\beta}{\sum_{l \in N_i^z} \tau_{il}^\alpha \cdot \eta_{il}^\beta}, \quad if \, j \in N_i^z \tag{10.4}$$

The m solutions generated by the ants, together with the *best-so-far* solution, are evaluated and compared via *F-Race*.

4. Experimental analysis

In the experimental analysis proposed here, we compare *ACO/F-Race* with *ACO-1*, *S-ACO* and *S-ACOa*. For convenience of the reader, we summarize here the main characteristics of the algorithms considered in this study.

ACO-1: Solutions produced at a given iteration are compared on the basis of single realization ω to select the *iteration-best* solution.

Again, on the basis of the same realization, the *iteration-best* solution is then compared with the *best-so-far* solution to select the new *best-so-far* solution.

S-ACO: Solutions produced at a given iteration are compared on the basis of a single realization ω to select the *iteration-best* solution. On the basis of a large sample of realizations, whose size is given by the equation 10.3, the *iteration-best* solution is then compared with the *best-so-far* solution.

S-ACOa: Solutions produced at a given iteration are compared on the basis of a single realization ω to select the *iteration-best* solution. On the basis of a large sample of realizations, the size of which is computed dynamically on the basis of a parametric statistical test, the *iteration-best* solution is then compared with the *best-so-far* solution.

ACO/F-Race: Solutions produced at a given iteration, together with the *best-so-far* solution, are evaluated and compared using the *F-Race* algorithm.

These four algorithms differ only for what concerns the technique used for comparing solutions and for selecting the *best-so-far* solution which is used for updating the pheromone. The implementations used in the experiments are all based on [15]. The problems considered are *homogeneous* PTSP instances obtained from TSP instances generated by the DIMACS generator [12]. We present the results of two experiments. In the first, cities are *uniformly distributed*, in the second they are *clustered*. For each of the two experiments, we consider 100 TSP instances of 300 cities. Out of each TSP instance we obtain 21 PTSP instances by letting the probability range in $[0, 1]$ with a step size of 0.05. computation time has been chosen as the stopping criterion: Each algorithm is allowed to run for 60 seconds on an AMD Opteron™ 244. These four algorithms were not fine-tuned. The parameters adopted are those suggested in [10] for *S-ACO* and are given in Table 10.1. This might possibly introduce a bias in favor of *S-ACO*. Also note that *S-ACOa* was not previously applied to PTSP. Furthermore, for PTSP, the expected cost of the objective function can be easily computed using an explicit formula given in [1]. Using this mathematical formula, the solutions selected by each algorithm on each instance were then evaluated.

In the plots given in Figures 10.1 and 10.2, the probability that cities require being visited is represented on the x-axis. The y-axis represents the expected length of the *a posteriori* tour obtained by *ACO/F-Race*,

Table 10.1. Value of the parameters adopted in the experimental analysis.

Parameter	Notation	Value
Number of ants	m	50
Pheromone exponent	α	1.0
Heuristic exponent	β	2.0
Pheromone evaporation factor	ρ	0.01
Best-so-far update constant	c	0.04

S-ACO and *S-ACOa* normalized by the expected length of the *a posteriori* tour obtained by *ACO-1*, which is taken here as a reference algorithm.

For each of the two classes of instances and for the probability values of 0.25, 0.50, 0.75, and 1.00, we study the significance of the observed differences in performance. We use the *Pairwise Wilcoxon rank sum* test [7] with p-values adjusted through Holm's method [17]. In our analysis, we consider a significance level of $\alpha = 0.01$. In Tables 10.2 and 10.3, the p-value reported at the crossing between row A and column B refers to the comparison between the algorithms A and B, where the null hypothesis is $A = B$, that is, the two algorithms produce equivalent results, and the alternative one is $A < B$, that is, A is better than B: A number smaller than $\alpha = 0.01$ in position (A, B) means that algorithm A is better than algorithm B, with confidence at least equal to $1 - \alpha = 0.99$.

From the plots, we can observe that the solution quality of *ACO-1* is better than *S-ACO*, *S-ACOa* and *ACO/F-Race* for probabilities larger than approximately 0.4, that is, when the variance of the *a posteriori* tour length is small. Under such conditions, an algorithm designed to solve TSP is better than one specifically developed for PTSP. This confirms the results obtained by Rossi and Gavioli [18]. This is easily explained: Using a large number of realizations for selecting the *best-so-far* solution is simply a waste of time when the variance of the objective function is very small.

On the other hand, for probabilities smaller than approximately 0.4, the problem becomes "more stochastic": Selecting the *best-so-far* solution on the basis of a large sample of realizations plays a significant role. The risk we run by following a single sample strategy, as in *ACO-1*, is that we might sample a particularly atypical realization which provides a misleading evaluation of solution. *S-ACO*, *S-ACOa* and *ACO/F-Race* by considering a large sample of realizations obtain better results than *ACO-1*.

Another important observation concerns the relative performance of *S-ACO*, *S-ACOa* and *ACO/F-Race*. Throughout the whole range of

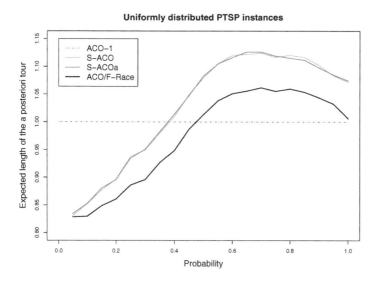

Figure 10.1. Experimental results on the uniformly distributed homogeneous PTSP. The plot represents the expected length of the *a posteriori* tour obtained by *ACO/F-Race*, *S-ACO*, and *S-ACOa* normalized by the one obtained by *ACO-1* for the computation time of 60 seconds.

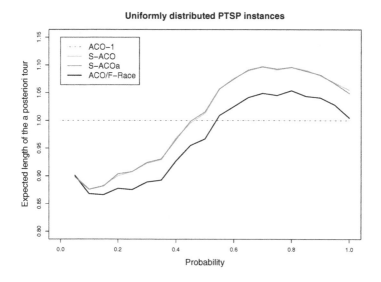

Figure 10.2. Experimental results on the clustered homogeneous PTSP. The plot represents the expected length of the *a posteriori* tour obtained by *ACO/F-Race*, *S-ACO*, and *S-ACOa* normalized by the one obtained by *ACO-1* for the computation time of 60 seconds.

probabilities, the solution quality obtained by *ACO/F-Race* is *significantly* better than the one obtained by *S-ACO* and *S-ACOa*. We can conclude that *ACO/F-Race*, with its nonparametric evaluation method, is more effective than *S-ACOa*, which uses parametric method, and than *S-ACO*, which adopts a linearly increasing sample size for selecting the *best-so-far* solution at each iteration.

In Figures 10.3 and 10.4, the average number of solutions explored by *ACO-1*, *S-ACO*, *S-ACOa* and *ACO/F-Race* is given. Since *ACO-1* uses a single realization to select the best solution, the average number of solutions explored by *ACO-1* is always larger than the those explored by *S-ACO*, *S-ACOa* and *ACO/F-Race*. Apparently a trade-off exists. The number of realizations considered should be large enough for providing a reliable estimate of the cost of solutions but at the same time it should not be too large otherwise too much time is wasted. The appropriate number of realizations depends on the stochastic character of the instance at hand. The larger the probability that a city is to be visited, the less stochastic an instance is. In this case, the algorithms that obtain the best results are those that consider a reduced number of realizations and therefore explore more solutions in the unit of time. On the other hand, when the probability that a city is to be visited is small, the instance at hand is highly stochastic. In this case, it pays off to reduce the total number of solutions explored and to consider a larger number of realizations for obtaining more accurate estimates.

In Figures 10.1 and 10.2, it should be observed that when the probability tends to 1 the curve of *ACO/F-Race* approaches 1 and therefore *ACO/F-Race* performs almost as well as *ACO-1*. This is due to the nature of the Friedman test adopted within *ACO/F-Race*. Indeed, in the deterministic case the Friedman test is particularly efficient and with a minimum number of realizations it is able to select the best solution. The computational overhead with respect to *ACO-1* is therefore relatively reduced. On the other hand, both *S-ACO* and *S-ACOa* adopt a number of realizations that is too large and therefore are able to explore only a limited number of solutions: In *S-ACO* the size of the sample does not depend on the probability and in *S-ACOa* the statistical test adopted is apparently less efficient than the Friedman test in detecting that the instance is deterministic and that a large sample is not needed. This can be observed on the far right hand side of Figures 10.1 and 10.2.

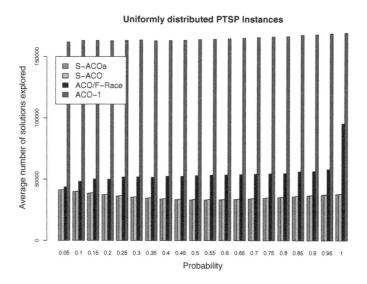

Figure 10.3. Experimental results on the uniformly distributed homogeneous PTSP. The plot represents the average number of solutions explored by *ACO-1*, *S-ACO*, *S-ACOa* and *ACO/F-Race* for the computation time of 60 seconds.

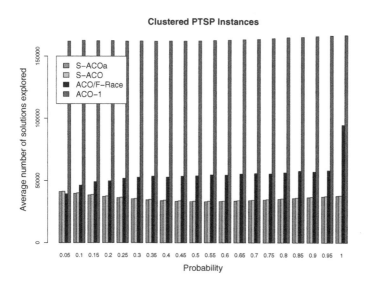

Figure 10.4. Experimental results on the clustered homogeneous PTSP. The plot represents the average number of solutions explored by *ACO-1*, *S-ACO*, *S-ACOa* and *ACO/F-Race* for the computation time of 60 seconds.

Table 10.2. The p-values of the paired Wilcoxon tests on uniformly distributed homogeneous PTSP instances. The quantities under analysis are the expected length of the *a posteriori* tour obtained by ACO/F-Race, S-ACO, S-ACOa and ACO-1.

Probability=0.25	ACO/F-Race	S-ACO	S-ACOa	ACO-1
ACO/F-Race	–	$< 2.2e - 16$	$< 2.2e - 16$	$< 2.2e - 16$
S-ACO	1	–	1	$< 2.2e - 16$
S-ACOa	1	$< 2.2e - 16$	-	$< 2.2e - 16$
ACO-1	1	1	1	–
Probability=0.5	ACO/F-Race	S-ACO	S-ACOa	ACO-1
ACO/F-Race	–	$< 2.2e - 16$	$< 2.2e - 16$	1
S-ACO	1	–	$< 2.2e - 16$	1
S-ACOa	1	1	–	1
ACO-1	$< 2.2e - 16$	$< 2.2e - 16$	$< 2.2e - 16$	–
Probability=0.75	ACO/F-Race	S-ACO	S-ACOa	ACO-1
ACO/F-Race	–	$< 2.2e - 16$	$< 2.2e - 16$	1
S-ACO	1	–	$< 2.2e - 16$	1
S-ACOa	1	1	–	1
ACO-1	$< 2.2e - 16$	$< 2.2e - 16$	$< 2.2e - 16$	–
Probability=1.0	ACO/F-Race	S-ACO	S-ACOa	ACO-1
ACO/F-Race	–	$< 2.2e - 16$	$< 2.2e - 16$	1
S-ACO	1	–	$< 2.2e - 16$	1
S-ACOa	1	1	–	1
ACO-1	$< 2.2e - 16$	$< 2.2e - 16$	$< 2.2e - 16$	–

Table 10.3. The p-values of the paired Wilcoxon tests on clustered homogeneous PTSP instances. The quantities under analysis are the expected length of the *a posteriori* tour obtained by ACO/F-Race, S-ACO, S-ACOa and ACO-1.

Probability=0.25	ACO/F-Race	S-ACO	S-ACOa	ACO-1
ACO/F-Race	–	$< 2.2e - 16$	$< 2.2e - 16$	$< 2.2e - 16$
S-ACO	1	–	0.6845	$< 2.2e - 16$
S-ACOa	1	< 0.3155	–	$< 2.2e - 16$
ACO-1	1	1	1	–
Probability=0.5	ACO/F-Race	S-ACO	S-ACOa	ACO-1
ACO/F-Race	–	$< 2.2e - 16$	$< 2.2e - 16$	$< 2.2e - 16$
S-ACO	1	–	$< 2.2e - 16$	1
S-ACOa	1	1	–	1
ACO-1	1	$< 2.2e - 16$	$< 2.2e - 16$	–
Probability=0.75	ACO/F-Race	S-ACO	S-ACOa	ACO-1
ACO/F-Race	–	$< 2.2e - 16$	$< 2.2e - 16$	1
S-ACO	1	–	$< 2.2e - 16$	1
S-ACOa	1	1	–	1
ACO-1	$< 2.2e - 16$	$< 2.2e - 16$	$< 2.2e - 16$	–
Probability=1.0	ACO/F-Race	S-ACO	S-ACOa	ACO-1
ACO/F-Race	–	$< 2.2e - 16$	$< 2.2e - 16$	1
S-ACO	1	–	1	1
S-ACOa	1	$< 2.2e - 16$	–	1
ACO-1	$< 2.2e - 16$	$< 2.2e - 16$	$< 2.2e - 16$	–

5.   Conclusions and future work

The preliminary experimental results proposed in Section 4 confirm the generality of the approach proposed by Gutjahr [9, 10], and show that the *F-Race* algorithm can be profitably adopted for comparing solutions in the framework of the application of *ant colony optimization* to combinatorial optimization problems under uncertainty.

Further research is needed for properly assessing the quality of the proposed *ACO/F-Race*. We are currently developing an *estimation*-based local search for PTSP. We plan to study the behavior of *ACO/F-Race* enriched by this local search on *homogeneous* and *non-homogeneous* problems.

In the experimental analysis proposed in Section 4, the goal was to compare the evaluation procedure based on *F-Race* with the one proposed in [10] and with the trivial one based on a single realization. For this reason, solution construction and pheromone update were implemented as described in [9, 10]. We plan to explore other possibilities, such as construction and update as defined in $\mathcal{MAX}$–$\mathcal{MIN}$ *ant system* [16]. Applications to other problems, in particular of the VEHICLE ROUTING class, will be considered too.

Acknowledgments. This research has been supported by COMP²SYS, a Marie Curie Early Stage Research Training Site funded by the European Community's Sixth Framework Programme under contract number MEST-CT-2004-505079, and by the ANTS project, an *Action de Recherche Concertée* funded by the Scientific Research Directorate of the French Community of Belgium. Marco Dorigo acknowledges support from the Belgian FNRS, of which he is a Research Director.

References

[1] D. J. Bertsimas, P. Jaillet, and A. Odoni. A priori optimization. *Operations Research*, 38:1019–1033, 1990.

[2] L. Bianchi. *Ant Colony Optimization and Local Search for the Probabilistic Traveling Salesman Problem: A Case Study in Stochastic Combinatorial Optimization*. PhD thesis, Université Libre de Bruxelles, Brussels, Belgium, 2006.

[3] L. Bianchi, L. M. Gambardella, and M. Dorigo. Solving the homogeneous probabilistic travelling salesman problem by the ACO metaheuristic. In M. Dorigo, G. Di Caro, and M. Sampels, editors, *Ant Algorithms, 3rd International Workshop, ANTS 2002*, volume 2463 of *LNCS*, pages 176–187, Berlin, Germany, 2002. Springer-Verlag.

[4] M. Birattari. Race. R package, 2003. http://cran.r-project.org.

[5] M. Birattari. *The Problem of Tuning Metaheuristics as Seen from a Machine Learning Perspective*. PhD thesis, Université Libre de Bruxelles, Brussels, Belgium, 2004.

[6] M. Birattari, T. Stützle, L. Paquete, and K. Varrentrapp. A racing algorithm for configuring metaheuristics. In W. B. Langdon, E. Cantú-Paz, K. Mathias, R. Roy, D. Davis, R. Poli, K. Balakrishnan, V. Honavar, G. Rudolph, J. Wegener, L. Bull, M. A. Potter, A. C. Schultz, J. F. Miller, E. Burke, and N. Jonoska, editors, *Proceedings of the Genetic and Evolutionary Computation Conference*, pages 11–18, San Francisco, CA, USA, 2002. Morgan Kaufmann.

[7] W. J. Conover. *Practical Nonparametric Statistics*. John Wiley & Sons, New York, NY, USA, third edition, 1999.

[8] M. Dorigo and T. Stützle. *Ant Colony Optimization*. MIT Press, Cambridge, MA, USA, 2004.

[9] W. J. Gutjahr. A converging ACO algorithm for stochastic combinatorial optimization In A. Albrecht, and T. Steinhöfl, editors, *Proc. SAGA 2003*, volume 2827 of *LNCS*, pages 10–25, Berlin, Germany, 2003. Springer-Verlag.

[10] W. J. Gutjahr. S-ACO: An ant based approach to combinatorial optimization under uncertainity. In M. Dorigo, M. Birattari, C. Blum, L. M. Gambardella, F. Mondada, and T. Stützle, editors, *Ant Colony Optimization and Swarm Intelligence, 4th International Workshop, ANTS 2004*, volume 3172 of *LNCS*, pages 238–249, Berlin, Germany, 2004. Springer-Verlag.

[11] P. Jaillet. *Probabilistic Travelling Salesman Problems*. PhD thesis, The Massachusetts Institute of Technology, Cambridge, MA, USA, 1995.

[12] D. S. Johnson, L. A. McGeoch, C. Rego, and F. Glover. 8th DIMACS implementation challenge. http://www.research.att.com/ dsj/chtsp/, 2001.

[13] O. Maron and A. W. Moore. Hoeffding races: Accelerating model selection search for classification and function approximation. In J. D. Cowan, G. Tesauro, and J. Alspector, editors, *Advances in Neural Information Processing Systems*, volume 6, pages 59–66, San Francisco, CA, USA, 1994. Morgan Kaufmann.

[14] A. W. Moore and M. S. Lee. Efficient algorithms for minimizing cross validation error. In *Proceedings of the Eleventh International Conference on Machine Learning*, pages 190–198, San Francisco, CA, USA, 1994. Morgan Kaufmann.

[15] T. Stützle. ACOTSP, version 1.0.
      http://www.aco-metaheuristic.org/aco-code/, 2002.

[16] T. Stützle and H. H. Hoos. $\mathcal{MAX}$–$\mathcal{MIN}$ ant system. *Future Generation Computer Systems*, 16(8):889–914, 2000.

[17] S. Holm. A simple sequentially rejective multiple test procedure. *Scandinavian Journal of Statistics*,volume 6, 65–70,1979.

[18] F. Rossi and I. Gavioli. Aspects of heuristic methods in the probabilistic traveling salesman problem. *Advanced School on Stochastics in Combinatorial Optimization.* pages 214–227. World Scientific, Hackensack, NJ, USA, 1987.

[19] S. M. Weiss and C. Kulikowski *Computer systems that learn. Classification and prediction methods from statistics neural nets machine learning and expert systems.* Morgan Kaufmann, San Mateo, CA, USA, 1991.

Chapter 11

ADAPTIVE CONTROL OF GENETIC PARAMETERS FOR DYNAMIC COMBINATORIAL PROBLEMS

Abdunnaser Younes
Systems Design Engineering
University of Waterloo
ayounes@uwaterloo.ca

Otman Basir
Systems Design Engineering
University of Waterloo
obasir@uwaterloo.ca

Paul Calamai
Systems Design Engineering
University of Waterloo
phcalama@solstice.uwaterloo.ca

Abstract   The idea of using diversity to guide evolutionary algorithms is gaining interest. However, it is mainly used in static problems or in dynamic continuous optimization problems. In this paper, we investigate the idea on dynamic combinatorial problems.

The paper uses a measure for population diversity based on distance from the population-best individual rather than distance between all possible pairs in the population. The measured diversity is used to adjust the mutation rate and the selection probability in a standard genetic algorithm whenever the diversity is found to be excessively low or excessively high.

This adaptive scheme aims to retain the algorithm ability to search the solution space even after the population converges prematurely around some suboptimal solution. This scheme also enables the algorithm to persevere after converging around solutions that become obsolete due to environmental changes. Tests on several benchmarks of dynamic travelling salesman problem show that the scheme is promising.

Keywords: Genetic algorithms, local search, adaptation, dynamic environments, combinatorial problems, TSP

Introduction

In recent years, there has been a growing interest in the use of evolutionary algorithms (EAs) in non-stationary (time-varying) environments, where the information is revealed progressively with time to the decision maker. However, most available work basically targets continuous optimization, while little work is directed to discrete optimization, although many real-world problems are both discrete and time-varying.

One promising subject of research is to hybridize schemes used for dynamic continuous optimization problems with techniques that proved successful on static combinatorial optimization problems (COPs). The goal of such hybrids is to develop EAs capable of tackling COPs in non-stationary environments. This paper introduces an adaptive scheme to enhance the performance of the standard Genetic Algorithm (GA) in dynamic environments by using population diversity as a guide to control mutation rate and selection probability.

The idea of using diversity to guide the evolutionary algorithms is adopted by several researchers in many recent publications: Zhu [22] presents an adaptive GA for vehicle routing problems. The population diversity is maintained at pre-defined levels by adapting rates of GA operators to the problem dynamics. Zhu and Liu [23] present an empirical study of population diversity measures and adaptive control of population diversity for a permutation-based genetic algorithm. Burke et al. [2] examine several measures of diversity in genetic programming. Ursem [17] measures population diversity as the sum of the distance to average point and uses it to guide the search process. Riget and Vesterstrom [15] use an approach similar to [17] but on particle swarm optimization. Sorensen and Sevaux [16] present a memetic algorithm with population management to control population diversity. They use edit distance between individuals in the population to measure its diversity.

These publications have targeted either static problems or dynamic continuous optimization problems—none for dynamic COPs. Furthermore, measuring diversity as the sum of the distances between each individual and the rest of the individuals in the population is computationally expensive. Computational cost can be reduced by limiting the population size to a few individuals [16], but this is not a sufficient reason for limiting the population size.

Another possibility is to use a single aggregation point as a representative of the whole population, and thus reducing computation requirements of the diversity by a factor of n, where n is the population size [15, 17]. Since it is often hard to define an "average" point for the population in COPs, the adaptive GA in this paper uses a diversity measure based on distance from the population-best rather than the population-average. The idea is to directly regulate the genetic parameters in response both to environmental changes and to the measured diversity. The regulated parameters will in turn alter the population diversity through the genetic operators.

This scheme is tested on dynamic Travelling Salesman Problem (TSP) in which edge lengths and number of cities change over time. The used benchmarks are similar to those given in [4, 7, 20, 21].

The paper is organized as follows. Section 1 highlights the grave consequences of diversity loss both for static and dynamic problems. Section 2 presents a linear model where the control parameters change abruptly in response to environmental changes then change linearly with time. Section 3 enhances the linear model by regulating the control parameters according to population diversity. Section 4 outlines the algorithm and its main components. Section 5 presents some selected results from the experimentation carried out on the two models of adaptation. Finally, Section 6 concludes the paper with some comments on how the models performed and possible future work.

1. Undesirable Population Convergence

Promoting population diversity is a central issue both in static and in dynamic optimization problems. In static optimization, loss of diversity is blamed for premature convergence, which leads to suboptimal solutions; in dynamic optimization, it is blamed for the algorithm's inability to further track the shifting optima. We refer to the latter situation as *obsolete convergence*, which can be defined as the convergence of a population around an optimal or suboptimal solution of an instance that is replaced by a newer one due to some change in the environment.

It is important to note that real-world problems seldom change completely at once, hence some of the information gathered from the past is re-usable in present and future instances. Therefore, for a GA to be successful in a dynamic environment, it should—without discarding information—be able to persevere after obsolete convergence and at the same time overcome premature convergence. The next sections of the paper introduce two models for tackling dynamic problems. Both models react to environmental changes by immediately increasing the explorative forces of the search. Once the environment becomes static again, the first model reduces the explorative forces gradually with time, while the second model regulates by monitoring population diversity.

2.   Linear Model

The idea of the linear model (LM) model is to increase population diversity when an environmental change is detected in order to diversify the search for newer optima, and to progressively reduce diversity during quiescent phases in order to fine-tune the search in new regions. A simple way to achieve this is by linearly changing the mutation rate repeatedly: When the environment changes (say at $t = t_m$), a cycle begins and the mutation rate $\mu(t)$ is set to an upper limit $\overline{\mu}$. Subsequently, it is reduced with time until, after a period ρ, it reaches a base value $\underline{\mu}$. The current cycle terminates at the next environmental change (at $t = t_{m+1}$), and a new cycle begins. The following formula gives the variation of mutation rate in the cycle between two consecutive environmental changes (i.e. $t_m \leq t < t_{m+1}$).

$$\mu(t) = \begin{cases} \overline{\mu} - \dfrac{\overline{\mu} - \underline{\mu}}{\rho}(t - t_m), & t_m \leq t < t_m + \rho \\ \underline{\mu}, & t_m + \rho \leq t < t_{m+1} \end{cases} \qquad (11.1)$$

In addition to changing the mutation rate, the LM model changes the probability of selection in a similar way. Tournament selection is often modified by injecting it with some degree of randomness that would lead to the selection of a less-fit individual from time to time (instead of selecting the fitter individual all the time). In the modified scheme, the fitter individual is selected at a fixed probability s in the range between 0.5 and 1.0. Thus, selection pressure and consequently the rate of population convergence can be controlled by changing the probability s. The proposed LM model takes advantage of this fact and explicitly changes the probability $s(t)$ in order to control diversity, in a manner analogous to that of $\mu(t)$.

3.   Adaptive Diversity Model

The problem with the LM model is that it addresses obsolete convergence but ignores premature convergence. In other words, the scheme is inflexible during the static phase that separates any two consecutive environmental changes.

One promising possibility is to make the search process adapt to changes in the population diversity. That is the population diversity is used to direct the search towards more exploration of new (unexplored) regions of the search space or towards partially tested areas that have shown promising results.

The proposed adaptive diversity model (ADM) adapts during the static phase by measuring the population diversity d relative to two fixed reference values d_l and d_h and then using the measured diversity to compute the GA parameters. This scheme is combined with the LM model in order to keep diversity under control all the time. In this scheme diversity is measured as the sum of the genotypic distances $dist(x_i, x^*)$ between individuals x_i in the population and the best found solution x^*. For the TSP problem considered in this paper, the genotypic distance $dist(x_i, x^*)$ is taken to be the number of edges that are part of x_i and not in x^*. The decision to use the population best solution as a reference point for measuring diversity is based on two reasons: First, the use of a single aggregation point to represent the population greatly reduces costs of computing diversity without imposing unnecessary limitations on the population size. Second, as GAs are designed to converge around the population-best, it is reasonable to measure the population diversity in terms of distances from the population best solution. Furthermore, we demonstrate the validity of the population-best based diversity measure by comparing it with the more commonly used population based diversity. The evolution of both measures is shown in Figure 11.1 for TSP instance kroA100 from [14]. The six subplots shown in the figure are the results of using three combinations of GA parameters for each diversity measure. Each subplot uses ten GA runs. The three parameter combinations are labelled L, M and H. The L combination uses a mutation rate of 0.001, a crossover rate of 0.0, and a selection probability of 1.0. The corresponding values for the M combination are 0.005, 0.3, and 0.55; for the H combination they are 0.1, 1.0, and 0.55. The parameter values were selected so that a wide range of diversity is considered. The figure clearly shows that diversity diminishes as the GA progresses, and that both measures of diversity give similar inferences on the behavior of the population.

Therefore, the proposed ADM model uses the population-best based diversity measure d to control the rate of mutation. Two corrections factors Z_l and Z_h are computed in proportion to the deviation of the measured diversity from the lower and the upper reference points d_l and d_h. These corrections are applied to the current rate of mutation to produce a new rate that brings the population diversity of the next

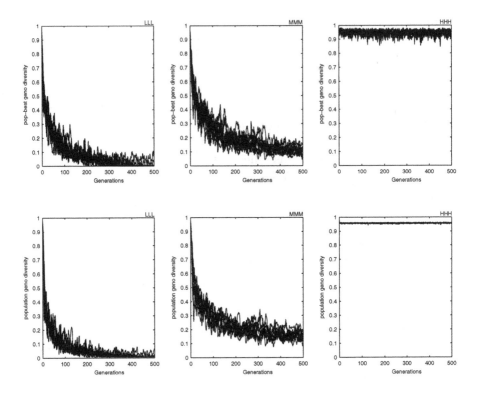

Figure 11.1. Evolution of genotypic measures of diversity

generation within an unbiased range D. Formally, the ADM model can be expressed as follows:

$$\mu(t) = \begin{cases} \overline{\mu}, & t = t_m \\ Z_l \cdot (\overline{\mu} - \mu(t-1)) + \mu(t-1), & t \neq t_m \,, d < d_l \\ \mu(t-1) - Z_h \cdot (\mu(t-1) - \underline{\mu}), & t \neq t_m \,, d > d_h \\ \mu(t-1), & t \neq t_m \,, d_l \leq d \leq d_h \end{cases} \quad (11.2)$$

$$\text{where} Z_l = min\left\{\frac{d_l - d}{D}, 1\right\}, Z_h = min\left\{\frac{d - d_h}{D}, 1\right\},$$

$$d = \sum_{i}^{popsize} dist(x_i, x^*) \,, D = d_h - d_l \text{ and } m = 1, 2, \cdots$$

The above ADM is extended to included adaptive selection probability as well, with similar formula for controlling the selection probability in response to measured diversity. Figure 11.2 illustrates how the adaptive model works on either parameter (mutation rate or selection probability).

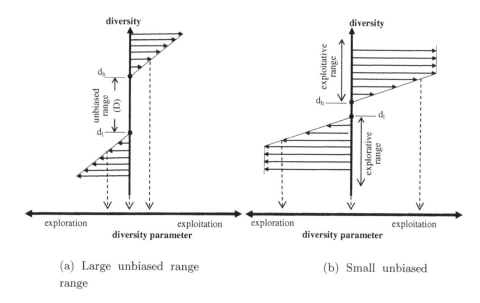

(a) Large unbiased range

(b) Small unbiased range

Figure 11.2. The horizontal arcs represent the adaptation forces (exploitation and exploration) initiated by the diversity measure. These forces are exerted on the genetic parameters and consequently on the entire evolutionary process. When diversity is above the upper reference point, it will map into an exploitative value of the genetic parameter. Similarly, very low diversity forces encourage explorative search. The farther the diversity from the unbiased range, the more is the adaptation response. Diversity values within the unbiased range tend to leave the search state unchanged since the genetic parameters will not be altered. The effect of size of the unbiased ranges is readily seen by comparing the two parts of the figure; reducing the unbiased range makes the search oscillate more frequently and more severely between exploitation and exploration.

4. Algorithm Structure

The two earlier models are used to convert a standard genetic algorithm into a dynamic solver shown in Figure 11.3. Chromosome representation in this algorithm is a straight forward path representation, where values of the genes are the city numbers, and the relative position of the genes reflect their order in the tour.

DyndamicGA

    Generate pop[0];
    t = 1;
    WHILE not terminating criteria
        ApplyDynamicStrategy;

        WHILE quiescent environment
            pop[t] = Select(pop[t-1]);
            pop[t] = Cross(pop[t]);
            pop[t] = Mutate(pop[t]);
            pop[t] = LocalSearch(pop[t]);
            parameters[t+1] = Adapt(parameters[t], diversity[t]);
            t = t+1;
        ENDWHILE

    ENDWHILE

Figure 11.3. Pseudocode for a GA-based dynamic solver

The algorithm is generational; that is, the whole population is replaced by the new offspring at every generation. The used tournament selection works in a manner very similar to ranking selection, which avoids the pitfalls of roulette wheel [12] and at the same time is easier to implement and less time consuming than ranking. The tournament selection used in this paper selects the better of any two competing candidates if a random number is less than a user defined selection parameter and selects the worst of the two solutions otherwise as shown in Figure 11.4. Selection pressure can thus be increased or decreased by changing the selection parameter.

Edge crossover [19] is employed to recombine solutions in the population. This operator has the advantage of preserving most edges of the parent solutions without producing infeasible child solutions. The mutation operator is a pairwise node swap. It sweeps down the list of bits in the chromosome, swapping each with a randomly selected bit if a probability test is passed. That is, for a problem of size l and a mutation rate of μ the expected number of swaps on an individual is μl. This operator produces little change in the individual, since no more than four edges are replaced in each swap. Nevertheless, more drastic changes can be applied by simply increasing the rate of mutation. The evaluation of individuals is integrated with the genetic operators so that newly created individuals are evaluated as soon as they are created.

The genetic algorithm is also hybridized with a local search heuristic. We employ the well-known 2-opt heuristic, which eliminates two edges from the tour and reconnects the two resulting paths in a different way to obtain a new tour. Some researchers [10, 11] reported very good results with exhaustive local search; that is when local search is continued till reaching local optima for all individuals in the population. However, this practice can be too time consuming especially for large problem instances, and it can also lead to a profound loss of diversity [9]. Therefore, although exhaustive local search might be acceptable in solving static problems, it is certainly unsuitable for dynamic problems, where both time and diversity are of utmost importance. For these reasons, we allow only a fraction of the population to undergo local search and do not require local optimality to terminate local search. Two input parameters LsRate and LsNeighbours are employed to respectively control the fraction of population undergoing local search and the fraction of the neighborhood tested during each application of local search.

5. Experimentation

This section describes experiments carried out to test the two models of adaptation presented in the paper. The section describes the setting of the test problems, the used genetic algorithm, the criterion of performance measure and the results obtained from the experimentations.

Test Problems

Developing suitable benchmark problems is one of the central issues in dynamic optimization. A dynamic test problem has to be addressed

```
Select(pop[t])

    i = 0;
    WHILE i < PopulationSize
        sequentially select indivA from pop[t-1];
        randomly select indivB from pop[t-1];

        IF rand(0,1) ≤ SelectionProbability;
            insert best(indivA, indivB) in pop[t+1];
        ELSE
            insert worst(indivA, indivB) in pop[t+1];
        ENDIF

        i = i+1;
    ENDWHILE
```

Figure 11.4. Pseudocode for tournament selection

on three levels: the base static instance, changes added to create the sequence of instances, and the problem dynamics. In continuous optimization, the base instance is usually a multi-modal function created from multiple Gaussian peaks [6]. In discrete optimization, base test problems can be borrowed from the available libraries of established COP benchmarks. In this paper, the base static problem is a 100-city problem, kroA100 from the TSP library [14]. Changes are introduced in three ways (modes): an edge change mode (ECM), an insert/delete mode (IDM) and a city swap mode (CSM). Problem dynamics are controlled by two parameters: frequency and severity. The frequency of change determines the number of generations between succeeding environmental changes, and the severity of change determines the number of elementary steps applied at every environmental change. An elementary step is the smallest possible change in the problem that causes the new instance to have different optimal solution from the previous one.

In the ECM mode, distance between cities is viewed as a time period or cost that may increase or decrease with time; hence the introduction and the removal of a traffic jam can be simulated respectively by the increase or decrease in the distance between cities [4]. The problem is changed by increasing the length of an edge randomly selected from the best found tour, or decreasing the length of an edge randomly selected but not from the best found tour. This scheme ensures that environmental changes affect optimal solutions. The elementary step of the change is the change in the cost of a single edge.

In IDM, environmental changes are imposed by adding new cities to the problem or by removing some of the existing cities. IDM reflects the addition of new assignments to the problem or the deletion of existing assignments [7]. The elementary step of the change in this mode is the addition (or the deletion) of a single city. This mode might prove to be the most difficult since it entails variable representation to reflect the changing number of cities.

In CSM, the labels of two randomly selected cities are interchanged in the mapping function that maps the chromosome into solutions. Although this mode does not reflect direct real-world scenarios, it is an efficient method to create dynamic problems with known optima without the need of re-optimization. A more detailed description of these benchmark modes is given in [21].

In the current experimentation, each benchmark problem includes 200 successive changes to the base problem; that is, there is a sequence of 200 static problems for each of the three modes of environmental shifts. Each sequence of static problems is translated into nine dynamic test problems by combining three degrees of severity (1, 10, 100 steps per shifts) and three periods of change (10,100, 1000 generations between shifts). In short, there are 27 dynamic test problems each made up from 200 static instances.

Algorithm Settings

The dynamic test problems are used to compare the performance the ADM and LM models of this paper against three well known models: *FM, Restart,* and *Rim.* FM is a fixed model that uses a standard GA with fixed operator rates. The Restart model utilizes a fixed model strategy but re-initiates the population whenever the environment changes. Rim is a random immigrant model which can be viewed as a partial restart model, since only a fraction of the population (20% in this paper) is replaced with random individual at every environmental change. All tested models use a generational GA with tournament selection and a two-point order crossover. Pairwise interchange is used as a mutation operator.

A population size of 50 and crossover rate of 0.3 are used throughout. Finally, since the five models depend in part on the underlying mutation rate, experiments are repeated for three values of the base mutation rate (0.0025,0.025 and 0.25). These rates are also the upper bounds for mutation in the LM and ADM models; the lower bound on mutation is 0.0025 for both models.

A rate of 0.1 is used for local search, and the number of neighbors tested at each local search iteration is limited to 10% of the problem size.

Results

The criterion of comparisons in this paper is based on the offline performance metric originally presented by De Jong [3]. The measure is a running average of the best solutions found in order to reflect the ultimate goal of the optimization process. In other words, it dynamically abstracts the search evolution in one value that at any time assesses the algorithm's performance up to that point. However, this measure is meaningless in dynamic environments, since the value of previously found solutions is irrelevant after an environmental change. Hence, many researchers use current-best performance [18], which is defined as the generation best fitness averaged over several runs. However, this measure is not suitable for comparing several algorithms nor for complex dynamic problems (in which the optimal solution is not necessarily monotonically changing). The resulting plots would be intermingled, prohibiting the determination of which of the competing algorithms is performing best overall (remember that because the problem is dynamic, we may need to consider as many solutions as the number of environmental changes). Branke [1] suggested a modified offline performance measure to overcome this shortcoming by resetting the computed best-so-far value at every environmental change. Nevertheless, the modified OFP is sensitive to extreme solution values, and instances with large optimal values will dominate others.

This shortcoming can be avoided by normalizing the evaluations by dividing them by their corresponding optimal solution values, which are usually known in advance in the case of test problems. Younes et al. [20] used a measure that takes into account changes in the optimal solutions and also averages the results over the considered runs. We present a normalized modified offline performance at a time step t and for a number of R runs considered (ten runs in this paper) as follows:

$$NormModOFP(t, R) = \frac{1}{R \cdot t} \sum_{r=1}^{R} \sum_{\tau=1}^{t} \left(\frac{\varepsilon_\tau^r}{f_\tau^*} \right) \quad (11.3)$$

where $\varepsilon_\tau^r = min\left\{ e_\theta^r, e_{\theta+1}^r, \ldots, e_\tau^r \right\}$, θ is the time step of the last environmental change occurred prior to τ, e_θ^r is the value of the evaluation at time step θ and run r, and f_τ^* is the value of the optimal solution (or of the best known solution) to the problem instance at time step τ.

In a TSP, $NormModOFP$ is based on the cost of solution rather than on the fitness, thus, the lesser is $NormModOFP$ the better is the performance. Moreover, since $NormModOFP$ is measured relative to the value of the best solutions found during benchmark construction, it will in general exceed unity. Less than unity values if encountered will indicate high performance of the relevant model—so high that it is exceeding expectation.

Such a measure was used by Younes et al. [20] $NormModOFP$ abstracts dynamics, runs, evaluations, variations in the value of optimal solution. Thus, it allows convenient evaluation of an EA and guarantees fair comparisons. We also limit the number of evaluations (i.e. time) between succeeding environmental changes. This will indirectly assess the ability of the algorithm to recover from influence of changes in environment.

Figure 11.5 shows the results of experiments when the period of changes is 1000 generations/shift. This is the longest of the three periods considered in this paper. Offline performance is plotted against the base mutation rate for several ranges of severity of environmental changes for the three modes of benchmarking (ECM, IDM, and CSM). The values shown in the figure are averaged from ten runs using different random initial populations. In every run, the environment is kept quiescent for the first 10000 generations, then allowed to change according to the specified severity and period of change. The initial static phase gives the GA sufficient time to reach initial convergence and thus make later performance less dependent on the initial population.

Although Figure 11.5 represents the longest of the three periods of environmental changes and thus the best case for the Restart model, the inferior performance of the Restart model is quite evident. Indeed, Restart results are so large that they do not show in some slides of the figure. Nevertheless, the Restart model does produce results comparative to other models when environmental changes are large. The results, in general, are in favor of schemes that can exploit knowledge from past solutions while retaining the diversity needed to track shifting optima.

The FM model shows relatively good behavior in slides (a) and (b) corresponding to the ECM mode. This seemingly unexpected performance is the result of *dynamically insignificant* changes associated with this mode [21]. Reducing the length of an edge on the optimal tour (or increasing the length of an edge not on the optimal tour) will not alter the optimal solution. Furthermore, increasing the length of an edge on the optimal tour (or reducing the length of an edge not on the optimal tour) will not alter the optimal solution unless the changes are sufficiently large. Such changes do not alter the optimal solution and hence do not necessarily induce any adaptation from the optimizing algorithm. However when more changes are added to the problem, newer instances become considerably different from previous ones, and the performance of the FM model starts to deteriorate, as shown in slide (c) of the same figure.

In ADM, the base rate of mutation is not as critical as in other models. The reason for this behavior is that base rate in ADM acts as an upper bound on the mutation rate, while the actual rate is controlled by diversity. This suggests that ADM can be used to reduce the number of GA parameters to tune. That is the user will be spared the time-consuming tuning process of the GA parameters.

Statistical Analysis

Statistical t-tests that are used to compare the means of two samples can be used to compare the performance of two algorithms. The typical t-test is performed to build a confidence interval that is used to either accept or reject a null hypothesis that both sample means are equal. In applying this test to compare the performance of two algorithms,

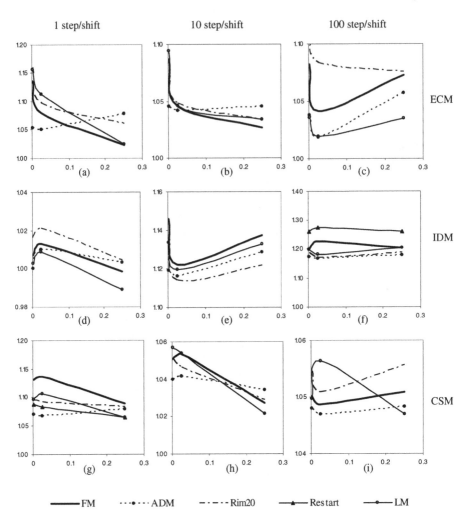

Figure 11.5. Modified offline performance against base mutation rate. Note that some results of the Restart model are too large to appear on the same slide with other results.

the measures of performance are treated as sample means, the required replicates of each sample mean are obtained by performing several independent runs of each algorithm, and the null hypothesis is that there is no significant difference in the performance of both algorithms.

However, when more than two samples are compared, the probability of multiple t-tests incorrectly finding a significant difference between a pair of samples increases with the number of comparisons [8]. Analysis of variance (ANOVA) overcomes this problem by testing the samples as a whole for significant differences. Therefore, ANOVA is performed to test the hypothesis that measures of performance of all the evolutionary models under considerations are equal. Then, a multiple post ANOVA comparison test, known as Tukey's test, is carried out to produce confidence intervals for the difference in the performance of each pair of models.

Statistical analysis reported in this section are obtained using a significance level of 5% to construct 95% confidence intervals on the difference in offline performance. Table 11.1, 11.2, and 11.3 show the results of multiple post ANOVA comparison test for the three modes of change (respectively, ECM, IDM, and CSM). Each table covers nine combinations of problem dynamics (three periods of change and three levels of severity of change). For the FM, Restart and Rim20 models, which use fixed rates of mutation, a mutation rate of $\frac{1}{100*4} = 0.0025$ is used. This value corresponds to the commonly used rate of mutation [5, 12, 13] (inverse of the chromosome length in a 100 city chromosome) and to the number of edges (four) replaced in each swap. For the LM and ADM models, the lower bound on mutation is 0.0025 and the upper bound is twice that value. The entries in the table are interpreted as follows. An entry of 1 signifies that the confidence interval for the difference in performance measures of the corresponding pair consists entirely of positive values, which indicates that the first model is inferior to the second model. Conversely, an entry of -1 signifies that the confidence interval for the corresponding pair consists entirely of negative values, which indicates that the first model is superior to the second one. An entry of 0 indicates that there is no significant difference between both models.

As can be seen in the three tables, there are significant differences between the performance of the ADM model and the others. The ADM outperforms other models in all cases except the K100-IDM problem with small severity of change (1 step/shift) and larger period of change (1000 generations/shift), where its performance is comparable to that of the LM model but still better than other models. We note here that these results are obtained using a "conventional" rate of mutation without attempting to fine tuning this parameter. Thus, an extended version of the ADM model that takes into account other input parameters may prove to be useful in reducing the usual cumbersome efforts of parameter tuning.

Table 11.1. Multiple comparison test (K100-ECM problem)

period →	10			100			1000		
severity →	1	5	25	1	5	25	1	5	25
FM vs Restart	-1	-1	-1	0	0	-1	0	0	0
FM vs Rim20	0	0	0	0	0	0	0	0	0
FM vs LM	1	1	1	1	1	1	1	1	1
FM vs ADM	1	1	1	1	1	1	1	1	1
Restart vs Rim20	1	1	1	0	0	0	0	0	0
Restart vs LM	1	1	1	1	1	1	1	1	1
Restart vs ADM	1	1	1	1	1	1	1	1	1
Rim20 vs LM	1	1	1	1	1	1	1	1	1
Rim20 vs ADM	1	1	1	1	1	1	1	1	1
LM vs ADM	1	1	1	1	1	1	1	1	1

Table 11.2. Multiple comparison test (K100-IDM problem)

period →	10			100			1000		
severity →	1	5	25	1	5	25	1	5	25
FM vs Restart	0	-1	0	0	0	0	1	0	0
FM vs Rim20	0	0	0	0	0	0	1	0	0
FM vs LM	1	1	0	1	1	1	1	1	1
FM vs ADM	1	1	1	1	1	1	1	1	1
Restart vs Rim20	0	1	0	0	0	0	0	0	0
Restart vs LM	1	1	0	1	1	1	1	1	1
Restart vs ADM	1	1	1	1	1	1	1	1	1
Rim20 vs LM	1	1	0	1	1	1	1	1	1
Rim20 vs ADM	1	1	1	1	1	1	1	1	1
LM vs ADM	1	1	1	1	1	1	0	1	1

Table 11.3. Multiple comparison test (K100-CSM problem)

period →	10			100			1000		
severity →	1	5	25	1	5	25	1	5	25
FM vs Restart	-1	-1	0	-1	-1	0	-1	-1	-1
FM vs Rim20	0	0	0	0	0	0	0	0	0
FM vs LM	1	1	0	1	1	1	1	1	1
FM vs ADM	1	1	1	1	1	1	1	1	1
Restart vs Rim20	1	1	0	1	1	0	0	1	1
Restart vs LM	1	1	0	1	1	1	1	1	1
Restart vs ADM	1	1	1	1	1	1	1	1	1
Rim20 vs LM	1	1	0	1	1	1	1	1	1
Rim20 vs ADM	1	1	1	1	1	1	1	1	1
LM vs ADM	1	1	1	1	1	1	1	1	1

6. Conclusions

This paper introduces a GA model that controls diversity throughout the search process in a dynamic environment. The idea is to maximize diversity when the environment changes and to control it during subsequent generations by changing the GA parameters in response to the diversity of the current population. In other words, diversity control in a dynamic problem is treated as an extension to that in the static problem.

Results and statistical analysis show that the idea is promising as it produces good results compared with other models. Currently, we are developing an enhanced ADM model that uses variable diversity limits instead of the fixed limits used in this paper.

Future work should investigate the use of ADM to manipulate crossover rate, local search rate, and fraction of the neighborhood tested during each application of local search. The model should also be tested on more complex dynamic COPs, such as job shop scheduling.

The ability of the ADM model to retain good performance over a wide range of mutation rate encourages investigating the use of this model as a substitute for parameter tuning for static problems as well.

Acknowledgments

Support of this work has been provided by the Natural Sciences and Engineering Research Council of Canada (NSERC). The authors would like to thank the anonymous reviewers for many valuable suggestions and comments.

References

[1] J. Branke. Evolutionary approaches to dynamic optimization problems: A survey. pp. 134–137, GECCO Workshops, A. Wu (ed.), 1999.

[2] E. K. Burke, S. Gustafson, and G. Kendall. Diversity in genetic programming: An analysis of measures and correlation with fitness. *IEEE Transactions on Evolutionary Computation*, 8(1):47–62, 2004.

[3] K. A. De Jong. *An Analysis of the Behavior of a Class of Genetic Adaptive Systems*. PhD thesis, University of Michigan, 1975.

[4] C. J. Eyckelhof and M. Snoek. Ant systems for a dynamic TSP. In *Ant Algorithms*, pages 88–99, 2002.

[5] M. Gen and R. Cheng. *Genetic Algorithms*. John Wiley & Sons, Inc., New York, NY, USA, 1999.

[6] J. J. Grefenstette. Evolvability in dynamic fitness landscapes: a genetic algorithm approach. In *1999 Congress on Evolutionary Computation*, pages 2031–2038, Piscataway, NJ, 1999. IEEE Service Center.

[7] M. Guntsch, M. Middendorf, and H. Schmeck. An ant colony optimization approach to dynamic tsp. In L. Spector, E. D. Goodman, A. Wu, W. Langdon, H.-M. Voigt, M. Gen, S. Sen, M. Dorigo, S. Pezeshk, M. H. Garzon, and E. Burke, editors, *Proceedings of the Genetic and Evolutionary Computation Conference (GECCO-2001)*, pages 860–867, San Francisco, California, USA, 7-11 July 2001. Morgan Kaufmann.

[8] Y. Hochberg and A. C. Tamhane. *Multiple Comparison Procedures*. Wiley, 1987.

[9] N. Krasnogor and J. Smith. A tutorial for competent memetic algorithms: model, taxonomy, and design issues. *IEEE Transactions on Evolutionary Computation*, 9(5):474– 488, 2005.

[10] P. Merz. Advanced fitness landscape analysis and the performance of memetic algorithms. *Evol. Comput.*, 12(3):303–325, 2004.

[11] P. Merz and B. Freisleben. Genetic local search for the TSP: New results. In *IEEECEP: Proceedings of The IEEE Conference on Evolutionary Computation, IEEE World Congress on Computational Intelligence*, 1997.

[12] M. Mitchell. *An Introduction to Genetic Algorithms*. The MIT Press, Cambridge, Massachusetts, 1996.

[13] C. R. Reeves and J. E. Rowe. *Genetic Algorithms: Principles and Perspectives: A Guide to GA Theory*. Kluwer Academic Publishers, Norwell, MA, USA, 2002.

[14] G. Reinelt. TSPLIB — a traveling salesman problem library. *ORSA Journal on Computing*, 3:376 – 384, 1991.

[15] J. Riget and J. Vesterstroem. A diversity-guided particle swarm optimizer - the arpso, 2002.

[16] K. Sörensen and M. Sevaux. MA—PM: Memetic algorithms with population management. *Computers and Operations Research*, 2004. In Press, available online 26 November 2004.

[17] R. K. Ursem. Diversity-guided evolutionary algorithms. In *Proceedings of Parallel Problem Solving from Nature VII (PPSN-2002)*, pages 462–471. Springer Verlag, 2002.

[18] K. Weicker. Performance measures for dynamic environments. In J. Merelo, P. Adamidis, H.-G. Beyer, J. Fernández-Villacañas, and

H.-P. Schwefel, editors, *Parallel Problem Solving from Nature*, volume 2439 of *LNCS*, pages 64–73. Springer, 2002.

[19] D. Whitley, T. Starkweather, and D. Shaner. The traveling salesman and sequence scheduling: Quality solutions using genetic edge recombination. In L. Davis, editor, *Handbook of Genetic Algorithms*, pages 350–372. Van Nostrand Reinhold, New York, 1991.

[20] A. Younes, O. Basir, and P. Calamai. A benchmark generator for dynamic optimization. In *the 3rd WSEAS International Conference on Soft Computing, Optimization, Simulation & Manufacturing Systems*, Malta, 2003. WSEAS.

[21] A. Younes, P. Calamai, and O. Basir. Generalized benchmark generation for dynamic combinatorial problems. In F. Rothlauf, M. Blowers, J. Branke, S. Cagnoni, I. I. Garibay, O. Garibay, J. Grahl, G. Hornby, E. D. de Jong, T. Kovacs, S. Kumar, C. F. Lima, X. Llorà, F. Lobo, L. D. Merkle, J. Miller, J. H. Moore, M. O'Neill, M. Pelikan, T. P. Riopka, M. D. Ritchie, K. Sastry, S. L. Smith, H. Stringer, K. Takadama, M. Toussaint, S. C. Upton, and A. H. Wright, editors, *Genetic and Evolutionary Computation Conference (GECCO2005) workshop program*, pages 25–31, Washington, D.C., USA, 25-29 June 2005. ACM Press.

[22] K. Q. Zhu. A diversity-controlling adaptive genetic algorithm for the vehicle routing problem with time windows. In *ICTAI*, pages 176–183, 2003.

[23] K. Q. Zhu and Z. Liu. Population diversity in permutation-based genetic algorithm. In *ECML*, pages 537–547, 2004.

Chapter 12

A MEMETIC ALGORITHM FOR DYNAMIC LOCATION PROBLEMS

Joana Dias[1], M. Eugénia Captivo[2] and João Clímaco[1]

[1]*Faculdade de Economia and Inesc-Coimbra, Universidade de Coimbra, Av. Dias da Silva, 165, 3004-512 Coimbra, Portugal;*[2]*Universidade de Lisboa, Faculdade de Ciências, Centro de Investigação Operacional, Campo Grande, Bloco C6, Piso 4, 1749-016 Lisbon, Portugal*

Abstract: In this paper a memetic algorithm integrating genetic procedures and local search, able to solve capacitated and uncapacitated dynamic location problems, is described. These problems are characterized by explicitly considering the possibility of a facility being opened, closed and reopened more than once during the planning horizon. It is also possible to explicitly consider different open and reopen fixed costs. The problems can be of single or multi-level nature. The computational results obtained show that the algorithm is capable of finding good quality solutions, but at the cost of large computational times, when compared with dedicated primal-dual heuristics and even with a general solver. Some changes are proposed to tackle this disadvantage.

Key words: location problems, genetic algorithms, local search

1. INTRODUCTION

When faced with a hard combinatorial optimization problem, the first obvious question that has to be answered is which algorithm(s) should be used to find the optimal solution. The use of exact methods becomes, most of the times, impracticable (in spite of the increase in computing power and hence speed of calculation). Heuristic methods try to calculate good solutions, without guaranteeing their optimality, but offering good compromises between solution quality, computational time and storage. Metaheuristics have shown their capability for calculating high quality solutions to complex problems. Genetic algorithms, for instance, have been widely used in solving combinatorial problems in general and location problems in particular. Hosage and Goodchild, 1986, study the possibilities of applying genetic algorithms to location problems. Houck *et al.*, 1996, study large location-allocation problems, where each possible location is a bidimensional point in a continuous space, and use genetic algorithms. Beasley and Chu, 1996, apply genetic algorithms to a covering problem, developing specific operators that guarantee the feasibility of the individuals. Kratica, 1999 and Kratica *et al.*, 2001, use genetic algorithms for the simple plant location problem, and propose hybridization with an *ADD* heuristic (at each iteration only the facility that contributes to the maximum reduction of the overall cost is open). Filipovic *et al.*, 2000, introduce the grained tournament selection operator. Lorena and Lopes, 1997, apply genetic algorithms to computationally hard covering problems. Abdinnour-Helm, 1998, describes a hybrid heuristic which includes genetic algorithms and tabu search, applying it to the uncapacitated hubs problem. Jaramillo *et al.*, 2002, study genetic algorithms as an alternative way of calculating good quality solutions to location problems, and conclude that these algorithms should not be used for capacitated location problems with fixed costs. Shimizu, 1999, mingles genetic algorithms and mathematical programming, solving the sanitary landfill for hazardous materials location problem, in a multi-objective environment. Correa *et al.*, 2001, apply genetic algorithms to a real problem that can be formulated as a p–median problem. Cheung *et al.*, 2001, describe a genetic algorithm, partially implemented in a parallel architecture, which is applied to several location and location-allocation problems in the oil industry. Cortinhal and Captivo, 2003, describe genetic algorithms applied to the capacitated location problem with total assignment. Salhi and Gamal, 2003, consider the problem of finding p facilities in the plane to serve n clients, minimizing the total transportation costs, using genetic algorithms with a real number representation. Domínguez-Marín *et al.*, 2005, solve location problems (p-centre and p-median) using genetic algorithms and variable neighborhood search.

In the present paper, we will describe an algorithm that hybridizes genetic algorithms and local search, and that is able to solve dynamic capacitated and uncapacitated location problems, with opening, closing and reopening of facilities. The authors have already studied this problem and developed several efficient primal-dual heuristics (Dias *et al.*, 2005a, b, 2006). The primal-dual heuristics developed calculate good quality solutions as well as lower bounds on the optimal objective function value. Nevertheless, their structure makes it difficult and time consuming to adapt these heuristics to even minor changes in the problem formulation. This paper is organized as follows: in the next section we present the problem, in section 3 the main characteristics of our algorithm are described, in section 4 some computational results are reported and, finally, in section 5 we point out some conclusions and possible future work.

2. THE DYNAMIC LOCATION PROBLEM

Consider that there are n clients, m facilities' possible locations, and T time periods considered in the planning horizon. A facility can be opened, closed and reopened more than once during the planning horizon. We develop a general model considering two different types of facilities: 1- uncapacitated facilities; 2- facilities with maximum and/or minimum capacity restrictions. The objective function considers the minimization of all fixed costs and transportation costs. The fixed costs account for the opening, closure and fixed operating costs during the operating time periods. The model assures that each client's demand in each time period has to be satisfied; each client can only be assigned to operating facilities. A facility can only be reopened at the beginning of period t if it has been opened earlier and can only be opened once during the planning horizon. Only one facility can be open at each location, in each time period. It is possible to extend this problem to the multi-level case, where a client has to be assigned to a path of facilities, instead of only one facility. A path of facilities can be constituted of, at most, k facilities (where k represents the maximum number of levels considered), and, at most, one facility of each level. Problems' formulations as Mixed Integer Linear Programming Problems can be found in Dias *et al.*, 2005a,b, 2006. Depending on the type of facilities present, after fixing a set of feasible values for the location variables, it is rather simple to calculate the optimal allocation variables in each time period: 1) if all facilities are uncapacitated, then each client is assigned to exactly one facility (or path of facilities), the cheapest one; 2) if there is at least one facility of type two, then it is necessary to solve a transportation (or a transshipment) problem.

Defining $J = \{1,..., j, ..., n\}$ as the set of indices corresponding to the clients' locations and $I = \{1,..., i,..., m\}$ as the set of indices corresponding to facilities' possible locations, the decision variables considered in the model are:

$$a_{it}^{\xi} = \begin{cases} 1 & \text{if facility } i \text{ is opened at the beginning of period } t \\ & \text{and stays open until the end of period } \xi \\ 0 & \text{otherwise} \end{cases}$$

$$r_{it}^{\xi} = \begin{cases} 1 & \text{if facility } i \text{ is reopened at the beginning of period } t \\ & \text{and stays open until the end of period } \xi \\ 0 & \text{otherwise} \end{cases}$$

If the problem is of multi-level nature, variables x_{pj}^{t} are considered, where p is a path of facilities, with at most one facility of each level. Otherwise variables x_{ij}^{t} are used.

x_{pj}^{t} = fraction of customer j demand assigned to path p during period t.

x_{ij}^{t} = fraction of customer j demand served by facility i during period t.

The objective functions' coefficients can be defined as follows:

c_{pj}^{t} or c_{ij}^{t} = cost of fully assigning client j to path p, or facility i, in period t;

FA_{it}^{ξ} = fixed cost of opening a facility i at the beginning of period t, and closing it at the end of period ξ (the facility will be in operation from the beginning of t to the end of ξ);

FR_{it}^{ξ} = fixed cost of reopening a facility i at the beginning of period t, and closing it at the end of period ξ (the facility will be in operation from the beginning of t to the end of ξ);

The location problems considered are generalizations of the simple location problem, so they are *NP*-hard.

3.   THE MEMETIC ALGORITHM

Our first experiences began with a simple genetic algorithm, without local search, but the results were far from being satisfactory, so a memetic algorithm was developed (Moscato and Cotta, 2003). According to Osman and Kelly, 1996, an evolutionary algorithm is composed of five basic components: a genetic representation of solutions; a way to create an initial population; an evaluation function and a selection operator; genetic operators that alter the genetic composition of children during reproduction and values for the parameters. In the following subsections, all these components will be described for our particular case. Algorithm 1 and Algorithm 2 describe the algorithm's functioning scheme.

Algorithm 1: Generation

x_{best} – represents the best individual in the preceding generation; $flag(x)$ – is equal to *true* if x has already passed through the local search procedure, *false* otherwise; $P_{current}$ – the current population; $f(x_j)$ – fitness of individual x_j; $npop$ – number of individuals in the population.

1. $x_1 \leftarrow x_{best}$; $x_{best} \leftarrow x_1$; $j \leftarrow 2$; $Newpop \leftarrow \{x_1\}$.
2. If $j > npop$ then $P_{current} \leftarrow Newpop$, else continue.
3. Select parents x_A and x_B using binary tournament selection.
4. Crossover to generate two children: x_j and x_{j+1}. $flag(x_j) \leftarrow false$; $flag(x_{j+1}) \leftarrow false$.
5. Apply the mutation operator to x_j.
6. If $x_j = x_A$ then $flag(x_j) \leftarrow flag(x_A)$; if $x_j = x_B$ then $flag(x_j) \leftarrow flag(x_B)$;[1]
7. Calculate the fitness of x_j: $f(x_j)$. If $f(x_j) = +\infty$, then apply the repair procedure to x_j. If $f(x_j) < f(x_{best})$ then $x_{best} \leftarrow x_j$.
8. Apply the change openings procedure to x_j.
9. If not $flag(x_j)$ then apply the local search procedure.
10. If $(j+1) \leq npop$ then repeat steps 5 to 9 with child x_{j+1}.
11. $Newpop \leftarrow Newpop \cup \{x_j\}$. If $(j+1) \leq npop$ then $Newpop \leftarrow Newpop \cup \{x_{j+1}\}$. $j \leftarrow j + 2$. Go to 2.

Algorithm 2: global memetic algorithm

$P_{current}$ – the current population; x_{best} – represents the best individual in the current population; $nimp$ - maximum number of generations without improving the best objective function value found; $Nger$ - Total maximum number of generations; β - Percentage of increase in the number of individuals.

1. Initialize $P_{current}$.
2. Initialize x_{best}, $best \leftarrow f(x_{best})$, $ngen \leftarrow 1$, $count \leftarrow 0$.
3. If $ngen > Nger$ or $count > nimp$ then stop. Else continue.
4. $ngen \leftarrow ngen + 1$. Call procedure *generation*.
5. If $f(x_{best}) \geq best$ then $count \leftarrow count + 1$. Else $count \leftarrow 0$.
6. If $count < nimp$ then $ngen \leftarrow ngen + 1$, then go to 3.
7. If $min\{\lceil npop(1 + \beta) \rceil, Nmaxpop\} > npop$, then $npop \leftarrow min\{\lceil npop(1 + \beta) \rceil, Nmaxpop\}$, initialize randomly the new individuals and $count \leftarrow 0$. Go to 3.

3.1 Representation and Evaluation of Solutions

Each individual is represented using two chromosomes, composed of genes that can only take values 0 or 1. Gene in position $(t-1)m+i$ of the *L*-chromosome is equal to 1 if facility i is open during time period t, and equal to 0 otherwise. This information is not sufficient to build an admissible solution for the problem, because it is necessary to determine the open and reopen periods: a facility i can be operating from τ to ξ but have been closed at the end of period η and reopened at the beginning of $\eta+1$, with $\tau \leq \eta \leq \xi$.

---

[1] x_j is considered equal to x_A if all the *L*-chromosome's genes are equal.

The other chromosome (*F*-chromosome) will give this information. Gene in position $(t-1)m+i$ will be equal to 1 if facility i is reopened at the beginning of period t, and 0 otherwise. The *F*-chromosomes complement the information provided by *L*-chromosomes. Their genes' values will only be taken into account when strictly needed. Consider the following example, with five possible locations and three time periods (a matrix notation is used for ease of understanding):

Chromosome *L*

i \ t	1	2	3	4	5
1	1	0	0	1	1
2	1	1	0	0	0
3	1	1	0	1	0

Chromosome *F*

i \ t	1	2	3	4	5
1	1	1	0	0	1
2	*0*	0	0	1	1
3	*1*	*0*	1	0	1

Figure 1: An individual's representation

In terms of location variables, these two chromosomes would be interpreted as all variables equal to zero except a_{11}^2, r_{13}^3, a_{22}^3, a_{41}^1, r_{43}^3, a_{51}^1. The three *F*-chromosome genes represented in bold italic are the only genes (from this chromosome) that really matter for building the solution.

This representation is *redundant* (Rothlauf and Goldberg, 2002): the number of genotypes exceeds the number of phenotypes[2]. The capacity restrictions are the only ones that can be violated. The fitness of each individual is given by the objective function value of the corresponding solution in the phenotype space. An unfeasible individual has fitness equal to $+\infty$, but it is not deleted from the current population, to increase diversity.

Definition 1: Consider two individuals that differ only in one $L(F)$-chromosome gene. If the solutions they represent in the phenotype space are different, then the $L(F)$-chromosome gene is called *determinant*, otherwise it is called *non-determinant*.

Proposition 1: All *L*-chromosome genes are determinant.

Proposition 2: The only *F*-chromosome genes that are determinant are genes in position $(t-1)mix$, for some i and $t>1$, such that the *L*-chromosome genes $(t-1)m+i$ and $(t-2)m+i$ are equal to one.

Proposition 3: It is possible to represent each and every admissible solution to the location problem using a pair of F and L-chromosomes.

---

[2] If there are p determinant genes within the *F*-chromosome, there can be $2^{(Tm-p)}$ different individuals codifying exactly the same solution.

3.2 Genetic Operators

The genetic algorithm uses selection, crossover and mutation. Selection is based on the binary tournament selection with sharing (Oei *et al.*, 1991, Deb, 2001). In each generation, individuals *i* and *j* are randomly selected. A sharing value $sh(i,j)$ is calculated. For each *i*, all values $sh(i,j)^3$ (considering all individuals *j* belonging to the new population) are summed up. If, at the moment of the selection, there are *num* individuals in the new population, then $nc_i = num - nc_i$. The individual's fitness value is divided by nc_i, and the resulting value ($f(i)$) is used in the binary tournament selection. In the presence of two randomly chosen individuals x_1 and x_2, if $f(x_1) < f(x_2)$ then individual x_1 wins the binary tournament with a given probability p_{bt} (usually closer to one).

$$d_{ij}^{\vartheta} = \begin{cases} 1, & \text{if the } \vartheta-\text{gene in the } L\text{-chromosome is different in } i \text{ and } j \\ 0, & \text{otherwise} \end{cases} \quad , d_{ij} = \sum_{\vartheta=1}^{mT} d_{ij}^{\vartheta}$$

$$nc_i = \sum_{j \text{ belongs to the new population}} sh(i,j), \quad sh(i,j) = \begin{cases} 1 - \left(d_{ij}/\alpha_{share}\right)^{\alpha}, & \text{if } d_{ij} \leq \alpha_{share} \\ 0, & \text{otherwise} \end{cases}$$

The crossover operator is an adaptation of the one-point crossover: two parent solutions are recombined yielding two children. A value $1 \leq \kappa \leq T$ is randomly chosen. The first child will have *L* and *F*-chromosome genes $(t-1)m+i$, with $t < \kappa$, equal to the first parent, and all the others equal to the second parent. The opposite happens with the second child. The children of two admissible parents are not guaranteed to be admissible. Two special operators were developed: a repair and a *change openings* procedure. An individual represents an unfeasible solution if it violates any maximum or minimum capacity restrictions, due to *L*–chromosome genes, that are changed, in a random but guided manner, by the repair procedure. If, for instance, maximum capacity restrictions are violated at period *t*, it randomly opens more facilities. If minimum capacity restrictions are violated, it randomly closes facilities. All changes are performed in a random manner[4], so a maximum number of tries had to be imposed. The *change openings* procedure studies the effect on fixed (re)open costs of changes in some of the determinant *F*-chromosome genes. It identifies situations such that a facility *i* is open from time period τ to time period ξ, $\xi > \tau$, and is reopened during that interval (in a time period $\tau \leq t \leq \xi$). The *F*-chromosome gene in

---

[3] The α_{share} value is calculated as described in Deb, 2001, and α is considered equal to one.
[4] To avoid the introduction of bias in the search (Coello Coello, 2002).

position $(t-1)m+i$ (equal to one) is changed to zero. If the total fixed costs diminish, then the gene's value is changed, else retains its original value.

3.3   Local Search

Local search plays a very important role in the algorithm developed[5]. Every individual in the new population is a potential starting solution for the local search procedure, ran with a given probability p_{ls}[6]. The procedure searches k-neighborhoods, from $k=1$ to $k=T$, where an individual x' is said to be in the k-neighborhood of individual x if and only if x' differs from x by the insertion or removal of at most k continuous operating time periods to a single facility i. Whenever the fitness function is improved, the genotype is changed, and the search continues from this new starting point. Algorithm 3 describes the local search procedure. This procedure is very time consuming (responsible for 95% or more of the algorithm's total computational time), so its performance has to be improved. If in presence of capacitated facilities, a sensitivity analysis based on the dual optimal solution of the assignment problems, helps to estimate if the present neighbor is or is not better than the current individual, decreasing the total computational time by 20% or more without a significant decrease in the final solution's quality.

Algorithm 3: Local Search Procedure

x – starting solution; $f(x)$ – fitness of individual x.

1. $k \leftarrow 1$.
2. If $k > T$ then stop. Else $count \leftarrow 0$.
3. $flag(i) \leftarrow false, \forall i$.
4. If $flag(i) = true, \forall i$ or $count > z$ then $k \leftarrow k + 1$ and go to 2. Else choose randomly a facility i such that $flag(i) = false$.
5. $flag(i) \leftarrow true; t \leftarrow 1$;
6. If $t > T - k + 1$ or $count > z$ then go to 4, else continue.
7. If facility i is operating during periods t to $t + k - 1$, then study k-neighbor x' obtained from x by the removal of operating periods t to $t + k - 1$ and go to 9. Else continue.
8. If facility i is not operating during periods t to $t + k - 1$, then study k-neighbor x' obtained from x by the insertion of operating periods t to $t + k - 1$ and go to 9. Else $t \leftarrow t + 1$ and go to 6.
9. If $f(x) > f(x')$ then $x \leftarrow x'$ and $count \leftarrow 0$, else $count \leftarrow count + 1$. $t \leftarrow t + 1$ and go to 6.

---

[5] Examples of genetic algorithms hybridized with local search can also be found in Huntley and Brown, 1996; Murata *et al.*, 1998; Reeves and Höhn, 1996; Yagiura and Ibaraki, 1996.
[6] If a child is equal to one of the parents that, in turn, resulted from local search, this probability is equal to zero.

3.4   Population

Our initial population is constituted by individuals randomly created that are modified by the *repair, change openings* and *local search* procedures (clones are allowed). We chose to work with small populations. As there is a risk of premature convergence to a poor quality solution, we overcame this disadvantage by changing on-line the total number of individuals in the current population. This number is increased whenever *nimp* generations are run without improving the best objective function value, until the maximum number of individuals in the population is reached (this is also the stopping criteria for the algorithm). The new individuals are randomly initialized. The population is initialized with *npop* individuals calculated as described in Reeves, 1993: it is the minimum value such that $\left(1-\left(1/2\right)^{npop-1}\right)^{l} \geq 0.99$ [7],

where l is the number of genes of each individual. Each individual has two chromosomes, each with mT genes, so l should be equal to $2mT$. As the F-chromosome has very few determinant genes, we chose to consider l equal to mT for the calculation of the initial value of *npop*, and equal to $2mT$ for the calculation of the maximum value *npop* can take. The child population replaces the parents population completely, with the exception of the best solution that passes from one generation to the next unchanged.

3.5   Parameters' Values

This algorithm has many parameters whose values have to be fixed and that can influence the algorithm's behavior. Aside from the number of individuals in the population, that can change during the execution, all other parameters are fixed before the run and do not change. Table 1 presents a list of all parameters, and the way in which they can influence its behavior.

4.   COMPUTATIONAL TESTS

The memetic algorithm was tested with a set of randomly generated problems. For each combination of (T, m, n), 5 problems were generated. Both uncapacitated and capacitated, single and multi-level location problems were solved. The test problems were generated as described in Dias *et al.*, 2005b. For multi-level problems, it is considered that the number of possible locations for facilities at level l is half the number at level $l-1$. All possible paths between clients and facilities are considered.

---

[7] The value 0.99 represents the probability of at least one allele being present at each locus in the initial population.

Table 12-1. Algorithm's parameters

Parameter	Description	Influence on the algorithm's behavior and Recommended Values[8]
$Nger$	Total maximum number of generations	It can be used to terminate the algorithm. If it is completely blind to the algorithm's performance, it can be responsible for premature terminations as well as for unnecessary generation runs. In our algorithm this parameter is not important, because it uses other termination rules.
$nimp$	Maximum number of generations without improving the best objective function value found	It indicates that the algorithm is converging. The number of individuals in the population is increased whenever there are no improvements in the objective function during $nimp$ generations, as a way of increasing the genetic diversity, and to avoid getting trapped in local minima. If the current number of individuals is equal to $Nmaxpop$, then the algorithm is terminated after $nimp$ generations without improving the objective function. It is fixed to 5.
β	Percentage of increase in the number of individuals	It controls, along with $Nmaxpop$, the number of times the population is increased. It is hard to predict how it will influence the quality of the solution or the total computational time: greater values will correspond to fewer generations but with longer computational times per generation. In our algorithm this value is equal to 25%.
p_{bt}	Probability of choosing the most fitted individual in the binary tournament selection	The greater the probability, the more difficult it is for less fit individuals to be passed on to the next generation. It can be used to influence the diversity of the population. If controlled on line, it could be increased as the number of generations increases, to ensure diversity in the beginning and convergence in the end. In our algorithm this value is fixed at 0.9.
$npop$	Number of individuals in the current population	It influences the computational time and the quality of the final solution: populations with more individuals will take longer to generate their children but are genetically more powerful. Small populations run the risk of under-covering the solution space (Reeves, 1993). The initial population is calculated as described in 3.4, and this value is increased whenever there are $nimp$ generations without improvement on the best objective function value.
$Nmaxpop$	Maximum number of individuals in the current population	This parameter influences the total execution time, and can influence the quality of the best solution found. It is calculated as described in 3.4.

---

[8] These "recommended values" are indicated according to the computational experiments made so far. These computational tests were executed using different instances of different location problems, both single and multi-level problems, of different sizes. They allowed us to assess, in a non-systematic way, the behavior of the algorithm due to changes in its parameters. The values indicated were the ones that presented better compromises throughout the preliminary tests.

p_{ls}	Probability of executing the local search procedure for each individual	It influences both the computational time and the quality of the best solution found. With values near 1, the algorithm will converge quicker and with good quality solutions. It is difficult to estimate how this parameter influences computational time because with smaller values each generation is executed in less computational time but the convergence towards a good solution is slower, so the total algorithm's computational time can increase. It is advised that p_{ls} should be equal to 1 at least in the last generation. In our algorithm this value is fixed to 1.
z	Maximum number of k-neighbors visited without improving the individual's fitness	It influences the algorithm's behavior in a way similar to the previous one. It should consider the total number of neighbors of a given solution which is hard to compute. In our algorithm we consider z equal to 10000.
p_v	Probability of visiting a neighbor expected to improve the individual's fitness	It influences the algorithm's behavior in a way similar to the two previous parameters. This probability should be always a value near to 1. In our algorithm it is fixed to one.
p_{nv}	Probability of visiting a neighbor not expected to improve the individual's fitness	It influences the computational time and also the quality of the final solution. To obtain a good compromise value, we recommend it should be fixed to a value between 0 and 0.1.
P_{μ}	Probability of changing one gene in the mutation operator	It can influence the diversity of the population. If controlled on-line, it could be decreased as the number of generations increases, or increased when the best fitness value does not improve in a given number of generations. In our algorithm this parameter is fixed at 0.002.

Experiments were carried out in a Pentium 4, 1.80 GHz, running under Windows 2000 operating system, with a maximum of 2000 MB of virtual memory and 260Mb of Ram.

The quality of the solutions calculated, as well as computational times, is compared with the results obtained with primal-dual heuristics and CPLEX (version 7.0 for the single-level problems and version 9.0 for the multi-level problems). Primal-dual heuristics and the memetic algorithm were programmed using C-language. CPLEX terminates without calculating the optimal solution whenever more than 2 100 000 000 nodes of the branch and bound tree are explored, or when the number of simplex iterations in a node exceeds 2 100 000 000, or when there is not enough memory to read the problem or when the execution times exceeds 200 000 seconds. After the execution of the primal-dual heuristic, a local search procedure was executed (Dias *et al.*, 2005b). Tables 2 to 9 present mean values (average and σ-standard deviation) of our computational results. Tables 2, 4, 7, and 8 present the quality of the best primal solution found and tables 3, 5, 6 and 9 present the computational time spent by the different approaches. For multi-level problems, m_k represents the number of services in the last level, and N represents the number of levels. The quality of the primal solution is calculated as $(Z\text{-}ZLB)/ZLB$, where Z is the objective function value of the final primal solution found and ZLB is the value of the best lower bound known. For memetic algorithms the computational time spent until the best solution is calculated is also shown. As can be seen, the memetic algorithm is capable of finding good quality solutions to the location problems solved, but at the expense of huge computational times, especially when compared with the primal-dual heuristics. The results for uncapacitated problems are better than the results for the capacitated ones, both for single and multi-level case. This can be justified by the increased complexity of the corresponding assignment problems.

Table 12-2. Quality of the primal solution (in percentage) – uncapacitated single level location problem

T	m	n	Primal-dual heuristic Average	σ	Memetic algorithm Average	σ	T	m	n	Primal-dual heuristic Average	σ	Memetic algorithm Average	σ
5	5	25	0.0	0.0	0.0	0.0	20	5	50	0.3	0.5	0.0	0.0
5	5	50	0.0	0.0	0.0	0.0	20	5	100	0.2	0.4	0.6	0.5
5	5	100	0.0	0.0	0.0	0.0	20	5	200	0.0	0.0	1.1	1.5
5	5	200	0.0	0.0	0.0	0.0	20	5	500	0.0	0.0	0.1	0.2
5	5	500	0.0	0.0	0.0	0.0	20	5	1000	0.0	0.0	0.0	0.0
5	5	1000	0.0	0.0	0.0	0.0	20	10	25	0.0	0.0	0.2	0.4
5	10	25	0.0	0.0	0.0	0.0	20	10	50	0.2	0.3	0.4	0.4
5	10	50	0.0	0.0	0.0	0.0	20	10	100	0.0	0.0	0.7	0.4
5	10	100	0.0	0.0	0.0	0.0	20	10	200	0.0	0.0	0.6	0.3
5	10	200	0.0	0.0	0.0	0.0	20	10	500	0.0	0.0	0.2	0.2
5	10	500	0.0	0.0	0.0	0.0	20	10	1000	0.0	0.0	0.0	0.0
5	10	1000	0.0	0.0	0.0	0.0	20	50	25	0.0	0.0	9.3	3.7
5	50	25	0.1	0.1	0.1	0.1	20	50	50	1.1	1.1	8.8	5.8
5	50	50	0.4	0.7	0.0	0.0	20	50	100	0.3	0.5	7.8	0.8
5	50	100	0.2	0.2	0.1	0.2	50	5	25	2.7	3.1	3.1	4.1
5	50	200	0.0	0.0	0.1	0.1	50	5	50	1.5	1.8	2.7	2.5
5	50	500	0.1	0.1	0.1	0.1	50	5	100	1.1	1.6	5.2	3.7
5	50	1000	0.1	0.1	0.0	0.0	50	5	200	0.4	0.4	1.5	0.8
5	100	25	0.5	1.1	0.4	0.8	50	5	500	0.6	0.7	0.7	0.3
5	100	50	0.4	0.6	0.0	0.0	50	5	1000	0.5	0.5	0.5	0.3
5	100	100	0.4	0.5	0.8	0.6	50	10	25	1.7	0.6	6.6	3.9
5	100	200	0.3	0.4	0.4	0.6	50	10	50	0.7	0.6	7.1	5.1
5	100	500	1.0	0.9	0.8	1.1	50	10	100	1.1	0.9	4.9	2.2
5	100	1000	2.2	0.4	2.0	0.2	50	10	200	2.2	1.6	3.5	2.2
20	5	25	0.0	0.0	0.0	0.1	50	10	500	0.6	0.8	2.3	0.6
							50	10	1000	0.9	1.1	0.9	0.3

Table 12-3. Computational times (in seconds) – uncapacitated single level location problem

T	m	n	Primal-dual heuristic Average	σ	CPLEX Average	σ	Memetic Algorithm Average	σ	Memetic Algorithm – best solution Average	σ
5	5	25	0.0	0.0	0.0	0.0	2.2	0.2	0.16	0.1
5	5	50	0.0	0.0	0.2	0.3	3.5	0.8	0.22	0.1
5	5	100	0.0	0.0	0.1	0.0	7.3	0.9	0.38	0.1
5	5	200	0.0	0.0	4.5	8.3	35.1	3.5	4.50	2.1
5	5	500	0.0	0.0	1.1	0.0	88.3	11.5	9.85	4.1
5	5	1000	0.0	0.0	2.9	0.1	170.2	13.2	12.34	2.4
5	10	25	0.0	0.0	0.1	0.0	11.3	0.8	0.67	0.0
5	10	50	0.0	0.0	0.4	0.5	22.1	2.0	1.41	0.2
5	10	100	0.0	0.0	1.1	1.6	65.2	7.1	3.98	0.9
5	10	200	0.0	0.0	2.5	3.4	184.8	5.4	48.18	28.7
5	10	500	0.1	0.0	28.4	30.8	493.7	33.8	48.94	21.6
5	10	1000	0.1	0.0	7.4	0.6	967.5	69.9	412.38	266.2
5	50	25	0.0	0.0	0.4	0.0	590.4	25.8	67.06	16.2
5	50	50	0.0	0.0	0.7	0.0	1511.6	57.5	184.89	70.8
5	50	100	0.0	0.0	110.5	170.4	3717.0	472.3	1099.40	910.7
5	50	200	0.2	0.0	146.1	217.7	7717.4	1476.1	3724.68	2733.0
5	50	500	1.2	0.2	2463.1	2256.0	16910.3	1296.4	6372.27	4333.8
5	50	1000	3.8	0.5	37328.2	58013.4	33957.0	4163.6	11978.96	10181.2
5	100	25	0.0	0.0	2.9	3.9	3479.2	379.8	1030.52	998.1
5	100	50	0.0	0.0	6.7	10.3	7737.8	600.4	1559.78	790.3
5	100	100	0.1	0.0	103.5	82.6	17925.6	2344.7	11200.10	3487.9
5	100	200	0.8	0.1	1128.5	1408.4	33449.2	2404.7	21422.60	4479.8
5	100	500	4.3	0.7	170075.4	156371.0	76531.5	8971.1	31624.03	24274.9
5	100	1000	11.7	1.8	---	---	147134.0	10427.1	58450.72	18559.0
20	5	25	0.0	0.0	1.2	0.1	95.1	20.6	15.77	14.6
20	5	50	0.1	0.0	2.3	0.0	237.0	21.7	46.21	33.1
20	5	100	0.2	0.1	28.7	47.2	605.1	38.6	216.80	95.6
20	5	200	0.3	0.1	15.7	3.6	1000.7	15.7	189.47	122.5
20	5	500	1.0	0.3	563.6	115.3	3589.9	1011.2	1284.81	1142.8
20	5	1000	2.1	0.6	7821.6	347.9	9014.2	1797.0	4276.58	3402.5
20	10	25	0.0	0.0	2.9	0.5	820.1	144.8	490.09	268.2
20	10	50	0.2	0.1	6.0	0.1	1843.2	203.1	835.54	497.0
20	10	100	0.6	0.1	14.2	0.4	4525.0	508.3	2163.48	736.1
20	10	200	1.3	0.2	271.3	140.0	9470.3	1777.7	4311.96	3051.7
20	10	500	3.4	0.6	9379.2	218.2	23506.7	2064.4	11443.03	4535.0
20	10	1000	8.0	1.8	---	---	30114.9	2475.7	11855.76	4360.1
20	50	25	0.9	0.2	118.2	114.1	19609.9	425.4	2748.19	781.1

20	50	50	5.0	0.5	1881.2	1094.0	42472.7	4313.7	7044.99	9003.2
20	50	100	15.5	0.4	9657.4	282.5	83202.3	2610.9	6581.03	1815.0
50	5	25	0.5	0.1	152.9	22.8	196.9	46.1	82.83	43.1
50	5	50	1.4	0.0	3994.2	465.7	533.9	53.2	207.49	92.1
50	5	100	2.6	0.1	---	---	931.2	166.3	322.54	226.0
50	5	200	5.8	0.8	---	---	2574.6	409.7	494.59	372.8
50	5	500	14.9	3.5	---	---	6270.2	1192.9	2056.53	1627.3
50	5	1000	33.6	2.2	---	---	13617.9	1520.4	2530.93	1606.0
50	10	25	2.2	0.2	4070.8	127.5	2841.4	242.0	1162.50	757.0
50	10	50	5.1	0.6	---	---	6983.3	1693.8	3948.66	2190.3
50	10	100	10.7	0.7	---	---	11666.1	1648.8	3925.25	2470.7
50	10	200	24.8	0.7	---	---	27557.8	4473.6	3857.47	1607.6
50	10	500	62.3	2.6	---	---	65213.5	10713.7	21151.65	24420.8
50	10	1000	126.6	12.1	---	---	149904.8	11332.1	55343.04	38990.3

Table 12-4. Quality of the primal solution (in percentage) - single-level capacitated location problem

T	m	n	Primal-dual heuristic		Memetic algorithm		T	m	n	Primal-dual heuristic		Memetic algorithm	
			Average	σ	Average	σ				Average	σ	Average	σ
5	25	5	0.1	0.1	0.0	0.0	5	500	10	1.2	1.2	0.7	1.3
5	25	10	1.8	2.2	0.8	1.4	5	500	20	0.8	0.3	4.1	1.5
5	25	20	1.7	0.8	2.6	1.3	10	25	5	2.4	1.5	0.0	0.0
5	50	5	0.1	0.1	0.0	0.0	10	25	10	0.6	0.7	0.0	0.1
5	50	10	0.9	1.0	0.0	0.1	10	25	20	2.6	1.9	1.1	0.5
5	50	20	3.3	1.4	2.7	1.5	10	50	5	1.3	1.8	0.0	0.0
5	100	5	0.6	1.1	0.0	0.0	10	50	10	1.9	0.8	0.1	0.2
5	100	10	1.0	1.2	0.5	1.1	10	50	20	1.5	0.5	1.8	0.8
5	100	20	1.7	1.2	2.6	0.5	10	100	5	1.5	1.8	0.0	0.0
5	100	50	1.4	1.1	6.2	1.1	10	100	10	0.2	0.4	0.2	0.2
5	200	5	0.3	0.5	0.0	0.1	10	100	20	3.5	1.3	1.5	1.2
5	200	10	0.6	0.9	1.7	1.0	10	100	50	1.3	0.4	3.6	0.8
5	200	20	1.9	1.0	2.4	1.0	10	200	5	0.9	0.5	0.0	0.0
5	200	50	2.1	0.6	5.9	1.3	10	200	10	1.3	1.1	0.3	0.3
5	500	5	0.6	1.1	1.1	1.7	10	200	20	1.3	1.0	0.6	0.5

In the uncapacitated multi-level case, the memetic algorithm is capable of finding better solutions than the primal-dual heuristic. The consideration of more than one level does not deteriorate the performance of the memetic algorithm, because the assignment problems are, in the case, very simple to solve. However, the primal-dual heuristic is much more complex in the multi than in the single-level case. It can also be seen that the memetic algorithm is wasting too much time on unfruitful generations, after finding the best solution.

Table 12-5. Computational times (in seconds) - single-level capacitated location problem

T	m	n	Primal-dual heuristic Average	σ	CPLEX Average	σ	Memetic Algorithm Average	σ	Memetic Algorithm – best solution Average	σ
5	25	5	0.0	0.0	0.2	0.1	25.7	5.8	0.7	0.3
5	25	10	0.1	0.0	1.1	0.8	147.2	46.0	16.0	13.9
5	25	20	0.4	0.3	4.1	2.2	552.5	97.4	312.2	121.2
5	50	5	0.1	0.0	1.2	1.2	48.3	21.4	1.5	0.2
5	50	10	0.2	0.1	25.9	37.4	323.0	72.9	68.9	51.6
5	50	20	0.7	0.1	10.9	9.7	1647.5	457.5	901.6	492.3
5	100	5	0.2	0.1	2.2	2.3	163.9	52.0	5.6	3.6
5	100	10	0.5	0.2	16.2	10.7	902.5	202.7	366.0	247.0
5	100	20	3.6	2.6	112.6	65.4	4072.5	649.1	1973.9	1037.6
5	100	50	26.0	18.9	360.0	130.1	25394.0	1672.4	15109.8	1271.7
5	200	5	0.3	0.3	7.4	6.6	416.1	171.3	45.2	71.5
5	200	10	1.8	1.2	113.5	120.8	1851.0	357.6	452.9	313.4
5	200	20	11.4	5.1	296.2	171.0	12640.7	2612.8	6124.6	4339.2
5	200	50	124.8	118.6	28346.6	36332.1	121565.0	23962.4	76304.3	33241.2
5	500	5	1.9	0.8	58.4	50.7	1277.0	528.9	58.5	35.5
5	500	10	5.1	2.3	228.3	231.9	9392.7	4832.1	3070.7	3666.5
5	500	20	70.5	30.7	7135.6	7897.5	53807.7	17463.3	25216.9	17827.1
10	25	5	0.2	0.1	2.4	1.9	95.0	17.0	3.7	3.1
10	25	10	0.5	0.2	6.1	4.4	926.6	265.8	111.7	69.8
10	25	20	2.2	1.3	119.1	120.1	3603.1	1266.7	2140.9	1439.1
10	50	5	0.5	0.2	2.7	1.9	292.1	87.6	57.2	57.2
10	50	10	2.1	0.7	19.1	14.2	1671.1	469.6	656.0	497.8
10	50	20	5.7	1.0	201.0	164.0	8257.6	2757.0	5130.2	2430.7
10	100	5	1.9	0.2	12.0	4.9	672.9	226.2	62.2	67.7
10	100	10	5.6	1.2	70.8	92.8	3802.3	371.5	1484.7	866.4
10	100	20	20.4	4.5	1681.9	1196.8	31651.2	7415.1	20313.1	9073.6
10	100	50	545.3	139.2	121381.2	96307.0	150728.3	37868.0	72466.1	46691.3
10	200	5	4.8	1.8	104.5	77.2	2426.0	785.2	635.6	753.7
10	200	10	17.0	3.1	278.0	157.3	13266.1	5280.5	5868.5	5953.2
10	200	20	96.9	26.2	14902.7	25460.5	76893.0	22831.8	26761.6	24415.7

Table 12-6. Computational time (in seconds) - multi-level uncapacitated location problem

T	m	n	m_k	Primal-dual heuristic Average	σ	CPLEX Average	σ	Memetic Algorithm Average	σ	Memetic Algorithm – best solution Average	σ
5	2	25	2	0.0	0.0	0.1	0.1	1.0	0.1	0.0	0.0
5	2	25	5	0.0	0.0	0.2	0.0	13.3	1.1	0.3	0.1
5	2	50	2	0.0	0.0	0.1	0.0	1.8	0.1	0.0	0.0
5	2	50	5	0.1	0.0	0.5	0.0	27.5	1.4	0.7	0.0
5	2	50	10	0.2	0.1	11.9	20.4	255.5	16.6	12.5	15.9
5	2	100	2	0.0	0.0	0.2	0.0	3.0	0.2	0.1	0.0
5	2	100	5	0.1	0.0	1.2	0.3	60.9	2.3	1.7	0.7
5	2	100	10	0.6	0.3	33.1	57.6	550.8	21.6	14.8	6.7
5	3	25	2	0.1	0.0	0.3	0.0	20.8	0.6	0.4	0.1
5	3	25	5	4.4	3.4	11.5	9.4	895.9	44.1	64.3	40.5
5	3	50	2	0.1	0.1	0.7	0.0	41.8	1.3	0.8	0.3
5	3	50	5	2.5	1.6	17.8	7.2	1626.5	96.2	35.4	13.4
5	3	100	2	0.2	0.2	1.4	0.1	20225.8	40254.1	282.5	561.7
5	3	100	5	4.8	3.0	117.2	20.8	3615.7	219.3	303.4	287.7
10	2	25	2	0.0	0.0	0.3	0.1	8.6	2.3	1.9	3.4
10	2	25	5	0.1	0.1	1.0	0.4	92.5	5.1	6.8	4.9
10	2	50	2	0.0	0.0	0.4	0.0	13.6	0.6	0.4	0.2
10	2	50	5	0.8	1.0	5.6	6.5	203.3	3.9	13.5	7.9
10	2	50	10	1.1	0.2	6.2	1.8	1922.6	336.9	515.5	578.9
10	2	100	2	0.1	0.0	0.9	0.0	26.6	1.9	2.0	1.8
10	2	100	5	0.7	0.2	7.3	4.9	470.3	34.3	49.4	50.0
10	2	100	10	5.9	5.5	49.6	45.5	3801.2	875.3	1187.1	1354.9
10	3	25	2	0.2	0.1	0.9	0.0	153.1	2.8	4.6	1.4

5. CONCLUSIONS

For this kind of problems, genetic algorithms have to be hybridized with local search procedures, otherwise they will have a weak performance. The computational results show that the memetic algorithm needs much longer computational times than primal-dual heuristics, and sometimes reaches better solutions. Nevertheless, the primal-dual heuristics present a better compromise between solution quality and computational time needed. The main advantage of the memetic algorithm relies on the fact that a single algorithm can solve several different problems, whereas a primal-dual heuristic has to be dedicated to a particular location problem. This is especially important if, in a given problem, there is a set of facilities to locate with different characteristics.

Table 12-7. Quality of the primal solution (in percentage) - multi-level capacitated location problem

T	m	n	m_k	Primal-dual heuristic Average	σ	Memetic Algorithm Average	σ	T	m	n	m_k	Primal-dual heuristic Average	σ	Memetic Algorithm Average	σ
5	2	25	2	0.0	0.0	0.3	0.4	5	3	50	2	0.4	0.8	1.7	0.9
5	2	25	5	1.7	2.2	1.7	0.9	5	3	100	2	1.2	1.7	2.4	0.7
5	2	50	2	0.0	0.0	1.0	0.7	10	2	25	2	0.4	0.5	2.4	1.1
5	2	50	5	0.3	0.3	0.9	0.7	10	2	25	5	1.5	1.0	4.5	2.8
5	2	50	10	3.1	1.3	4.4	1.8	10	2	50	2	0.5	0.7	1.4	0.7
5	2	100	2	0.2	0.3	0.9	1.2	10	2	50	5	1.1	0.7	4.1	0.7
5	2	100	5	1.9	1.8	2.0	2.5	10	2	50	10	3.6	0.5	8.2	0.8
5	2	100	10	4.5	2.0	6.1	1.9	10	2	100	2	1.0	1.6	1.1	0.5
5	3	25	2	1.1	1.5	2.1	1.8	10	2	100	5	1.8	0.6	3.4	1.8

Table 12-8. Computational Times (in seconds) - multi-level capacitated location problem

T	m	n	m_k	Primal-dual heuristic Average	σ	CPLEX Average	σ	Memetic Algorithm Average	σ	Memetic Algorithm – best solution Average	σ
5	2	25	2	0.2	0.1	0.3	0.3	4.9	2.4	0.7	0.5
5	2	25	5	1.2	0.2	38.1	14.0	137.6	88.2	100.4	73.6
5	2	50	2	0.4	0.0	1.3	0.5	33.7	21.5	12.8	16.6
5	2	50	5	2.7	0.6	517.8	355.1	301.4	113.0	205.6	83.4
5	2	50	10	22.6	2.8	206468.8	12932.7	1614.1	385.4	829.5	490.8
5	2	100	2	0.8	0.1	10.5	8.0	54.4	43.7	22.1	25.0
5	2	100	5	6.6	1.5	618.9	462.1	948.9	207.8	628.8	167.9
5	2	100	10	70.1	4.0	184026.7	31963.0	6139.6	1964.6	4543.9	1856.1
5	3	25	2	1.3	0.2	11.0	6.6	72.4	43.6	46.9	32.8
5	3	25	5	159.6	18.9	30412.0	11380.6	1413.1	2826.1	344.7	689.3
5	3	50	2	3.6	1.4	109.2	78.1	104.4	69.9	56.6	42.4
5	3	100	2	15.9	4.4	385.4	111.4	182.4	54.0	125.3	53.6
10	2	25	2	0.6	0.1	3.9	3.3	57.4	53.5	23.5	14.5
10	2	25	5	4.5	1.2	1571.0	2613.3	870.2	539.4	475.4	375.8
10	2	50	2	1.1	0.2	13.1	10.5	146.3	163.4	103.4	116.8
10	2	50	5	6.0	1.8	3152.9	4061.9	3434.8	2218.4	2568.3	2077.1
10	2	50	10	57.9	10.4	120585.3	59943.9	11647.2	3484.6	7119.0	3266.5
10	2	100	2	2.9	0.5	29.5	25.1	510.2	460.3	260.0	299.3
10	2	100	5	18.7	4.0	16433.4	15863.2	14402.6	9964.6	3835.1	2368.4

Several things can be done to improve the performance of the memetic algorithm. It is possible to explore different definitions of neighborhoods to use during local search, improving the procedure that is responsible for most of the computational time spent. The repair procedure presently relies only on randomness. However, the algorithm could incorporate some of the primal procedures developed for primal-dual heuristics. The use of solutions calculated using the primal-dual heuristics in the initialization of the population can also decrease the importance of the local search procedure, decreasing the total computational time (if the local search procedure needs to be executed fewer times). The crossover operator could also be changed, so that children can be generated in a different way. With the present operator, children have the pattern of opened and closed facilities in each time period equal to one of their parents. If an adaptation of the three point crossover is considered instead, then each matrix will be divided into four sub-matrixes. There will be more degrees of freedom to create children, increasing the population's diversity and helping the algorithm to converge to a good quality solution more quickly. Some kind of tournament between all the generated children could be considered. This kind of crossover can also be used to increase the number of individuals in the population, instead of randomly generating all of them. Last, but not least, we should look carefully to the codification of solutions. If, at a first glance, the option of not codifying the assignment variables seems to be a good approach, and does not seem to have any disadvantage (apart from the fact that we need to solve an assignment problem for each individual), when additional restrictions are introduced, things are not that simple. It is necessary to continue studying possibilities of different representations for solutions, or choose to work simultaneously with two different populations. The results obtained thus far encourage us to follow this line of work and extend the use of this memetic algorithm to multi-objective dynamic facility location problems, as well as to situations with several decision makers.

6.   REFERENCES

Abdinnour-Helm, S. (1998). "A Hybrid Heuristic for the Uncapacitated Hub Location Problem." European Journal of Operational Research 106: 489-499.

Beasley, J. E. and Chu, P. C. (1996). "A Genetic Algorithm for the Set Covering Problem." European Journal of Operational Research 94: 392-404.

Cheung, B. K.-S., Langevin, A. and Villeneuve, B. (2001). "High Performing Techniques for Solving Complex Location Problems in Industrial System Design." Journal of Intelligent Manufacturing 12: 455-466.

Coello Coello, C. A. (2002). "Theoretical and Numerical Constraint-Handling Techniques used with Evolutionary Algorithms: A Survey and the State of the Art." Computer Methods in Applied Mechanics and Engineering 191: 1245-1287.

Correa, E. S., Steiner, M. T. A., Freitas, A. A. and Carnieri, C. (2001). A Genetic Algorithm for the P-Median Problem. Proceedings 2001 Genetic and Evolutionary Computation GECCO20011268-1275.

Cortinhal, M. J. and Captivo, M. E. (2003). Genetic Algorithms for the Single Source Capacitated Location Problem. Metaheuristics: Computer Decision-Making. M. Resende and J. P. d. Sousa, Kluwer Academic: 187-216.

Deb, K. (2001). Multi-objective Optimization using Evolutionary Algorithms, John Wiley & Sons.

Dias, J., Captivo, M. E. and Clímaco, J. (2004). Capacitated Dynamic Location Problems with Opening, Closure and Reopening of Facilities, Inesc-Coimbra.

Dias, J., Captivo, M. E. and Clímaco, J. (2005a). Dynamic Multi-Level Capacitated and Uncapacitated Location Problems: an approach using primal-dual heuristics. INOC'05, LisbonB1.227-B1.234.

Dias, J., Captivo, M. E. and Clímaco, J. (2005b). "Efficient Primal-Dual Heuristic for a Dynamic Location Problem." Computers & Operations Research to appear.

Dias, J., Captivo, M. E. and Clímaco, J. (2006). "Capacitated Dynamic Location Problems with Opening, Closure and Reopening of Facilities." IMA Journal of Mathematical Management, Special Issue on Location Analysis: Applications and Models, S. Salhi e Zvi Drezner (eds) (to appear)

Domínguez-Marín, P., Nickel, S., Hansen, P. and Mladenovic, N. (2005). "Heuristic Procedures for Solving the Discrete Ordered Median Problem." Annals of Operations Research 136: 145-173.

Filipovic, V., Kratica, J., Tosic, D. and Ljubic, I. (2000). Fine Grained Tornament Selection for the Simple Plant Location Problem. Proceedings of the 5th Online World Conference on Soft Computing Methods in Indsutrial Applications WSC5, September 2000152-158.

Hosage, C. M. and Goodchild, M. F. (1986). "Discrete Space Location-Allocation Solutions from Genetic Algorithms." Annals of Operations Research 6: 35-46.

Houck, C., Joines, J. and Kay, M. (1996). "Comparison of Genetic Algorithms, Random Restart and Two-Opt Switching for Solving Large Location-Allocation Problems." Computers & Operations Research 23: 587-596.

Huntley, C. and Brown, D. (1996). "Parallel Genetic Algorithms with Local Search." Computers & Operations Research 23: 559-571.

Jaramillo, J., Bhadury, J. and Batta, R. (2002). "On the Use of Genetic Algorithms to Solve Location Problems." Computers & Operations Research 29: 761-779.

Kratica, J. (1999). "Improvement of Simple Genetic Algorithm for Solving the Uncapacitated Warehouse Location Problem." Advances in Soft Computing, Engineering Design and Manufacturing: 390-402.

Kratica, J., Tosic, D., Filipovic, V. and Ljubic, I. (2001). "Solving the Simple Plant Location Problem by Genetic Algorithm." RAIRO Operation Research 35: 127-142.

Lorena, L. and Lopes, L. D. S. (1997). "Genetic Algorithms Applied to Computationally Difficult Set Covering Problems." Journal of the Operational Research Society 48: 440-445.

Moscato, P. and Cotta, C. (2003). A Gentle Introduction to Memetic Algorithms. Handbook of Metaheuristics. F. Glover and G. Kochenberger. Boston, Kluwer Academic Publishers.

Murata, T., Ishibuchi, H. and Gen, M. (1998). "Neighborhood Structures for Genetic Local Search Algorithms." IEEE Transactions on Systems Man and Cybernetics: 259-263.

Oei, C. K., Goldberg, d. E. and Chang, S.-J. (1991). Tournament Selection, Niching and the Preservation of Diversity. Illinois Genetic Algorithms Laboratory (IlliGAL) Report.

Osman, I. H. and Kelly, J. P. (1996). Meta-Heuristics: Theory & Applications, Kluwer Academic Publishers.

Reeves, C. and Höhn, C. (1996). "Integrating Local Search into Genetic Algorithms." Modern Heuristic Search Methods: 99-115.

Reeves, C. R. (1993). Using Genetic Algorithms With Small Populations. Proceedings of the Fifth International Conference on Genetic Algorithms, Morgan Kaufmann, San Mateo, CA.

Rothlauf, F. and Goldberg, D. (2002). Redundant Representations in Evolutionary Computation. Illinois Genetic Algorithms Laboratory (IlliGAL) Report.

Salhi, S. and Gamal, M. D. H. (2003). "A Genetic Algorithm Based Approach for the Uncapacitated Continuous Location-Allocation Problem." Annals of Operations Research 123: 203-222.

Shimizu, Y. (1999). "Multi-Objective Optimization for Site Location Problems through Hybrid Genetic Algorithm with Neural Networks." Journal of Chemical Engineering of Japan 32: 51-58.

Yagiura, M. and Ibaraki, T. (1996). "The Use of Dynamic Programming in Genetic Algorithms for Permutation Problems." European Journal of Operational Research 92: 387-401.

Chapter 13

A STUDY OF CANONICAL GAs FOR NSOPs

Panmictic versus Decentralized Genetic Algorithms for Non-Stationary Problems

Enrique Alba,[1] Juan F. Saucedo Badia,[2] and Gabriel Luque[3]

[1] *Departamento de Lenguajes y Ciencias de la Computación,*
E.T.S. Ingeniería Informática
Campus de Teatinos, 29071 Málaga (SPAIN)
eat@lcc.uma.es

[2] *Departamento de Lenguajes y Ciencias de la Computación,*
E.T.S. Ingeniería Informática
Campus de Teatinos, 29071 Málaga (SPAIN)
juanfsb@andaluciajunta.es

[3] *Departamento de Lenguajes y Ciencias de la Computación,*
E.T.S. Ingeniería Informática
Campus de Teatinos, 29071 Málaga (SPAIN)
gabriel@lcc.uma.es

Abstract   In order to solve a *Non-Stationary Optimization Problem (NSOP)* it is necessary that the used algorithms have a set of suitable properties for being able to dynamically adapt the search to the changing fitness landscape. Our aim in this work is to improve our knowledge of existing canonical algorithms (steady-state, generational, and structured –cellular– genetic algorithms) in such a scenario. We study the behavior of these algorithms in a basic Dynamic Knapsack Problem, and utilize quantitative metrics for analyzing the results. In this work, we analyze the role of the mutation operator in the three algorithms and the impact of the frequency of dynamic changes in the resulting difficulty of the problem. Our conclusions outline that the steady-state GA is the best in fast adapting its search to a new problem definition, while the cellular GA is the best in preserving diversity to finally get accurate solutions. The generational GA is a tradeoff algorithm showing performances in between the other two.

Keywords: Non-Stationary Problem, Dynamic Knapsack Problem, Cellular GA.

Introduction

In recent years we have seen an increasing number of applications of Evolutionary Algorithms to *Non-Stationary Optimization Problems (NSOPs)* [Branke, 2001]. Many systems can be modeled as a changing search landscape, such as the optimization of the car traffic flow by optimally synchronizing red light controllers, or the optimal service of a multielevator system to minimize the waiting time of passengers in a building. However, we feel that much work has been put into developing new algorithms while maybe it should be first necessary to know more on the comparative advantages of existing solvers for this kind of problems.

We will proceed in this study by analyzing evolutionary algorithms (EAs) [Bäck et al., 1997] including panmictic as well as structured ones. This article is an extension of [Alba and Saucedo, 2005] which now provides a more thorough analysis of the dynamic scenarios for the problem under study.

Defining and using quantitative metrics becomes also important with NSOPs to avoid extracting conclusions by simple visual inspection, and since most existing metrics are devoted to static optimization. We will also include a statistical significance study since we are dealing with non deterministic algorithms.

Our contributions will allow to rank three kinds of EAs according to the used metrics to guide further research. We also study the impact of the mutation rate and several features of the dynamic problem to solve (e.g., the period of change and the severity of change).

The work is structured as follows: Section 1 introduces the NSOP concept, and a (very) brief state of the art on the *Dynamic Knapsack Problem*. In Section 2 we describe our algorithms. Section 3 contains a discussion of the metrics used and the details of the different configurations of our approaches used in the experimentation section. We perform an experimental analysis and explain the results in Section 4. Finally, Section 5 summarizes the main conclusions and draws some future research lines.

1.   Dynamic Knapsack Problem (DKP)

A non-stationary optimization problem is known to change over time the optimized function, the problem instance, and/or the restrictions.

There is a wide variety of possible types of environmental dynamics, therefore, a formal description for such problems is necessary. A mathematical definition was given by K. Weicker in [Weicker, 2000] but because it is too general and we want to study the basic behavior of the standard GAs (a subclass of EAs), we now delimit the used NSOP subclass.

In our case, the non-stationary function shifts in a regular (cyclic) manner among different versions of the same basic problem. In consequence, we can state this function as a shifting among n subfunctions (Equation 13.1), in our case $n = 2$, 3, or 8, with a certain period of change of p generations. As a consequence, the optimization process becomes a process of learning several optima. The technique showing the highest velocity of adaptation with the lower distance to the optimum between each period of change is usually determined as the best.

$$
f\left(\vec{x}, t_i\right) =
\begin{cases}
f_1\left(\vec{x}\right), & \text{for } i \bmod n = 0 \\
f_2\left(\vec{x}\right), & \text{for } i \bmod n = 1 \\
\quad \vdots & \qquad \vdots \\
f_n\left(\vec{x}\right), & \text{for } i \bmod n = (n-1)
\end{cases}
\tag{13.1}
$$

The *Dynamic Knapsack Problem (DKP)* falls in this class of problems. DKP is a variant of the well-known knapsack problem which consists in maximizing the total value of a subset of objects (selected from a set of N possible objects) that are placed into a knapsack which has a maximum weight constraint. Each object has an associated value v_i and weight w_i. Therefore, the objective is

$$
\max \sum_{i=1}^{N} v_i x_i \text{ subject to } \sum_{i=1}^{N} w_i x_i \leq W
$$

where W is the weight limit, which in this study changes in time. Here, the x_i variables take on the values 1 or 0 to indicate that the object is in or out of the knapsack, respectively.

One of the first works using this problem can be found in [Goldberg and Smith, 1987]. They used a knapsack problem with 17-objects whose values, weights, and optimal solutions are shown in Table 13.1. Therefore, their problem encodes the x_i values contiguously to form a 17-bit string. The constraint inequality is managed as an external penalty method, where weight violations are squared and multiplied by a penalty coefficient ($\lambda = 20$). The resulting fitness function for unfeasible solutions can be defined as

$$
f(x) = \sum_{i=1}^{N} v_i x_i - \lambda \left(\sum_{i=1}^{N} w_i x_i - W \right)^2
\tag{13.2}
$$

Table 13.1. The 17-Object, 0-1 knapsack problem plus its two optimal solutions

Object Number i	Object Value v_i	Object Weight w_i	Optimum= 71 $(W_1 = 60)$	Optimum= 87 $(W_2 = 104)$
1	2	12	0	0
2	3	5	1	1
3	9	20	0	1
4	2	1	1	1
5	4	5	1	1
6	4	3	1	1
7	2	10	0	0
8	7	6	1	1
9	8	8	1	1
10	10	7	1	1
11	3	4	1	1
12	6	12	1	1
13	5	3	1	1
14	5	3	1	1
15	7	20	0	1
16	8	1	1	1
17	6	2	1	1
Total:	**91**	**122**	**13 Objects**	**15 Objects**

In [Goldberg and Smith, 1987], the non-stationary effect is produced by the change on the maximum capacity from W_1 to W_2, and vice versa. When the weight value changes from $W_2 = 104$ to $W_1 = 60$ the current solutions also change their fitness evaluation, due to some feasible solutions for W_2 can become unfeasible with the new weight, and therefore their fitness value will be penalized according to Equation 13.2. The previous solutions must be usually removed in some manner, and the algorithm is forced to search anew. In our case, we are going to use the same parameters as Goldberg and Smith but we change the weight limit between more than two weights, and we make a wider study by analyzing different periods of change.

This problem has been widely used in the literature [Goldberg and Smith, 1987; Dasgupta and McGregor, 1992]. Sometimes it has been used in order to compare the Diploid/Dominance mechanism, [Goldberg and Smith, 1987; Ryan, 1997]. In [Lewis et al., 1998], the authors used also a knapsack problem with 14-objects and random periods of change.

Besides, this problem has been useful for comparing algorithms, e.g., steady state genetic algorithm with aging of individuals in [Ghosh et al., 1998], the feedback thermodynamical GA in [Mori et al., 1998] the replacement strategies in steady state GA in [Smith and Vavak, 1999], several approaches for maintaining diversity in [Andrews and Tuson,

2003], and so on. Recently, the Dynamic Knapsack Problem has been labeled as a "Suitable Benchmark Problem" in [Branke, 2001].

2. The Algorithms

The canonical algorithms whose behavior we analyze in this article are: (1) a steady-state GA (ssGA), (2) a generational GA (genGA), and (3) a cellular GA (cGA). There exist partial studies with similar goals in the literature [Vavak and Fogarty, 1996; Smith and Vavak, 1999; Salomon and Eggenberger, 1998]. In particular, in [Vavak and Fogarty, 1996] the authors compare ssGA versus genGA for a single non-stationary bit-matching instance. However, the cellular GA is not considered at all, and the scope of the study is very narrow (no quantitative measures, which have appeared more recently). The well known high capacity of cellular EAs for maintaining diversity is the main reason for including them in a study on non-stationary problems. One (isolated) leading work on comparing a cellular GA versus more traditional ones can be found in [Sarma and De Jong, 1999]; again, the same narrow scope can be imputed to this interesting study, since measures are not used, nor different classes of panmictic EAs are considered. In order to unambiguously report the algorithms, we proceed to a detailed (but brief) description of them.

In steady-state algorithms [Syswerda, 1991] only a few individuals are created and replaced in each iteration (one individual in our case). With ssGA, the least fit individual is replaced by the offspring resulting from crossover and mutation of the selected individuals, as presented in the pseudocode reported in Figure 13.1.

The pseudocode of a genGA [Whitley, 1989] can also be seen in Figure 13.1. Note that, at each iteration, the new population consists entirely of offspring computed after parents in the previous generation. The basic step creates a whole new population that replaces the old one (the best individual is preserved –elitism–). This algorithm is expected to maintain diversity for a larger number of generations.

The ssGA is expected to have a fast convergence to an optimum with respect to genGA. The question is whether this is a good feature for a given non-stationary problem or not.

Finally, a cGA replaces the population at each iteration (like genGA). However, the genetic operators are always applied inside neighborhoods of 5 individuals. Each individual is iteratively considered as the central point of the neighborhood composed by it plus its north-south-east-west neighboring individuals. This means that an individual may only interact with its nearby neighbors in the breeding loop.

ssGA	genGA

proc Reproductive_Cycle(ga):
  for $s = 1$ **to** MAX_STEPS **do**
    parent1=Select(ga.pop);
    parent2=Select(ga.pop);
    Crossover(ga.Pc,
      parent1, parent2, ind_aux.chrom);
    Mutate(ga.Pm, ind_aux.chrom);
    ind_aux.fitness=
      ga.Evaluate(Decode(ind_aux.chrom));
    Insert_New_Ind(ga,
      ind_aux, [if_better | worst]);
  end_for;
end_proc Reproductive_Cycle;
	proc Reproductive_Cycle(ga):
	  for $s = 1$ **to** MAX_STEPS **do**
	    p_list = Select(ga.pop);
	    for $i = 1$ **to** POP_SIZE / 2 **do**
	      Crossover(ga.Pc, p_list[i],
	        p_list[POP_SIZE/2 + i], ind_aux.chrom);
	      Mutate(ga.Pm, ind_aux.chrom);
	      ind_aux.fitness=
	        ga.Evaluate(Decode(ind_aux.chrom));
	      Insert_New_Ind(pop_aux, ind_aux);
	    end_for;
	    ga.pop=pop_aux; [elitist | non elitist]
	  end_for;
	end_proc Reproductive_Cycle;

cGA

proc Reproductive_Cycle(ga):
  for $s = 1$ **to** MAX_STEPS **do**
    for $x = 1$ **to** WIDTH **do**
      for $y = 1$ **to** HEIGHT **do**
        n_list= Calculate_neigbors(ga , position (x,y));
        parent1=Select(n_list);
        parent2=Select(n_list);
        Crossover(ga.Pc, n_list[parent1], n_list[parent2], ind_aux.chrom);
        Mutate(ga.Pm, ind_aux.chrom);
        ind_aux.fitness=ga.Evaluate(Decode(ind_aux.chrom));
        Insert_New_Ind(position(x,y),ind_aux,[if_better | always], ga, pop_aux);
      end_for;
    end_for;
    ga.pop=pop_aux;
  end_for;
end_proc Reproductive_Cycle;

Figure 13.1. Pseudocode of the Algorithms.

The new offspring replaces the central individual of the neighborhood only if it has a higher fitness (assuming maximization). The basic algorithm is synchronous: the computed new population is stored in a temporary population that replaces the old one at the end of each step. This algorithm is similar to that of [Sarma and De Jong, 1996] and is described in Figure 13.1. Its population is structured in a toroidal 2D grid. More details can be found in [Manderick and Spiessens, 1989; Alba and Troya, 2000]. Since an individual belongs to several neighborhoods at the same time, any change in its contents affects its neighbors in a smooth manner, being a good tradeoff between convergence and exploration of the search space. In fact, one outstanding feature of cGAs is that they can be easily tuned to match any desired exploration/exploitation tradeoff as shown in [Alba and Troya, 2000].

All our algorithms implement a fitness proportionate sampling mechanism for parents selection. Also, they use one point crossover yielding only one child: the one having the largest portion of the best parent. The applied mutation is a standard bit-flip operation.

3.   Experimental Setup

In the NSOP literature, the algorithms are usually compared by means of human observation of the graphs containing the average/best fitness curves along the time, which is too subjective. We use here quantitative values by adhering to a metric and the graphs are just used as summary of objective data. Let us begin by pointing out the ideas developed in [Branke, 2001; Weicker, 2002; Morrison, 2003] for comparing the performance of the algorithms in NSOPs. As stated in [Weicker, 2002]

> "The goal of an evolutionary search process in a dynamic environment is not only to find an optimum within a given number of generations but rather a perpetual adjustment to changing environmental conditions".

So, we will consider, here, his three metrics (1) accuracy, (2) stability, and (3) ε−reactivity (see [Weicker, 2002] for more details).

In order to measure the degree of accuracy, we have used Equation 13.3. The stability (Equation 13.4) measures the change in the accuracy values at time t (best value is 0). The reactivity (Equation 13.5) measures the time in which accuracy is similar to that of time t, and its optimum value is 1. Our summaries show the mean over all generations.

$$\text{accuracy}(t) \quad = \quad \frac{f(\text{best}, t)}{\text{Optimum}(t)} \tag{13.3}$$

$$\text{stability}(t) \quad = \quad \max\{0, \text{accuracy}(t) - \text{accuracy}(t-1)\} \tag{13.4}$$

$$\varepsilon - \text{reactivity}(t) \quad = \quad \min\{D \cup \text{maxgenerations} - t\} \tag{13.5}$$

$$D = \left\{ t' - t / \frac{\text{accuracy}(t')}{\text{accuracy}(t)} \geq (1 - \varepsilon), t < t' \leq \text{maxgenerations}; t, t' \in \mathbb{N} \right\}$$

Because of the fact that we use numeric metrics, we can include different periods of change in order to study how well the three algorithms adapt their search depending of the frequency of problem changes. In each one, we use several mutation rates to find general conclusions, from $\frac{1}{5*L}$ to $\frac{5}{L}$, where $L = 17$ is the length of binary string. The parameters used in each algorithm are summarized in Table 13.2.

Table 13.2.   Parameters of the algorithms.

Popsize	String Length	Parent's Selection	Crossover	Bit Mutation Reference	Replacement
225 $\begin{pmatrix} 15 \times 15 \\ \text{for cGA} \end{pmatrix}$	17	Roulette Wheel + Roulette Wheel	SPX, $p_c = 1.0$	$p_m = 1/L$ $= 0.0588235$	Rep_Least_Fit

Firstly, we compare the algorithms using the same parameters as in [Goldberg and Smith, 1987]. Then, we do a comparison of the algorithms using the DKP with the same parameters but with different periods of change. We shift the period of change from 1 to 31 generations in steps of 2 (in [Goldberg and Smith, 1987], a single period of 15 generations are used). As the Hamming distance between two optima, which is 2, is low, we include a third change which is more disruptive. We use a new weight limit, $W = 30$, whose associated optimum is 50. This optimum has a Hamming distance of 4 with respect to optimum 71 ($W_2 = 60$) and 6 from optimum 87 ($W_1 = 104$). Finally, we study a progressive change among 8 optima (i.e., the optimum value is smoothly increased from W_1 to W_8, and then a drastic change occurs between W_8 and W_1, and so on), where the Hamming distance changes in the sequence shown in Table 13.3.

Table 13.3.   Weight Limit and Hamming Distance for the problem with eight optima.

Weight Limit							
W_1	W_2	W_3	W_4	W_5	W_6	W_7	W_8
104	91	79	67	60	54	42	30

Hamming Distance between two consecutive optimal solutions							
W_1-W_2	W_2-W_3	W_3-W_4	W_4-W_5	W_5-W_6	W_6-W_7	W_7-W_8	W_8-W_1
3	4	1	3	2	1	3	7

We always perform 100 independent runs of the problem for each experiment, and compute the mean of each of the three metrics in every case. In order to compare these means, we first check if data are normally distributed using the Kolmogorov-Smirnov test, and if so, we carry out an ANOVA test in order to compare the means. If data are not normally distributed then we do a Kruskal-Wallis test in order to compare the medians. We perform a multiple comparison test in order to determine which means (or medians) are significantly different. All these tests use a level of confidence of 95%.

4.   Computational Experiments and Analysis

In this section we describe and discuss the results after running our three approaches over the dynamic problem. We will proceed in four phases. First, we analyze the problem as in [Goldberg and Smith, 1987]. In the second phase, we study the impact of the period. Later, we extend this scenario by considering three optima. The last phase, we examine the effect of a progressive change of optima over the behavior of our algorithms.

Comparing to Golberg and Smith

We begin the experimental part by using the same parameter as in [Goldberg and Smith, 1987] to compare the three basic algorithms.

We show the impact of the mutation rate over the behavior of the algorithms in Figure 13.2. For low values of mutation rate we can distinguish two different behaviors: generational algorithms (genGA and cGA) achieve their best accuracy value while ssGA is not able to generate diversity enough to adapt the dynamic problem and it obtains a very poor results. For high values of mutation rate all the algorithms have a similar performance; as this parameter is increased, a smooth loss of accuracy is provoked. This is due to algorithms maintain a high diversity but they are not able to explote the good regions of the search space found.

As a first result (Figure 13.2), ssGA (with mutation rate $\frac{8}{5*L}$) and genGA $(\frac{2}{5*L})$ are the most accurate algorithms (0.991 for both). Although cGA is not so accurate (0.987), its behavior is quite stable for any value of mutation rate, and in fact, cGA obtains the best mean accuracy out of all the algorithms (Table 13.4). Also, its interquartile range (IQR, i.e. the difference between the 25th and the 75th percentiles of the data) is the lowest, indicating the mean is a very representative value of the global behavior of the method. Therefore, cGA appears as the most robust algorithm, while ssGA and genGA are the most accurate but only for one concrete mutation rate.

Table 13.4. Mean accuracy values and IQR for all the mutation rates and algorithms.

(for all mutation rates)	ssGA	genGA	cGA
Mean Accuracy Values	0.9506	0.9527	0.9711
IQR	0.0307	0.0370	0.0221

If we concentrate in other metrics (only using the optimal mutation rates found), then ssGA $(\frac{8}{5*L})$, and genGA $(\frac{2}{5*L})$ with *mean_stab* = 0.004 and *mean_stab* = 0.003 respectively (which are not significantly different each other ($p\_value > 0.05$)), are more stable than cGA $(\frac{1}{L})$ with *mean_stab* = 0.006. The ssGA is also the one reacting fastest ($\varepsilon - reactivity$ = 1.058) whereas there are not significant differences between genGA and cGA (1.086, 1.081 respectively). So, while stability seems to be a meaningful characteristic, reactivity does not seem to have influence on accuracy which is the main characteristic. This is the reason why in the next analysis, we mainly pay attention to the accuracy and the stability metrics.

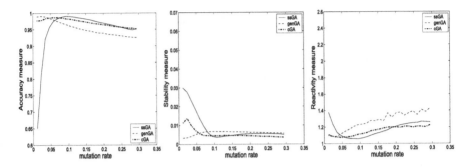

Figure 13.2. Results for DKP with same parameters as in [Goldberg and Smith, 1987].

Summarizing, in this case, for concrete mutation rates the ssGA and genGA are the best, but the most robust algorithm is cGA, independently of the mutation rate used.

Impact of the period of change

Now, we analyze the effect of the period of change over the behavior of the algorithms. For sake of clarity, we only show the accuracy for three (representative) mutation rates (Figure 13.3). In general, we can observe that for periods of change very low, the algorithms are not able to achieve good accuracy, since the objective changes very frequently. Using higher values of period, all the algorithms improve their performance, due to they have time enough to adapt the search to the new optimum.

As we stated in the previous section, mutation rate plays an important role in the behavior of ssGA and genGA (cGA is quite robust with respect to this parameter). For a very low rate $(\frac{2}{5*L})$ and a period of change is close to 7 generations (Figure 13.3), ssGA has a poor accuracy (0.745), while genGA obtains the best accuracy values (0.995). On contrary, for high values of mutation, the behavior of this two algorithms is the opposite, i.e., ssGA is the best one, and genGA is the worst method.

For a wide range of periods of dynamism, we can notice that cGA is the algorithm which has the smallest variation in the accuracy value, and therefore, it is again the most robust algorithm out of the three studied. However, using a concrete mutation (the optimal value found for each algorithm), we can observe that ssGA is the most accurate in general (left graphic of Figure 13.5). The genGA is the least accurate algorithm when the period of change is low (the objective function changes frequently),

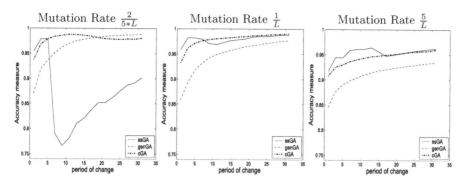

Figure 13.3. Results for DKP with several periods of change (two optima).

and there are not significant differences among the algorithms when the period of change is high.

Extension to three optima

In order to observe the behavior of the algorithm when the change in the environment is drastic. Now, the optima of two contiguous periods are far each other with regard to Hamming distance, we include a third optimum (Figure 13.4). Similarly to the previous case, mutation rate plays an important role since it is the standard form of algorithms of introducing diversity. This is confirmed in the actual results: with low mutation rates ($\frac{2}{5*L}$), the algorithms have low accuracy (lower than 0.9), whereas with high mutation rates ($\frac{5}{L}$), they are more accurate. Again, ssGA for low mutation rates has a loss of accuracy when the period of change is close to 8 generations (this fact should need further study in the future). As in the previous experiments, genGA is more accurate than ssGA for low mutation rates, while ssGA outperforms genGA for high values of mutation rate. The genGA is the most robust algorithm in spite of in several concrete configurations it is outperformed by our other methods.

If we only concentrate in the best accuracy values of all mutation rates for each period of change (Figure 13.5) then we can observe that ssGA and genGA have a similar behavior if period of change is high. However, when the change is produced in few generations then ssGA is the most accurate. The cellular genetic algorithm is the least accurate in general.

Summarizing, genGA is the most accurate for the best rate and it also is the most robust. This is because the abrupt change between

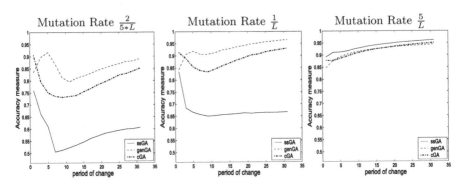

Figure 13.4.   Results for DKP with several periods of change (three optima).

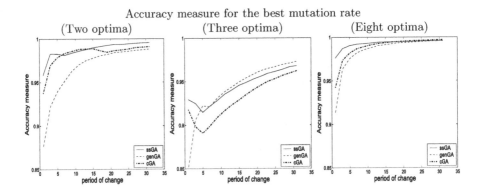

Figure 13.5.   Results for DKP with the optimal mutation rates.

two consecutive optima makes the algorithms need a larger speed of convergence and more diversity, characteristics that can be found in genGA.

Progressive change (8 optima)

Finally, we study the problem with eight optima with a progressive change. In this case, the first observation is that the accuracy values of all the algorithms are very high (Figure 13.6). This fact can be due to the progressive change between two optima, that makes easy the adaptation of the algorithms.

In this case, the behavior is similar to the problems with two optima. So, ssGA, is less accurate for low mutation rates and genGA is less accurate for high mutation rates. cellular GA is the most accurate in general, exceeding in most cases an accuracy of 0.95. Observing the algorithms analyzed at their best mutation rates (right graphic of Figure 13.5), ssGA is the best when the period of change is less than 15 due to it is very fast exploiting the search space. While cGA is the most accurate algorithm when the period of change is more than or equal to 15, because it is able to maintain a high diversity during a large number of generations. So, when the change happens between two or three optima, or the Hamming distance between optima is high then the algorithm needs to be fast in order to adapt to the change, and the ssGA is the best, and if the change is progressive then population needs more diversity, and the cGA is the best.

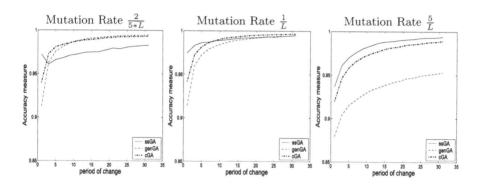

Figure 13.6. Results for DKP with several periods of change (eight optima).

5. Conclusions

The main purpose of this paper is to compare canonical versions of panmitic GAs (steady-state and generational models) and structured GAs (cellular model) in order to know their suitability for an important class of non-stationary problems showing different dynamics. In this way, we have used the latest advances in metrics (Weicker's measures), and a rigorous statistical significance study in order to have quantitative conclusions. In this full analysis, we have considered the mutation rate, the period of change, and the severity of change.

Our conclusions are that the accuracy of the algorithms depend on the severity of change (in this paper, *severity of change* is the Hamming distance between optima) and the mutation rate chosen. So, if we choose an appropriate mutation rate for each case then ssGA is the best choice when the problem needs fast adaptation, but, on the contrary, this algorithm is very dependent on the choice of mutation rate. On the one hand, we notice that genGA obtains accurate results in NSOPs, maybe better than what could have been expected after some works on static problems [Alba and Troya, 2002]. While cGA is the most robust, almost as accurate as the best algorithm in each case, and it is the clear best when the problem needs actual diversity in the adaptation. With regard to the mutation rate, genGA has a constant behavior for all the problems with its best mutation rate close to $1/L$, being L the length of the individual. However, for ssGA, the optimal mutation rate is very dependent on the difficulty of the problem. The summary conclusion for the case of having very different optima in two consecutive periods (intense change in the seeked solution) is that a higher mutation rate would be needed for the three algorithms.

Important future research lines include using specific problems generators and proposing a new cellular model for NSOPs.

Acknowledgments

This work has been partially funded by the Ministry of Science and Technology and FEDER under contract TIN2005-08818-C04-01 (the OPLINK project, http://oplink.lcc.uma.es).

References

Alba, E. and Saucedo, J. (2005). Panmictic versus decentralized genetic algorithms for non-stationary problems. In *Procs. the Sixth MIC*. Electronic publication.

Alba, E. and Troya, J. M. (2000). Cellular evolutionary algorithms: Evaluating the influence of ratio. In Schoenauer, Marc, Deb, K., and et al., editors, *Proc. of PPSN VI*, volume 1917 of *LNCS*, pages 29–38, France. Springer.

Alba, E. and Troya, J. M. (2002). Improving flexibility and efficiency by adding parallelism to genetic algorithms. *Statistics and Computing*, 12(2):91–114.

Andrews, M. and Tuson, A. (2003). Diversity does not necessarily imply adaptability. In Barry, Alwyn M., editor, *GECCO 2003: Proceedings of the Bird of a Feather Workshops, Genetic and Evolutionary Computation Conference*, pages 118–122, Chigaco. AAAI.

Bäck, T., Fogel, D. B., and Michalewicz, Z., editors (1997). *Handbook of Evolutionary Computation*. Oxford University Press.

Branke, J. (2001). *Evolutionary Optimization in Dynamic Environments*. Klüwer Academic Publishers.

Dasgupta, D. and McGregor, D. R. (1992). Nonstationary function optimization using the structured genetic algorithm. In Männer, Reinhard and Manderick, B., editors, *Proc. of PPSN II*, pages 145–154, Amsterdan, Holand. Elsevier Science Publishers, B. V.

Ghosh, A., Tsutsui, S., and Tanaka, H. (1998). Function optimization in nonstationary environment using steady state genetic algorithms with aging of individuals. In *Proc. of CEC'98*, pages 666–671. IEEE Press.

Goldberg, D. E. and Smith, R. E. (1987). Nonstationary function optimization using genetic algorithms with dominance and diploidy. In Grefenstette, J. J., editor, *Proc. of ICGA'87*, pages 59–68. Lawrence Erlbaum Associates.

Lewis, J., Hart, E., and Ritchie, G. (1998). A comparison of dominance mechanisms and simple mutation on non-stationary problems. In Eiben, Agoston E., Bäck, T., and et al., editors, *Proc. of PPSN V*, volume 1498 of *LNCS*, pages 139–148, Berlin, Germany. Springer.

Manderick, B. and Spiessens, P. (1989). Fine-grained parallel genetic algorithms. In Schaffer, J. D., editor, *Proc. of ICGA'89*, pages 428–433, San Mateo, CA, USA. Morgan Kaufmann.

Mori, N., Kita, H., and Y, Nishikawa (1998). Adaptation to a changing environment by means of the feedback thermodynamical genetic algorithm. In Eiben, Agoston E., Bäck, T., and et al., editors, *Proc. of PPSN V*, volume 1498 of *LNCS*, pages 149–158, Berlin, Germany. Springer.

Morrison, R. W. (2003). Performance measurement in dynamic environments. In Barry, Alwyn M., editor, *GECCO 2003: Proceedings of the Bird of a Feather Workshops, Genetic and Evolutionary Computation Conference*, pages 99–102, Chicago. AAAI.

Ryan, C. (1997). Diploidy without dominance. In Alander, Jarmo T., editor, *Proc. of the Third Nordic Workshop on Genetic Algorithms and their Applications*, pages 63–70, Vaasa, Finnland. Department of Information Technology and Production Economics, University of Vaasa.

Salomon, R. and Eggenberger, P. (1998). Adaptation on the evolutionary time scale: A working hypothesis and basic experiments. In Hao, J.-K., Lutton, Evelyne, and et al., editors, *Proc. of the Third AE'98*, volume 1363 of *LNCS*, pages 251–262, France. Springer.

Sarma, J. and De Jong, K. A. (1996). An analysis of the effect of the neighborhood size and shape on local selection algorithms. In Voigt, Hans-Michael, Ebeling, Werner, and et al., editors, *Proc. of PPSN IV*, volume 1141 of *LNCS*, pages 236–244, Berlin, Germany. Springer.

Sarma, J. and De Jong, K. A. (1999). The behavior of spatially distributed evolutionary algorithms in non-stationary environments. In Banzhaf, Wolfgang, Daida, J. M., , Eiben, Agoston E., Garzon, Max H.,

Honavar, Vasant, Jakiela, Mark, and Smith, R. E., editors, *Proc. of GECCO'99*, pages 572–578, Orlando, FL, USA. Morgan Kaufmann.

Smith, J. E. and Vavak, F. (1999). Replacement strategies in steady state genetic algorithms: dynamic environments. *Computing and Information Technology*, 7(1):49–60.

Syswerda, G. (1991). A study of reproduction in generational and steady-state genetic algorithms. In Rawlins, Gregory J., editor, *Proc. of FOGA'91*, pages 94–101, San Mateo, CA, USA. Morgan Kaufmann.

Vavak, F. and Fogarty, T. C. (1996). Comparison of steady state and generational gas for use in nonstationary environments. In *Proc. of CEC'96*, pages 192–195. IEEE Press.

Weicker, K. (2000). An analysis of dynamic severity and population size. In Schoenauer, Marc, Deb, K., and et al., editors, *Proc. of PPSN VI*, volume 1917 of *LNCS*, pages 159–168, France. Springer.

Weicker, K. (2002). Performance measures for dynamic environments. In Merelo Guervós, Juan Julián, Adamidis, Panagiotis, Beyer, Hans-Georg, Fernández-Villacañas, José-Luis, and Schwefel, Hans-Paul, editors, *Parallel Problem Solving from Nature – PPSN VII*, pages 64–73, Berlin. Springer.

Whitley, D. (1989). The GENITOR algorithm and selection pressure: Why rank-based allocation of reproductive trials is best. In Schaffer, J. D., editor, *Proc. of ICGA'89*, pages 116–121, San Mateo, CA, USA. Morgan Kaufmann.

Chapter 14

PARTICLE SWARM OPTIMIZATION AND SEQUENTIAL SAMPLING IN NOISY ENVIRONMENTS

Thomas Bartz-Beielstein

Chair of Algorithm Engineering and Systems Analysis,
Department of Computer Science, Dortmund University, Germany

thomas.bartz-beielstein@udo.edu

Daniel Blum

Chair of Algorithm Engineering and Systems Analysis,
Department of Computer Science, Dortmund University, Germany

daniel.blum@udo.edu

Jürgen Branke

Institute AIFB, University of Karlsruhe
76128 Karlsruhe, Germany

branke@aifb.uni-karlsruhe.de

Abstract   For many practical optimization problems, the evaluation of a solution is subject to noise, and optimization heuristics capable of handling such noise are needed. In this paper we examine the influence of noise on particle swarm optimization and demonstrate that the resulting stagnation can not be removed by parameter optimization alone, but requires a reduction of noise through averaging over multiple samples. In order to reduce the number of required samples, we propose a combination of particle swarm optimization and a statistical sequential selection procedure, called optimal computing budget allocation, which attempts to distribute a given number of samples in the most effective way. Experimental results show that this new algorithm indeed outperforms the other alternatives.

Keywords: Particle Swarm Optimization, Noise, Sequential Sampling

Introduction

In many real-world optimization problems, solution qualities can only be estimated but not determined precisely. Falsely calibrated measurement instruments, inexact scales, scale reading errors, etc. are typical sources for measurement errors. If the function of interest is the output from stochastic simulations, then the measurements may be exact, but some of the model output variables are random variables. The term "noise" will be used in the remainder of this article to subsume these phenomena.

This article discusses the performance of *particle swarm optimization* (PSO) algorithms on functions disturbed by Gaussian noise. It extends previous analyses by also examining the influence of algorithm parameters, considering a wider spectrum of noise levels, and analyzing different types of noise (multiplicative and additive). Furthermore, we integrated a recently developed sequential sampling technique into the particle swarm optimization method. Similar techniques have been integrated into other metaheuristics like evolutionary algorithms, but their application to the PSO algorithm is new.

The paper is structured as follows. First, we briefly introduce PSO in Section 1. Then, the effects of noise and sequential sampling techniques are discussed in Section 2. A sequential selection procedure is introduced in Section 3. Section 4 presents several experimental results, including the effect of parameter tuning, some algorithmic variants with perfect local and global knowledge, and the integration of sequential sampling. The paper concludes with a summary and an outlook.

1.   Particle swarm optimization

PSO uses a population (*swarm*) of *particles* to explore the search space. Each particle represents a candidate solution of an optimization problem and has an associated position, velocity, and memory vector. The main part of the PSO algorithm can be described formally as follows: Let $S \subseteq \mathbb{R}^n$ be the n-dimensional search space of a (fitness) function $f : S \to \mathbb{R}$ to be optimized. Without loss of generality, throughout the rest of this article, optimization problems will be formulated as minimization problems. Assume a swarm of s particles. The ith particle consists of three components. The first one, x_i, is its position in the search space, the second component, v_i, describes the velocity, and the third component, p_i^*, is its memory, storing the best position encountered so far. This vector is often referred to as *personal best* in the PSO literature. Finally, the term p_g^* denotes the best position found so far by the whole swarm, and is generally referred to as *global best*. Let t

denote the current generation. Velocities and positions are updated for each dimension $1 \leq d \leq n$ as follows:

$$v_i(t+1) = \underbrace{wv_i(t)}_{\text{momentum}} + \underbrace{c_1 u_{1i} \{p_i^*(t) - x_i(t)\}}_{\text{local information}} + \underbrace{c_2 u_{2i} \{p_g^*(t) - x_i(t)\}}_{\text{global information}},$$

(14.1)

$$x_i(t+1) = x_i(t) + v_i(t+1).$$

Before optimization can be started, several parameters or factors have to be specified. These so-called *exogenous* factors will be analyzed in more detail below. Parameters that are used inside the algorithm are referred to as *endogenous*. The endogenous factors u_{1i} and u_{2i} are realizations of uniformly distributed random variables U_{1i} and U_{2i} in the range $[0, 1]$. The exogenous factors c_1 and c_2 are weights that regulate the influence of the local and the global information. The factor w in the *momentum* term of Eq. 14.1 is called *inertia weight*. It was introduced in (Shi and Eberhart, 1998) to balance global and local search abilities.

2. The effect of noise

Noise makes it difficult to compare different solutions and select the better ones. In PSO, noise affects two operations: In every iteration (i) each particle has to compare the new solution to its previous best and retain the better one and (ii) the overall best solution found so far has to be determined. Wrong decisions can cause a *stagnation* of the search process: Over-valued candidates—solutions that are only seemingly better—build a barrier around the optimum and prevent convergence. The function value at this barrier will be referred to as the *stagnation level*. Or, even worse, the search process can be *misguided:* The selection of seemingly good candidates moves the search away from the optimum. This phenomenon occurs if the noise level is high and the probability of a correct selection is very small.

There is very little research on how strongly the noise affects the overall performance of PSO, and what measures are suitable to make PSO more robust against noise. (Parsopoulos and Vrahatis, 2001) were probably the first to present some results regarding the behavior of the PSO algorithm in noisy and continuously changing environments. In (Parsopoulos and Vrahatis, 2002), they focused on noise alone and concluded that "... in the presence of noise the PSO method is very stable and efficient." In both papers, fitness proportional noise models were used. (Krink et al., 2004) compared differential evolution, evolutionary algorithms, and PSO on noisy fitness functions. The noise was independent of the solution's fitness. To reduce the effect of the noise, they suggested to average over a number of samples. Finally, in (Liu et al., 2005), PSO is combined with local simulated annealing and a hypothesis test to tackle flow shop problems with noisy processing times. Thereby, the hypothesis tests are used to decide whether the new solution should replace a particle's personal best position.

The influence of noise on *evolutionary algorithms* (EAs) has received much more attention. EAs have been shown to work quite well in the presence of noise. Also, it has been proven analytically that under certain conditions, increasing the population size may help an evolution strategy to cope even better with the noise (Beyer, 2001). Several papers report on the successful integration of statistical tests or selection procedures into evolutionary algorithms, see, e.g., (Rudolph, 1997; Bartz-Beielstein and Markon, 2004; Branke and Schmidt, 2004; Buchholz and Thümmler, 2005; Schmidt et al., 2006). A comprehensive overview on the topic of evolutionary algorithms in the presence of noise is given in (Jin and Branke, 2005). Based on the good performance of evolution strategies in noisy environments, one might hope that also PSO can somehow cope with the noise, and that it is sufficient to adjust its parameters.

Alternatively, one may attempt to reduce the effect of noise explicitly. The simplest way to do so is to sample a solution's function value n times, and use the average as estimate for the true expected function value. While this reduces the standard deviation of the mean by a factor of $\sqrt{n}$, it also increases the running time by a factor of n, which is often not acceptable.

3.   Optimal computing budget allocation

We consider statistical selection procedures that use only a small number of samples to identify the best out of a set of solutions with a high probability. There is a multitude of selection procedures in the literature. (Bechhofer et al., 1995) give a comprehensive introduction into statistical selection methods. *Two-stage procedures* use the samples of a first stage to estimate means and variances. In the second stage, an additional amount of samples is drawn for each candidate solution, each amount depending on the variance and the overall required probability of correct selection. *Sequential procedures* allow even more than two stages. Such methods use either an elimination mechanism to reduce the number of alternatives considered for sampling, or they assign additional samples only to the most promising alternatives. The intuition is to use all available information as soon as possible to adjust the further process in a promising way. The most sophisticated sampling approaches include the information about variances and the desired probability of a correct selection and adjust the overall number of samples accordingly. A comparison of three state-of-the-art selection procedures can be found in (Branke et al., 2005).

In this paper, we use a procedure that assigns a fixed total number of samples to candidate solutions, but sequentially decides how to allocate the samples to different candidate solutions. A recently suggested sequential approach that falls into this category is the *optimal computing budget allocation* (OCBA) (Chen et al., 2000). OCBA is based on Bayesian statistics and aims at maximizing the *Approximate Probability of Correct Selection* (APCS), i.e., a lower bound for the probability of correct selection $P(CS)$. It is defined as

$$\text{APCS} = 1 - \sum_{i=1,i\neq b}^{k} P[\bar{X}_b > \bar{X}_i] \le P(CS),$$

where k is the number of solutions considered and b denotes the solution with the smallest sample mean performance, and $\bar{X}_i$ denotes the sample mean for solution i.

(Chen et al., 2000) show that for a fixed total number of samples $T = \sum_{i=1}^{k} N_i$, the APCS can be asymptotically maximized if

$$\frac{N_i}{N_j} = \frac{\sigma_i/(\bar{X}_i - \bar{X}_b)}{\sigma_i/(\bar{X}_j - \bar{X}_b)}, i,j \in 1,2,\ldots,k, \text{ and } i \neq j \neq b \qquad (14.2)$$

and

$$N_b = \sigma_b \sqrt{\sum_{i=1,i\neq b}^{k} \frac{N_i^2}{\sigma_i^2}} \qquad (14.3)$$

with σ_i being the standard deviation of the samples for solution i.

Based on these propositions, OCBA draws samples iteratively until the computational budget is exhausted.

1 Initialization: Draw n_0 initial samples for each solution. Set $l = 0, N_1^l = N_2^l = \cdots = N_k^l = n_0$, and $T = T - kn_0$

2 WHILE $T > 0$ DO

    (a) Set $l = l + 1$. Increase the computing budget by Δ_l (i.e., number of additional simulations in this iteration) and compute the new budget allocation to approximate Eq. 14.2 and Eq. 14.3.

    (b) Draw additional $\max(0, N_i^l - N_i^{l-1})$ samples for each solution $i, i = 1, 2, \ldots, k$

    (c) $T = T - \Delta_l$.

3 Return the index b of the system with the lowest mean $\bar{X}_b$, where $\bar{X}_b = \min_{1 \le i \le k} \bar{X}_i$.

For a more detailed description of OCBA, the reader is referred to (Chen et al., 2000).

4.   Experiments on the noisy sphere

Experimental setup

Sequential parameter optimization (SPO) is an algorithmical procedure to adjust the exogenous parameters of an algorithm, the so-called *algorithm design*, and to determine good *tuned* parameter settings for optimization algorithms (Bartz-Beielstein, 2006). It combines methods from computational statistics, *design and analysis of computer experiments* (DACE), and exploratory data analysis to improve the algorithm's performance and to understand why an algorithm performs poorly or well. SPO provides a means for reasonably fair comparisons between algorithms, allowing each algorithm the same effort and mechanism to tune parameters. Table 14.1 presents an algorithm design for PSO algorithms. The seven parameters were tuned during the SPO procedure. Details of this tuning procedure are presented in (Blum, 2005) and (Bartz-Beielstein, 2006).

In our experiments, we use the 10-dimensional sphere ($\min y = \sum_i^{10} x_i^2$) as test problem, because in this unimodal environment, the algorithm can easily find the optimum, and if it does not, this can be directly attributed to the noise. We consider additive ($\tilde{y} = y + \epsilon$) and

Table 14.1. Algorithm design of the PSO algorithm. Similar designs were used in (Shi and Eberhart, 1999) to optimize well-known benchmark functions.

Symbol	Parameter	Range	Default
s	Swarm size	$\mathbb{N}$	40
c_1	Cognitive parameter	$\mathbb{R}_+$	2
c_2	Social parameter	$\mathbb{R}_+$	2
w_{max}	Starting value of the inertia weight w	$\mathbb{R}_+$	0.9
w_{scale}	Final value of w in percentage of w_{max}	$\mathbb{R}_+$	0.4
$w_{iterScale}$	Percent iterations with reduced w_{max}	$\mathbb{R}_+$	1.0
v_{max}	Max. value of the step size (velocity)	$\mathbb{R}_+$	100

multiplicative ($\tilde{y} = y(1 + \epsilon)$) noise, where ϵ denotes a normally distributed random variable with mean zero and standard deviation σ. A broad range of noise levels σ was used to analyze the behavior of the algorithms and to detect the highest noise level PSO was able to cope with. For example, the experiments with additive noise used σ values from the interval $[10^{-4}, 10^2]$. All particle positions were initialized to $x_i = 10$ in all dimensions.

Unless specified otherwise, the best function value found after 10,000 function evaluations was used as a performance measure, because (i) a fixed number of evaluations is a quite fair and comparable criterion, as it does not depend on programming skills, hardware, etc., and (ii) many real-world optimization problems require simulation runs that are computationally expensive compared to the computational effort of the optimization algorithm itself.

Performance in noisy environments

For the experiments in this subsection we used the following fixed parameter settings which have proven reasonable in preliminary tests and have also been used, e.g., in (Shi and Eberhart, 1999): $s = 20, c_1 = c_2 = 2, w\text{max} = 0.9, w_{\text{scale}} = 4/9, v_{\text{max}} = 100$. Figure 14.1 shows the convergence curves (fitness over time) of the standard PSO algorithm for different levels of additive noise. As can be seen, while the PSO without noise keeps improving, the algorithm stagnates in noisy environments, with the fitness level reached depending on the noise level. It is interesting to note that the performance of PSO before reaching the stagnation level is almost unaffected by the noise, indicating that a certain noise level is tolerated before the system breaks down.

While for the case of additive noise, the influence of the noise on the performance was quite regular, the effect of multiplicative noise was quite different. Figure 14.2 shows the final solution quality obtained depending on the noise level. For low levels of noise, the algorithm is basically unaffected, as the noise scales with the fitness values and most decisions can be made correctly throughout the run. This confirms the observations made in (Parsopoulos and Vrahatis, 2002) regarding the robustness of PSO with respect to proportional noise. On the other hand, if the noise exceeds a certain threshold, the algorithm may actually diverge and end up with solutions worse than the initial solutions. This may happen if the worse solutions have a much higher noise, making it likely that the worse fitness is accidentally compensated by an overvaluation.

When comparing the performance of PSO to a simple $(1+1)$-*evolution strategy* (ES), we observed a much faster progress of the $((1+1)$-ES) on

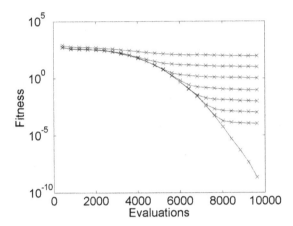

Figure 14.1.   Convergence curves for the standard PSO for different levels of additive noise (from *top to bottom*, $\sigma = 100, 10, 1, 0.1, 10^{-2}, 10^{-3}, 10^{-4}, 0$).

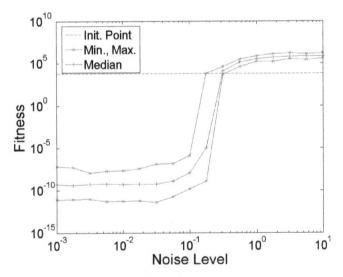

Figure 14.2.   Fitness after 10,000 evaluations for various noise levels. *Horizontal line* indicates initial fitness.

the Sphere function during the first phase of the optimization. However, the algorithm encountered the same problem as the PSO: the stagnation on a certain level. Having tuned the parameters of both algorithms using SPO, we observed that the (1+1)-ES stagnated on a higher fitness level than the PSO, possibly due to an inherent advantage of population-based approaches in noisy environments.

Global versus local certainty

As described above, noise can affect PSO in two steps: when each particle chooses its local best, and when the global best is determined. In order to find out which of the two steps is more critical for the algorithm's performance, we tested two special variants of the PSO algorithm.

Variant PSO_{pc} was given the correct information whenever a particle decided between its new position and the old personal best. This variant was able to find significantly better results than the variant $PSO_{default}$. Moreover, it did not stagnate at certain fitness levels and showed a progress during the whole optimization process. Similar results were observed for multiplicative noise. The search was not misguided by the noise and for all noise levels the solutions obtained were better than the initial point.

Variant PSO_{gc} was provided with the information which of the particles' presumed best was the true best, i.e., it could correctly select the global best from all the presumed local best. In the presence of additive noise the variant could find better solutions than the $PSO_{default}$, but the optimization stagnated and could not converge to the minimum. Experiments with multiplicative noise also showed an improvement compared to $PSO_{default}$, but again the basic problem remained: the search was misled.

Overall, in our experiments with additive and multiplicative noise models, PSO_{pc} showed clearly superior performance compared to the PSO_{gc} variant. However, we have to keep in mind that variant PSO_{pc} received in each iteration the knowledge for a number of decisions, which was equal to the swarm size. In contrast, variant PSO_{gc} could decide once per iteration correctly. Furthermore, PSO_{gc} could potentially loose the true global best, namely when the decision on the first level was wrong. This could not happen with PSO_{pc}.

Parameter tuning vs. multiple evaluations

Next, we examined whether parameter tuning is sufficient for PSO to cope with the noise, or whether multiple evaluations are necessary. Results are summarized in Table 14.2. Thereby, the PSO_{rep} variant uses a fixed number of repeated function evaluations (5 in the default setting, and this parameter is optimized by SPO as well). While parameter tuning improves the final solution quality for the single-evaluation and multiple-evaluation case, better results can be obtained when allowing multiple evaluations rather than only optimizing parameters and letting the algorithm cope with noise.

Integrating OCBA into PSO

Now we examine how PSO can benefit from integrating a sequential sampling procedure like OCBA (see Section 2). The variant PSO_{OCBA} uses the OCBA procedure to search for the swarm's global best among the set of positions considered in the iteration (i.e., all new and all local best positions). With the design of the PSO_{OCBA} algorithm, we aim at two objectives: (i) an increased probability to select the swarm's global best correctly, (ii) an increased probability to select the particles' personal bests correctly (as a byproduct of the repeated function evaluations of candidate positions by the OCBA method). In accordance with the OCBA technique, the position with the resulting lowest mean of the function values is selected as the swarm's global best. The new personal bests of the particles result from the comparison of the function value means of their old personal best and new positions. Function values from previous generations were stored for re-use in the next generation.

The results in Table 14.2 indicate that variant PSO_{OCBA} with an improved algorithm design generated by SPO significantly outperformed the other algorithm variants optimizing the Sphere function disturbed by additive noise. OCBA enables the algorithm to distinguish smaller function value differences than the other variants. This seems obvious with respect to $PSO_{default}$, as it has no noise reduction mechanism, but it also reaches a lower stagnation level than PSO_{rep}. OCBA's flexible assignment of samples seemed to be an advantage in the selection process. Furthermore, as OCBA allocates more samples to promising solutions, which are more likely to survive, the total number of function evaluations used for decisions in one iteration (new and preserved) is higher for the OCBA variant than if each position had received the same amount of function evaluations. In fact, in our experiments we observed this number to be about twice as high, allowing more accurate decisions.

5.   Summary and outlook

We have examined the influence of noise on the performance of PSO, and compared several algorithmic variants with default and tuned parameters. Based on our results, we make the following conclusions:

Table 14.2. Comparison of the standard algorithm *(standard)*, a variant with multiple evaluations *(rep)* and the new variant with integrated OCBA *(OCBA)*. Each variant has been tested with default *(default)* and optimized *(SPO)* parameter settings. Results ± standard error for the sphere with additive noise ($\sigma = 10$). Varying the noise level σ led to similar results that are reported in (Blum, 2005). Smaller values are better.

Algorithm	Parameter settings	
	Default	SPO
$\text{PSO}_{\text{standard}}$	9.08 ± 0.43	6.94 ± 0.30
PSO_{rep}	7.59 ± 0.35	4.99 ± 0.27
PSO_{OCBA}	6.81 ± 0.67	$\mathbf{1.98 \pm 0.11}$

- Additive noise leads to a stagnation of the optimization process, multiplicative noise can even lead to divergence.

- Parameter tuning alone cannot eliminate the influence of noise.

- Sequential selection procedures such as OCBA can significantly improve the performance of particle swarm optimization in noisy environments. Local information plays an important role in this selection process and cannot be omitted.

Why did sequential selection methods improve the algorithm's performance? First, the selection of the swarm's best of one iteration was correct with a higher probability compared to reevaluation approaches. Second, as more samples were drawn for promising positions, positions that remained and reached the next iteration were likely to have received more samples than the average. Samples accumulated and led to a greater sample base for each iteration's decisions. These two advantages might be transferable to other population-based search heuristics, in which individuals can survive several generations. Summarizing, we can conclude that it was not sufficient to only tune the algorithm design (e.g., applying SPO), or to only integrate an advanced sequential selection procedure (e.g., OCBA). The highest performance improvement was caused by the combination of SPO and OCBA.

However, our experiments were restricted to artificial test functions only. The application of the PSO_{OCBA} variant to real-world problems will be the next step. In such problems, the noise might not be normally distributed. First experiments, which analyzed the applicability of OCBA to an elevator group control problem proposed in (Markon et al., 2006) produced promising results. Noise-dependent, variable swarm sizes as proposed in (Bartz-Beielstein and Markon, 2004) for evolution strategies, might improve the convergence velocity. Furthermore, it might be interesting to replace the current OCBA by sampling techniques with variable stopping rules, where the number of samples allocated per generation is not fix but depends on the confidence in the decisions.

References

Bartz-Beielstein, T. (2006). *Experimental Research in Evolutionary Computation–The New Experimentalism.* Natural Computing Series. Springer, Berlin, Heidelberg, New York.

Bartz-Beielstein, T. and Markon, S. (2004). Tuning search algorithms for real-world applications: A regression tree based approach. In Greenwood, G. W., editor, *Congress on Evolutionary Computation*, volume 1, pages 1111–1118. IEEE Press.

Bechhofer, R. E., Santner, T. J., and Goldsman, D. M. (1995). *Design and Analysis of Experiments for Statistical Selection, Screening, and Multiple Comparisons.* Wiley, New York.

Beyer, H.-G. (2001). *The Theory of Evolution Strategies.* Springer, Berlin, Heidelberg, New York.

Blum, D. (2005). Particle Swarm Optimization für stochastische Probleme. Interner Bericht der Systems Analysis Research Group SYS–2/05, Universität Dortmund, Fachbereich Informatik.

Branke, J., Chick, S., and Schmidt, C. (2005). New developments in ranking and selection: An empirical comparison of the three main approaches. In Kuhl, N., Steiger, M. N., Armstrong, F. B., and Joines, J. A., editors, *Winter Simulation Conference*, pages 708–717. IEEE.

Branke, J. and Schmidt, C. (2004). Sequential sampling in noisy environments. In *Parallel Problem Solving from Nature*, volume 3242 of *LNCS*, pages 202–211, Berlin, Heidelberg, New York. Springer.

Buchholz, P. and Thümmler, A. (2005). Enhancing evolutionary algorithms with statistical sselection procedures for simulation optimization. In Kuhl, N., Steiger, M. N., Armstrong, F. B., and Joines, J. A., editors, *Winter Simulation Conference*, pages 842–852. IEEE.

Chen, C.-H., Lin, J., Yucesan, E., and Chick, S. E. (2000). Simulation budget allocation for further enhancing the efficiency of ordinal optimization. *Discrete Event Dynamic Systems: Theory and Applications*, 10(3):251–270.

Jin, Y. and Branke, J. (2005). Evolutionary optimization in uncertain environments—a survey. *IEEE Transactions on Evolutionary Computation*, 9(3):303–317.

Krink, T., Filipic, B., Fogel, G. B., and Thomsen, R. (2004). Noisy optimization problems - a particular challenge for differential evolution? In *Congress on Evolutionary Computation*, pages 332–339. IEEE Press.

Liu, B., Wang, L., and Jin, Y. (2005). Hybrid particle swarm optimization for flow shop scheduling with stochastic processing time. In Hao, Y. et al., editors, *Computational Intelligence and Security*, volume 3801 of *LNAI*, pages 630–637, Berlin, Heidelberg, New York. Springer.

Markon, S., Kita, H., Kise, H., and Bartz-Beielstein, T., editors (2006). *Modern Supervisory and Optimal Control with Applications in the Control of Passenger Traffic Systems in Buildings*. Springer, Berlin, Heidelberg, New York.

Parsopoulos, K. E. and Vrahatis, M. N. (2001). Particle swarm optimizer in noisy and continuously changing environments. In Hamza, M., editor, *Artificial Intelligence and Soft Computing*, pages 289–294. IASTED/ACTA Press.

Parsopoulos, K. E. and Vrahatis, M. N. (2002). Particle swarm optimization for imprecise problems. In Fotiadis, D. and Massalas, C., editors, *Scattering and Biomedical Engineering, Modeling and Applications*, pages 254–264. World Scientific.

Rudolph, G. (1997). Reflections on bandit problems and selection methods in uncertain environments. In Bäck, T., editor, *International Conference on Genetic Algorithms*, pages 166–173. Morgan Kaufmann.

Schmidt, C., Branke, J., and Chick, S. (2006). Integrating techniques from statistical ranking into evolutionary algorithms. In Rothlauf, F. et al., editors, *Applications of Evolutionary Computation*, number 3907 in LNCS, pages 752–763, Berlin, Heidelberg, New York. Springer.

Shi, Y. and Eberhart, R. (1998). Parameter selection in particle swarm optimization. In Porto, V., Saravanan, N., Waagen, D., and Eiben, A., editors, *Evolutionary Programming*, volume VII, pages 591–600. Springer, Berlin, Heidelberg, New York.

Shi, Y. and Eberhart, R. (1999). Empirical study of particle swarm optimization. In Angeline, P. J. et al., editors, *Congress on Evolutionary Computation*, volume 3, pages 1945–1950.

Part VI

Distributed and Parallel Algorithms

Chapter 15

EMBEDDING A CHAINED LIN-KERNIGHAN ALGORITHM INTO A DISTRIBUTED ALGORITHM

Thomas Fischer
University of Kaiserslautern
Department of Computer Science
Distributed Algorithms Group
fischer@informatik.uni-kl.de

Peter Merz
University of Kaiserslautern
Department of Computer Science
Distributed Algorithms Group
pmerz@informatik.uni-kl.de

Abstract

The Chained Lin-Kernighan algorithm (CLK) is one of the best heuristics to solve Traveling Salesman Problems (TSP). In this paper a distributed algorithm is proposed, where nodes in a network locally optimize TSP instances by using the CLK algorithm. Within an Evolutionary Algorithm (EA) network-based framework the resulting tours are modified and exchanged with neighboring nodes. We show that the distributed variant finds better tours compared to the original CLK given the same amount of computation time. For instance f13795, the original CLK got stuck in local optima in each of 10 runs, whereas the distributed algorithm found optimal tours in each run requiring less than 10 CPU minutes per node on average in an 8 node setup. For instance sw24978, the distributed algorithm had an average solution quality of 0.050% above the optimum, compared to CLK's average solution of 0.119% above the optimum given the same total CPU time (10^4 seconds). Considering the best tours of both variants for this instance, the distributed algorithm is 0.033% above the optimum and the CLK algorithm 0.099%.

Keywords: Traveling salesman problem, combinatorial optimization, distributed algorithm, evolutionary algorithm

1.   Introduction

The Traveling Salesman Problem (TSP) is one of the best-known *combinatorial optimization problems*. Given a number of cities (or customers), a salesman has to find a cost-minimal route visiting each city exactly once. The problem can be represented by a fully connected graph $G = (V, E)$ and a function $d_{i,j}$ denoting the distance between two vertices $v_i, v_j \in V$. For *symmetric* TSPs (STSP) $d_{i,j} = d_{j,i}$ holds for all i and j. For *asymmetric* TSPs (ATSP) $d_{i,j} \neq d_{j,i}$ holds for at least one pair (v_i, v_j). An optimal solution for a problem instance, which is a *Hamiltonian* cycle of minimal length on G, is a permutation π of its cities that has a minimum value for the cost function $C(\pi)$:

$$C(\pi) = \sum_{i=1}^{n-1} d_{\pi(i), \pi(i+1)} + d_{\pi(n), \pi(1)} \tag{15.1}$$

For an instance with n cities ($n = |V|$), there are $\frac{(n-1)!}{2}$ different tours. Although having a simple setup, the TSP is a *NP-hard* problem [Cook, 1971, Garey and Johnson, 1979].

To solve TSP instances, different types of algorithms have been developed [Applegate et al., 1999]. *Exact algorithms* enumerate implicitly each possible solution and find a provable optimal solution (see e.g. [Dantzig et al., 1954, Lawler and Wood, 1966, Applegate et al., 1995]). *Approximation algorithms* construct a valid tour providing a guarantee for the worst-case time complexity and the expected tour length (see e.g. [Christofides, 1976, Arora, 1998]). *Heuristic algorithms* perform a non-complete search in the solution space and do not guarantee to find an optimal solution. Their main advantage is that they can find a good sub-optimal solution in much shorter time than exact algorithms. Heuristic algorithms performing a *local search* usually exploit neighborhood relations between nodes. One example is the *k-opt neighborhood* [Flood, 1956, Lin, 1965], where two tours are called neighbors if one tour can be transformed to the other by exchanging k edges. A tour is called *k-optimal* (*k*-opt) if the tour cannot be improved any further by exchanging k edges. Increasing k and performing an exhaustive search for possible exchange moves increases the tour quality, but also requires a fast growing amount of computation time. So, for most applications k is limited to $k \leq 3$.

Lin and Kernighan [Lin and Kernighan, 1973] approached the problem of finding a trade-off between tour quality and computation cost by introducing an algorithm (LK) where k is variable. Here, a complex move is built by a sequence of simple edge removal and insertion moves. Initially, the tour is transformed into a 1-tree by removing one edge and inserting one new, shorter edge, which introduces a sub-cycle into the tour. In each iteration, the algorithm removes one edge from the current sub-cycle and has to choose whether to repair the tour by connecting both cities with degree 1 or to continue searching by adding a new edge creating a new sub-cycle. The search may terminate

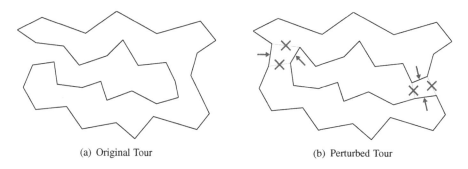

(a) Original Tour (b) Perturbed Tour

Figure 15.1. Double-bridge move

if the repair step results in a better tour than the original one. Alternatively, the algorithm may backtrack and explore different edges for removal and insertion. The LK algorithm can be improved utilizing nearest neighbor lists, don't look bits or sophisticated data structures. Furthermore, it can be embedded into a meta-heuristic such as the *Chained Lin-Kernighan algorithm* (CLK) introduced by Martin, Otto and Felten [Martin et al., 1991]. Instead of iteratively restarting a LK search to get better tours, Martin et al. suggested to *kick* the current tour before applying the LK algorithm again.

They proposed to use the *double-bridge move* (DBM, Figure 15.1), which is a cheap way to perform 4-exchange move and is quite unlikely reversible by the LK. In a double-bridge move, four edges are removed from the tour and four new edges are inserted to connect the four sub-tours again. Depending on the graphical representation, the new edges shape the form of two crossing bridges. By applying this move, the current tour leaves the local optimum allowing the LK algorithm to continue exploring the search space, while maintaining most properties of the previous solution.

In the search for optimal solutions the most recent achievement was the proof of optimality for a tour of instance sw24978 in May 2004. Here, 96 dual processor machines needed a total of over 80 CPU years to prove optimality with an exact algorithm. For heuristic algorithms, in contrast, distributed computation has not yet become very common. This might be due to the fact that heuristic algorithms require less computation time and therefore there is no exigent need for distributed computation compared to exact algorithms. To solve large TSP instances with today's heuristic algorithms, however, it is inevitable from our point of view to distribute computation.

In this paper we present a distributed algorithm for solving Traveling Salesman Problems. This algorithm utilizes the existing CLK implementation from Applegate et al. [Cook, a, Applegate et al., 1999] and embeds it into an evolutionary algorithm that is running distributed over several nodes in a network. Therefore, it resembles memetic algorithms (MA) [Merz and Freisleben, 2001], which combine evolutionary algorithms with local search techniques.

With the approach presented here, our algorithm finds both better tours given a computation time limit and it converges faster towards an optimal solution compared to the original CLK algorithm.

Previously, other distributed algorithms have been proposed of which some will be presented in the remainder of this section.

Baraglia et al. introduce a genetic algorithm (GA) [Baraglia et al., 2001] using an island model [Grosso, 1985], where each node represents an island with a sub-population. Here, the tours are encoded in a compact form [Harik et al., 1999] storing only the probability values in a $k \times k$ triangular matrix P (k is the number of cities). The matrix element $p_{i,j}$ represents the probability that the edge (i, j) is part of an individual's tour. In each generation a tour L is constructed using the probability values. This tour L is refined to tour W by the CLK algorithm. The matrix elements' values are increased if the corresponding edge occurs only in W, but not in L, decreased in the opposite case and remain unchanged if the edge occurs in both or none of the tours. Although the paper's conclusions are meager with respect to numerical data, the supplied plots show that the more processes cooperate the less generations are required for each one to find the optimal solution for an instance. The instance sizes analyzed in this paper range from 532 to 1002.

Nguyen et al. describe a GA-based algorithm [Hung, 2004], which uses their LK implementation (applying a 5-opt basic move) for local tour improvement. This algorithm can be parallelized by settling sub-populations on cluster machines. The GA algorithm performs in each generation a mutation or crossover operation on one member of each sub-population. For mutation, a selected tour is mutated by a Random-walk kick and optimized by the LK algorithms for a number of iterations (thus reproducing an Iterated LK). The best intermediate tour will replace the original tour if it was better than the original one. For the crossover operation (MPX3), two parents are selected from the sub-population and merged. Common sub-tours from both parents are fixed for the LK algorithm to follow. The resulting tour will replace the worst parent if better. The GA will terminate, if no improvement has been found for a number of iterations. Nguyen et al.'s algorithm can compete with Helsgaun's LK regarding the tour quality, but requires significantly less time for larger instances. E.g. for instance d18512, the GA-based algorithm requires about 5000 seconds in a 10 node setup for an average tour quality of 645 323.8, whereas LKH requires over 100 000 seconds for a worse tour (645 332.2) in on a single computing node.

2.   Architecture & Algorithm

The system is a structured network of eight computing nodes arranged in a hypercube topology. Each node in the distributed system consists of an evolutionary algorithm (EA) that uses a network module to communicate with other nodes and a CLK module for solving problem instances locally. The

CLK module taken from the Concorde package (031219) [Cook, a, Applegate et al., 1995] is well-known in the TSP community and is used by a variety of researchers (e.g. Walshaw in [Walshaw, 2000]).

During an initial setup phase the nodes connect to a dedicated bootstrapping node B (called "hub") which constructs the structure (see Figure 15.2) by supplying neighborhood lists to each node. Initially, the nodes know only how to connect to B, which is contacted for a list of neighbors. The hub determines the node's position within the hypercube and assembles the node's neighbor list based on nodes that are already known to the hub. As the first nodes will receive a sparse list of neighbors, to build the connected hypercube a node contacts each neighbor after receiving the list. If the contacting node is unknown to the contacted node, the contacting node is added to the contacted node's neighbor list.

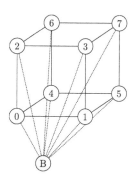

Figure 15.2. Architecture of the network

As the nodes communicate directly, the hub is the only central component in the network and is only used during initialization. For systems with a small network size, this approach appears to be feasible. In this work, however, we focus on the effects of distributed (population-based) optimization which is independent of the utilization of centralized or decentralized protocols for network setup.

Simplified, the main algorithm is structured as follows. In each iteration the node's only tour is perturbed by one or several random double-bridge moves. This perturbed tour is optimized by the Chained Lin-Kernighan algorithm. In the next step the new tour is compared with all the tours received meanwhile from other nodes. The best tour is stored as the node's new tour. If this tour was the result of the local CLK optimization it is multicasted to all neighboring nodes. The pseudo-code for this algorithm is shown in Figure 15.3.

The strength of the perturbation has to be chosen carefully. A perturbation that is too weak might not be able to leave the current local optimum, while a too strong perturbation might damage the tour heavily causing a loss of quality. Therefore the used strategy begins with a weak perturbation and increases its strength if no better tours are found.

Whenever the CLK function does not find a better tour than the previous best tour, a dedicated counter is increased. This counter gets reset when a better tour has been found or received from another node. The counter's value determines the number of random double-bridge moves applied to a tour; a higher number of perturbation moves leads to stronger perturbation. Eventually, perturbation moves will modify the tour in a way that it will leave the current local optimum. When the number of iterations without improvements reaches the value of a

```
function DISTRIBUTEDALGORITHM
    s := INITIATETOUR();
    s_best := CHAINEDLINKERNIGHAN(s);
    s_prev := s;
    while not TERMINATIONDETECTED do
        if NumNoImprovements > c_r then
            NumNoImprovements := 0;
            s := INITIATETOUR();
        else
            NumPerturbations := ⌊ NumNoImprovements / c_v ⌋ + 1;
            s := PERTURBETOUR(s, NumPerturbations);
        end if
        s := CHAINEDLINKERNIGHAN(s);
        S_received := ALLRECEIVEDTOURS;
        s_best := SELECTBESTTOUR(S_received ∪ {s} ∪ {s_prev});
        if LENGTH(s_best) = LENGTH(s_prev) then
            NumNoImprovements + +;
        else
            NumNoImprovements := 0;
        end if
        if s_best = s then
            MULTICASTTONEIGHBORS(s_best);
        end if
        s_prev := s_best;
    end while
```

Figure 15.3.   Pseudo-code for the distributed algorithm.

parameter c_r, the current tour is discarded and a new tour is constructed (the ILK is restarted thereby).

The termination criterion as represented by the function TERMINATION-DETECTED (not shown) is triggered when a node's solution quality equals an already known optimum (if available) or when a given time bound is hit. The notification on the occurrence of an optimal solution is propagated through the network terminating all nodes.

3.   Experimental Setup

For our analysis a set of instances from well-known sources have been selected. From Reinelt's TSPLIB [Reinelt, 1991] the instances fl1577, pr2392, pcb3038, fl3795, fnl4461, usa13509, pla33810 and pla85900 were used, from the collection of national TSPs [Cook, b] instances fi10639 and sw24978 and from the 8th DIMACS challenge [Johnson,] the random instances C1k.1 (randomly clustered city distribution) and E1k.1 (randomly uniform city distribution). See Table 15.1 for details.

Each simulation setup was performed 10 times, from which the average values were used. The number of runs was limited due to time constraints. The program linkern that is part of the concorde package [Cook, a] has been used as CLK engine. As the Random-walk kicking strategy performed good for most instances in initial tests and is the default kicking strategy in linkern,

Instance	Size	Held-Karp Bound	Optimal Tour Length
C1k.1	1000	11330836	11376735
E1k.1	1000	22839568	22985695
fl1577	1577	21886	22249
pr2392	2392	373490	378032
pcb3038	3038	136588	137694
fl3795	3795	28477	28772
fnl4461	4461	181569	182566
fi10639	10639	520527	*520 383**
usa13509	13509	19851464	19982859
sw24978	24978	855528	855597
pla33810	33810	66050535	*66 005 185**
pla85900	85900	142383704	*142 307 500**

Table 15.1. Testbed Instances. Instances marked with a star are not solved to optimum yet, the values represent the length of the best known tour. The Held-Karp Bound is a lower bound estimation for optimal tour lengths (see [Held and Karp, 1970, Held and Karp, 1971]).

simulations were performed primarily with this kicking strategy. For this strategy, the first of four relevant cities for the double-bridge move is chosen randomly. Starting there, three independent random walks terminate at the other three cities.

For the first part of the analysis, instances were solved by the original CLK code. The number of iterations (termination criterion of the algorithm) was set to a very high value to make time bounds the primary termination criterion. The time limit was set to 10^4 CPU seconds for instances with less than 10^4 cities and 10^5 CPU seconds for larger instances. The resulting values were used for comparison with later results from the distributed algorithm. The distributed algorithm itself was tested with different setup values. The number of CLK calls (termination criterion of the algorithm) has been set to infinity to make time bounds the only termination criterion here, too. The time limit was set to 10^3 CPU seconds per node for instances with less than 10^4 cities and 10^4 CPU seconds per node for larger instances, which is a tenth of the values for the original CLK. As for the distributed algorithm eight nodes were working in parallel, this is more than fair for CLK. The number of iterations per CLK call within the distributed algorithm was set to the number of cities (default for the original CLK).

The parameters were set to $c_v = 64$ and $c_r = 256$. Other parameters have been changed in different simulations to observe effects of different values. Primary simulations were performed using a hypercube topology with 8 nodes. Additionally, setups with only 1 node were performed to check the influence of parallelization.

The cluster used for this analysis consisted of eight identical computer nodes with one 3.0 GHz SMT processor (Pentium 4) and 512 MB RAM each running

Instance	CLK	DistCLK
C1k.1	9/10	**10/10**
E1k.1	3/10	**10/10**
fl1577	0/10	**8/10**
pr2392	4/10	**10/10**
pcb3038	0/10	**7/10**
fl3795	0/10	**10/10**
fnl4461	0/10	**1/10**

Table 15.2. Number of CLK and Dist-CLK runs that known the optimum within a given time bound. For CLK, the limit was set to 10^4 seconds and to 10^3 seconds per node for the distributed variant with 8 nodes solving in parallel. Larger instances were omitted as both algorithms did not find optimal solutions for them.

Linux (Kernel 2.6). The nodes were connected by a switched Gigabit Ethernet network.

4. Experimental Results

The default setup in the following discussion was a distributed algorithm variant running on 8 nodes and using the double-bridge move for perturbation as described before. For comparison, several instances were analyzed using a distributed algorithm that was (1) restricted on one single node, (2) running without variable strength perturbation or (3) running with both restrictions.

As shown in Table 15.2, for instances with a size above 3000 cities CLK could not find an optimum at all in any run. The distributed algorithm (abbreviated "DistCLK") finds the optimal solution for most instances up to fnl4461 in at least one run. In cases where not all runs were successful within 1000 CPU seconds, the results were already close to the optimum. The distributed algorithm can handle instances like fl3795 very well (successful in each run), which the standard CLK fails to solve every time within its time bound.

The approximation towards the optimum is faster with the distributed algorithm compared with the original algorithm. As shown for instances fl1577 and sw24978, the distributed version is better than CLK for these instances. For the instance fl1577, CLK gets stuck after about 150 seconds in local optima (9 runs in 22 395, 1 run in 22 256) that are not left within the time bound. The distributed variant, however, finds the optimum in 8 out of 10 runs in less than 300 CPU seconds per node, the other two runs need about 2000 CPU seconds (Figure 15.4).

For instance sw24978 (see Figure 15.5) the distributed algorithm has an average tour quality of 0.050% over the optimum after reaching its time bound (best is 0.033%), whereas CLK has an average tour quality of 0.088% (best is 0.064%). The CLK algorithm's final average tour length is already reached after 2377.6 CPU seconds per node by the distributed algorithm. Considering that 8 nodes were cooperating on that problem, the original CLK algorithm requires 5 times more total CPU time to reach this tour quality compared to the distributed variant.

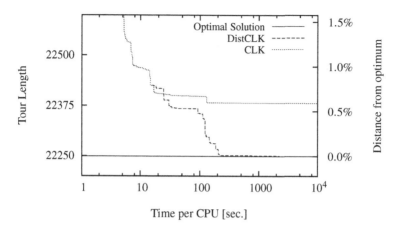

Figure 15.4.   Relation between tour length and CPU time for the Distributed Chained Lin-Kernighan algorithm (DistCLK) compared with the the original CLK for instance fl1577

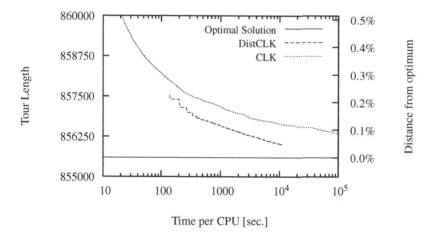

Figure 15.5.   Relation between tour length and CPU time for the Distributed Chained Lin-Kernighan algorithm (DistCLK) compared with the the original CLK for instance sw24978

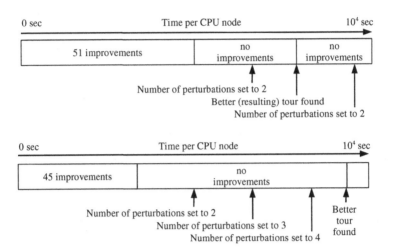

Figure 15.6. Time scale of the occurrence of improvements and perturbations during two example runs.

The perturbation and restarting strategy can effectively help the CLK to leave local optima. The following two example runs are selected out of ten simulation runs with instance fil10639 (see also Figure 15.6).

For run *A*, only a weak perturbation was required to find better tours. During the first 4952 CPU seconds 51 improving tours were found by the nodes. As after about 6600 seconds no new improvements were made, within a small time frame all eight nodes increased *NumPerturbations* to 2. Before requiring any further increase, a better tour was found (7858 seconds) by a node. As this tour was multicasted in the net and improved the local best tours, the local *NumNoImprovements* variables were reset, too. After about 9500 seconds *NumPerturbations* increased again as no new tour was found in the meantime. Finally the best tour's length was 520627, which is 0.047% above the Held-Karp bound.

Run *B* showed that strong perturbations can be necessary. For the first 3396 CPU seconds, 45 improving tours were found by the nodes. Hereafter *NumPerturbations* was increased sequentially: After about 5020 seconds to level 2, after about 6700 seconds to level 3 and after 8370 seconds to level 4. A better tour was found by a node after 9337 seconds preventing a further increase of *NumPerturbations*. This tour was improved four more times resulting in a final tour of length 520584 (0.039% above Held-Karp bound).

Tour qualities of all runs with the same parameters were between 520563 (0.035%) and 521002 (0.119%).

To compare the effects of parallelization a subset of the instances was run in setups with both 1 and 8 nodes, while keeping other setup parameters constant. Between two local CLK search steps the already described variable strength

| Algorithm | CLK | | DistCLK | |
Instance	100 sec	10^4 sec	10 sec	10^3 sec
E1k.1	0.005%	0.002%	OPT	OPT
C1k.1	0.024%	0.016%	◇	OPT
f11577	0.670%	0.594%	◇	0.006%
pr2392	0.237%	0.093%	0.152%	OPT
pcb3038	0.103%	0.060%	◇	0.004%
f13795	0.643%	0.524%	◇	OPT
fnl4461	0.098%	0.041%	◇	0.013%
fi10639	0.217%	0.106%	◇	0.116%
usa13509	0.204%	0.112%	◇	0.062%
sw24978	0.307%	0.122%	◇	0.116%
pla33810	0.519%	0.287%	◇	0.126%
pla85900	0.334%	0.160%	◇	0.182%

Table 15.3. Distance of the average tour length compared to known optimum (Held-Karp bound if not available) after 100 and 10^4 (for CLK) and 10 and 1000 CPU seconds per node (for DistCLK), respectively. For cells marks with ◇, there is no data available as the algorithm did not return any tour at this point of time.

double-bridge move perturbation was performed. In case of the 8 node variant the locally improved tours were exchanged between neighboring nodes. Simulation results show that the distributed algorithm can scale well with the number of nodes.

For the following discussion, "speed-up factor" (Equation 15.2) is the relation between the original CLK algorithm and the distributed algorithm regarding the total CPU time summed over all CPU nodes. For a setup with eight nodes, a speed-up factor of more than 8 means that the distributed algorithm required less total CPU time than the original CLK algorithm.

$$f_{\text{speed-up}} = \frac{t_{\text{1-node}}}{t_{\text{8-nodes}}} \tag{15.2}$$

In a comparison of instance pr2392 (see Figure 15.7) between the original CLK algorithm and the distributed algorithm running on 1 or 8 nodes, respectively, the variant with 8 nodes is more than twice as fast as expected from parallelization (regarding median values). It reaches a tour quality level of 0.1% above the optimum after 10.7 CPU seconds per node compared to 246.2 seconds for the distributed algorithm's single node variant (speed-up factor 23.01). The original CLK algorithm reaches this level after 8510.7 CPU seconds. For the quality level of 0.05% above the optimum the 8 node variant is still two times faster than the single node variant (speed-up factor 17.4), in respect to CPU seconds. Here, CLK does not reach this level as well as the optimum within the given 10^4 second time limit. The parallel variant with 8 nodes requires about a quarter of the time of the single node variant (speed-up factor 3.57) to find an optimal solution. This behavior is caused by three runs

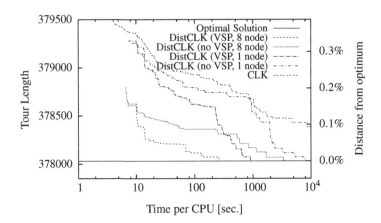

Figure 15.7. Effects of parallelization running the distributed algorithms on a different number of nodes and optional variable strength perturbation (VSP) for instance `pr2392`.

Distance to Optimum	CPU time per node [sec]			Speed-up Factor
	CLK	1 node	8 nodes	
Instance `pr2392`				
0.10%	8510.7	246.2	10.7	23.01
0.05%	–	421.1	24.2	17.40
0.00%	–	937.1	262.2	3.57
Instance `fl3795`				
0.50%	–	336.9	78.4	4.30
0.25%	–	1153.3	199.8	5.77
0.00%	–	4223.7	569.0	7.42
Instance `fi10639`				
0.12%	3912.6	1183.4	188.8	6.27
0.10%	15183.3	2671.7	350.6	7.62
0.08%	–	6960.5	723.0	9.63

Table 15.4. Speed-up with several instance. Average over 10 runs each.

in the parallel variant, that need between 110 and 260 seconds, while the other seven runs require less than 43 seconds to find the optimum. Thereby the medians over the optimum finding times for both variants are 71.2 seconds versus 596.5 seconds (factor 8.38) which suits the expectations from parallelization.

The required time to find a tour for instance `fl3795` that is 0.5% above the optimum the single node variant requires 337 CPU seconds, whereas the parallel variant requires 78 CPU seconds. Here, the speed-up factor is only about 4 for using 8 nodes. The speed-up factor gets better the closer the tour qualities get to the optimum. For a quality level of 0.25% above the optimum, the required CPU seconds are 1153 versus 200 (factor 5.77). To reach the optimal solution, the single node variant requires 4224 CPU seconds on average.

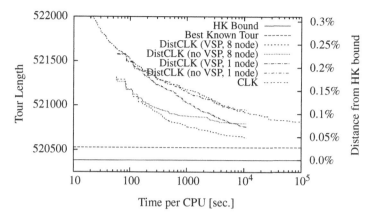

Figure 15.8. Effects of parallelization running the distributed algorithms on a different number of nodes and optional variable strength perturbation (VSP) for instance fi10639.

Having again a good speed-up factor of 7.42, the parallel variant requires 569 seconds per node.

As for instance fi10639 (Figure 15.8) no optimal solution is known, the Held-Karp bound was used to measure tour qualities. The first distance level of 0.12% above the Held-Karp bound for this instance was reached after 1183 CPU seconds in the one node variant, compared to the eight node variant requiring 189 seconds. This is a speed-up of 6.27, which is improved subsequently. The tour quality of 0.10% is reached on average after 2672 seconds versus 351 seconds (speed-up factor 7.62). Finally, the quality level of 0.08% required a computation time of 6961 seconds for the single node variant and 723 seconds for the parallel variant resulting in a speed-up factor of 9.63. As shown parallelization works for this distributed algorithm when comparing single versus multiple node variants. Especially for larger instances with long running times the speed-up factor may be optimal regarding used CPU time.

The variable strength perturbation (VSP) in the distributed algorithm may be seen redundant, as the Chained Lin-Kernighan algorithm already applies DBMs on the tours. However, as shown here with instances pr2392 (Figure 15.7, not discussed below) and fi10639 (Figure 15.8), this VSP actually improves the results of the distributed algorithm. In a setup where the distributed algorithm runs on only one node and no VSP gets applied, this algorithm obtains the same performance as the original CLK algorithm. In Figure 15.8 the performance of both simulations is represented by the two lines labeled with "DistCLK (no VSP, 1 node)" and "CLK". For comparison, the distributed algorithm was executed with VSP (labeled "DistCLK (VSP, 1 node)") in a third setup. Right from the start, this setup performs clearly better than both the original CLK and the distributed algorithm without VSP. After 10^4 CPU seconds the third setup is only 0.073% away from the Held-Karp lower bound.

Instance	Distance	Helsgaun LK LKH	DistCLK
C1k.1	0.12%	8.89	< 944.43
E1k.1	0.08%	9.78	< 9059.94
pr2392	0.24%	34.87	< 205.37
f13795	6.73%	74.06	< 914.73
fnl4461	0.07%	129.23	978.12
usa13509	0.21%	1133.81	< 2272.18
pla33810	0.96%	7982.09	< 2785.89
pla85900	1.25%	48 173.84	< 9350.55

Instance	Distance	Johnson&McGeoch ILK ILK-JM	DistCLK
C1k.1	0.00%	1292.40	944.43
E1k.1	0.05%	65.14	< 9059.94
pr2392	0.05%	220.54	681.95
f13795	0.00%	20 597.78	16 402.12
fnl4461	0.11%	722.42	674.93
usa13509	0.11%	8640.36	5418.11
pla33810	0.68%	47 599.30	10 662.38
pla85900			

Instance	Best of 10 runs Distance	DistCLK
C1k.1	0.000%	198.29
E1k.1	0.000%	65.07
pr2392	0.000%	575.90
f13795	0.000%	4283.36
fnl4461	0.000%	14 536.58
usa13509	0.008%	179 213.99
pla33810	0.561%	171 839.09
pla85900	0.468%	189 023.53

Table 15.5. Normalized computation time compared with other algorithms. "Distance" is the distance to the optimum or Held-Karp Lower Bound (for instances pla33810 and pla85900) as listed for the corresponding instance in the DIMACS challenge [Johnson,]. The two columns next to the distance are the CPU times for the two algorithms mention in the columns' header. For cells marked with "<", the distributed algorithm's intermediate results included only tours of better quality, so the value given is the point of time when an average value was available for the first time.

In contrast, the distributed algorithm without VSP is 0.110% above this bound and the original CLK 0.106% above. The variant with VSP required only about 1700 CPU seconds to reach this quality level. Finally, the distributed algorithm was run with two 8 node setups with VSP (labeled "DistCLK (VSP, 8 nodes)") and without a VSP (labeled "DistCLK (no VSP, 8 nodes)"). Soon it turns out that the VSP variant performs better in approximating the optimal solution. The final tour lengths after 10^4 CPU seconds per node are 0.080% for the variant without VSP and 0.050% above the Held-Karp bound for the VSP variant. The latter required only 753.4 CPU seconds per node on average to reach the first variant's final tour quality.

For comparison with other TSP solvers, the running times of selected instances have been normalized to a 500 MHz Alpha processor as standardized for the 8th DIMACS Implementation Challenge for the TSP [Johnson and Mc-Geoch, 2002, Johnson,], which generated by running a greedy algorithm on a testbed of random Euclidean instances. The normalization factor is calculated by comparing running times for the testbed instances to the known values for the Alpha machine. The computational data for the following presentation and comparison of other TSP solvers has been taken from the same source. Different distance values are due to the fact that for each algorithm different pairs of tour quality and CPU times were available.

Helsgaun's LK LKH by Helsgaun [Helsgaun, 2000] is a LK-based algorithm modifying the original LK algorithm. LKH uses a sequential 5-exchange step operating on neighborhood restricted on 5 members and is based on a α-nearness. The α-values are calculated by using one-trees on a modified weight matrix (π-values). Johnson and McGeoch report [Johnson and McGeoch, 2002] that LKH finds better tours than their own LK implementation (LK-JM) for most instances in their testbed, but LKH requires significantly more time to reach these tour qualities.

Johnson & McGeoch's ILK In their comparison of ILK algorithms in [Johnson and McGeoch, 2002] the authors use their own algorithm [Johnson and McGeoch, 1997] as reference. Here, the data of a variant with $10N$ iterations, 20 quadrant neighbors, don't-look-bits and maximum depth of 50 is compared to the results of the distributed algorithm. This variant from the DIMACS challenge [Johnson,] is the one with the longest running time and the best tour qualities over all ILK variants by Johnson and McGeoch.

For most of the compared instances, the distributed algorithm has on average better tours compared to the final tour quality levels of Helsgaun's LK (LKH) already after the first iteration, which is due to the underlying CLK algorithm. The current tour quality of a distributed algorithm is the best single node tour quality within the network. But for this initial tour quality, the distributed algorithm requires significantly more time than LKH to reach its final tour quality level for smaller instances (up to usa13509). For the two largest instances, however, less time is required. The ratio between the computation times for both algorithms shifts towards the distributed algorithm for increasing instances size: It grows from 0.13 for instance fnl4461 and 0.50 for usa13509 to 2.87 (instance pla33810) and 4.46 for instance pla85900.

Compared to Johnson & McGeoch's Iterated LK the distributed algorithm performs better for most instances. Except for instances pr2392 and E1k.1 the distributed algorithm requires significantly less time, up to the factor of 4.5 for instance pla33810.

Finally, the last block of Table 15.5 contains the distributed algorithm's best results out of 10 runs and the normalized CPU time until the first occurrence of this result.

5.   Conclusion

The proposed distributed algorithm improves the quality and performance of the original CLK algorithm in different ways. By exchanging tours between nodes, nodes with worse tours can leave their neighborhood to enter more promising areas of the search space. To increase the effectiveness of the distributed algorithm even further, a perturbation move with variable strength was introduced. Our approach therefore converges faster towards good solutions and finds better tours within a time bound summed over all nodes compared to the original CLK algorithm. In our experiments, the distributed algorithm finds optimal solutions for instances pr2392 and fl3795, where the plain CLK algorithm fails to find optimal solutions. The comparison with other heuristic TSP solvers indicates that the distributed variant is best suited for large instances. Due to the fact that 8 machines were running in parallel the *absolute* time to find a good solution makes the distributed algorithm competitive to existing heuristics for real-world applications.

There are different aspects of the distributed algorithm that are subject for improvements. A peer-to-peer (P2P) network will be used for communication between nodes. Here, nodes will be organized in a Chord-like ring using an Epidemic Algorithm for information propagation (see [Merz and Gorunova, 2005]). The CLK implementation will be replaced by a pure Java implementation allowing both better customization and portability in heterogeneous systems. The Evolutionary Algorithm will be substituted by a more sophisticated variant integrating recombination of different tours and supporting larger populations. Large TSP problems (10000 cities and more) will be processed by applying various problem reduction operators.

References

[Applegate et al., 1995] Applegate, David, Bixby, Robert, Chvátal, Vašek, and Cook, William (1995). Finding Cuts in the TSP (a Preliminary Report). Technical report, Rutgers University, Piscataway NJ.

[Applegate et al., 1999] Applegate, David, Bixby, Robert, Chvátal, Vašek, and Cook, William (1999). Finding tours in the TSP. Technical Report 99885, Forschungsinstitut für Diskrete Mathematik, Universit Bonn.

[Arora, 1998] Arora, Sanjeev (1998). Polynomial Time Approximation Schemes for Euclidean TSP and other Geometric Problems. *Journal of the ACM*, 45(5):753–782.

[Baraglia et al., 2001] Baraglia, Ranieri, Hidalgo, José Ignacio, and Perego, Raffaele (2001). A Parallel Hybrid Heuristic for the TSP. In *Proceedings of EvoCOP2001, the First European Workshop on Evolutionary Computation in Combinatorial Optimization*, pages 193–202.

[Christofides, 1976] Christofides, N. (1976). Worst-case analysis of a new heuristic for the traveling salesman problem. In Traub, J. F., editor, *Symposium on New Directions and Recent Results in Algorithms and Complexity*, page 441, Orlando, Florida. Academic Press.

[Cook, 1971] Cook, Stephen A. (1971). The complexity of theorem-proving procedures. In *Proceedings of the third annual ACM symposium on Theory of computing*, pages 151–158. ACM Press.

[Cook, a] Cook, William. Concorde TSP Solver. See: http://www.tsp.gatech.edu/concorde.html.

[Cook, b] Cook, William. National Traveling Salesman Problems. http://www.tsp.gatech.edu/world/countries.html.

[Dantzig et al., 1954] Dantzig, George, Fulkerson, Ray, and Johnson, Selmer (1954). Solution of a large-scale traveling-salesman problem. *Operations Research*, 2:393–410.

[Flood, 1956] Flood, M. M. (1956). The Travelling Salesman Problem. *Operations Research*, 4:61–75.

[Garey and Johnson, 1979] Garey, Michael R. and Johnson, David S. (1979). *Computers and Intractability: A Guide to the Theory of NP-Completeness.* W. H. Freeman, New York.

[Grosso, 1985] Grosso, P. B. (1985). *Computer Simulation of Genetic Adaptation: Parallel Subcomponent Interaction in a Multilocus Model.* PhD thesis, University of Michigan.

[Harik et al., 1999] Harik, G., Lobo, F., and Goldberg, D. (1999). The compact genetic algorithm. *IEEE Transactions on Evolutionary Computation*, 3:287–297.

[Held and Karp, 1970] Held, M. and Karp, Richard M. (1970). The Travelling Salesman Problem and Minimum Spanning Trees. *Operations Research*, 18:1138–1162.

[Held and Karp, 1971] Held, M. and Karp, Richard M. (1971). The Travelling Salesman Problem and Minimum Spanning Trees: Part II. *Mathematical Programming*, 1(1):6–25.

[Helsgaun, 2000] Helsgaun, Keld (2000). An effective implementation of the Lin-Kernighan traveling salesman heuristic. *European Journal of Operational Research*, 126(1):106–130.

[Hung, 2004] Hung, Nguyen Dinh (2004). *Hybrid Genetic Algorithms for Combinatorial Optimization Problems.* PhD thesis, Department of Information Systems Engineering, Graduate School of Engineering, University of Miyazaki, Miyazaki, Japan.

[Johnson,] Johnson, David S. 8th DIMACS Implementation Challenge: The Traveling Salesman Problem. http://www.research.att.com/~dsj/chtsp/.

[Johnson and McGeoch, 1997] Johnson, David S. and McGeoch, Lyle A. (1997). *Local Search in Combinatorial Optimization*, chapter The Traveling Salesman Problem: A Case Study in Local Optimization, pages 215–310. John Wiley and Sons, Ltd.

[Johnson and McGeoch, 2002] Johnson, David S. and McGeoch, Lyle A. (2002). Experimental Analysis of Heuristics for the STSP. In Gutin and Punnen, editors, *The Traveling Salesman Problem and its Variations*. Kluwer Academic Publishers.

[Lawler and Wood, 1966] Lawler, E. L. and Wood, D. E. (1966). Branch-and-Bound Methods: A Survey. *Operations Research*, 14(4):699–719.

[Lin, 1965] Lin, S. (1965). Computer Solutions Of The Traveling Salesman Problem. *Bell System Technical Journal*, 44:2245–2269.

[Lin and Kernighan, 1973] Lin, S. and Kernighan, B. W. (1973). An effective heuristic algorithm for the traveling salesman problem. *Operations Research*, 21:498–516.

[Martin et al., 1991] Martin, Olivier, Otto, Steve W., and Felten, Edward W. (1991). Large-Step Markov Chains for the Traveling Salesman Problem. *Complex Systems*, 5:299–326.

[Merz and Freisleben, 2001] Merz, Peter and Freisleben, Bernd (2001). Memetic Algorithms for the Traveling Salesman Problem. *Complex Systems*, 13(4):297–345.

[Merz and Gorunova, 2005] Merz, Peter and Gorunova, Katja (2005). Efficient Broadcast in P2P Grids. In *Proceedings of the IEEE/ACM International Symposium on Cluster Computing and the Grid (CCGrid 2005)*, Cardiff, UK.

[Reinelt, 1991] Reinelt, Gerhard (1991). TSPLIB – A Traveling Salesman Problem Library. *ORSA Journal On Computing*, 3:376–384.

[Walshaw, 2000] Walshaw, Chris (2000). A Multilevel Approach to the Travelling Salesman Problem. Mathematics Research Report 00/1M/63, Computing and Mathematical Sciences, University of Greenwich, London SE10 9LS, UK.

Chapter 16

EXPLORING GRID IMPLEMENTATIONS OF PARALLEL COOPERATIVE METAHEURISTICS

A Case Study for the Mirrored Traveling Tournament Problem

Aletéia P.F. Araújo[1,3], Cristina Boeres[2], Vinod E.F. Rebello[2], Celso C. Ribeiro[1,2], and Sebastián Urrutia[1]

[1] *Department of Computer Science, Catholic University of Rio de Janeiro,*
Rio de Janeiro, Brazil
{aleteia, celso, useba}@inf.puc-rio.br

[2] *Instituto de Computação, Universidade Federal Fluminense,*
Niterói, Brazil
{boeres, vinod, celso}@ic.uff.br

[3] *Department of Computer Science, Catholic University of Brasília,*
Brasília, Brazil

Abstract   Metaheuristics are general high-level procedures that coordinate simple heuristics and rules to find good approximate solutions to computationally difficult combinatorial optimization problems. Parallel implementations of metaheuristics appear quite naturally as an effective approach to speedup the search for approximate solutions. Besides the accelerations obtained, parallelization also allows solving larger problems or finding better solutions. We present in this work four slightly differing strategies for the parallelization of an extended GRASP with ILS heuristic for the mirrored traveling tournament problem, with the objective of harnessing the benefits of grid computing. Computational experiments on a dedicated cluster illustrate the effectiveness and the scalability of the proposed strategies. In particular, we show that the parallel strategy implementing cooperation through a pool of elite solutions scales better than the others and is able to find solutions that cannot be reached by the others. Computational grids are distributed high latency environments which offer significantly more computing power than traditional clusters. The best parallel strategy was also implemented and tested using a true grid platform. We report original results from pioneer computational experiments on a shared computational grid formed by 82 machines distributed over four clusters in three cities, illustrating the potential of the application of computational grids in the fields of metaheuristics and combinatorial optimization.

Keywords: Parallel metaheuristics, grid computing, computational grids, traveling tournament problem, GRASP, iterated local search

1.   Introduction

The organization and management of sporting events and championships is a worldwide multi-billion dollar industry. Schedules with minimum traveling times and offering similar costs and conditions to all teams taking part in a competition are of major interest to teams, leagues, sponsors, fans, and the media. In the case of the Brazilian national soccer championship, a single trip from Porto Alegre to Belém takes almost a full day's journey, with numerous connections due to the absence of direct flights, to cover a distance of approximately 4,000 kilometers. The total distance traveled becomes a key issue to be minimized, so as to reduce costs and to give the players more time to train and time off along the season that lasts for approximately eight months.

Several authors in different contexts (see e.g. [4, 5, 8, 20, 27, 35–37, 40, 41]) have tackled the problem of tournament scheduling for a variety of leagues and sports including soccer, basketball, hockey, baseball, rugby and cricket, using different techniques such as integer programming, tabu search, genetic algorithms, simulated annealing, and constraint programming.

The Traveling Tournament Problem is an inter-mural championship timetabling problem that abstracts certain characteristics of scheduling problems in sports [11]. It combines tight feasibility constraints with a difficult objective function to be optimized. The objective is to minimize the total distance traveled by the teams, subject to the constraint that no team can play more than three consecutive games at home or away. Since the total distance traveled is a major issue for every team taking part in the tournament, solving a traveling tournament problem may be a starting point for the solution of real timetabling applications in sports, in general.

Metaheuristics are general high-level procedures that coordinate simple heuristics and rules to find good approximate (often optimal)

solutions to computationally difficult combinatorial optimization problems. Among them, we find simulated annealing, tabu search, Greedy Randomized Adaptive Search Procedure (GRASP), genetic algorithms, scatter search, Variable Neighborhood Search, ant colonies, and others. They are based on distinct paradigms and offer different mechanisms to escape from locally optimal solutions. Metaheuristics are among the most effective strategies for solving hard combinatorial optimization problems. The customization (or instantiation) of a metaheuristic to a given problem yields a heuristic for that problem.

Recent years have witnessed huge advances in computer technology and communication networks. Cung et al. [9] noted that parallel implementations of metaheuristics not only appear as quite natural alternatives to speed up the search for good approximate solutions, but also facilitate solving larger problems and finding improved solutions, with respect to their sequential counterparts, due to the partitioning of the search space and to the increased possibilities for search intensification and diversification. As a consequence, parallelism can improve the effectiveness and robustness of metaheuristic-based algorithms. The latter are less dependent on sophisticated parameter tuning and their success is not limited to a few or small classes of problems.

The growing computational power requirements of large scale applications and the high costs of developing and maintaining supercomputers has fuelled the drive for cheaper high performance computing environments. With the considerable increase in commodity computers and network performance, cluster computing and, more recently, grid computing [15, 16] have emerged as a real alternatives to traditional supercomputing environments for executing parallel applications that require significant amounts of computing power.

A computing cluster generally consists of a fixed number of homogeneous resources, interconnected on a single administrative network, which together execute one parallel application at a time. Grids in some sense are just the opposite, aiming to harness sufficient computing power from a diverse pool of resources, available on the internet, to execute a number of applications simultaneously. Grids aggregate geographically distributed collections (or sites) of resources which typically belong to different owners and thus are shared between multiple users. Each of these sites could consist of one or more uni-processor machines, a symmetric multiprocessor cluster, a distributed memory multicomputer system, or a massively parallel supercomputer. Clearly, the physical nature of the resources and the computing power available are both heterogeneous. Unlike local area network environments, grids are more susceptible to resource and network failures. Additionally, since the resources and network are being shared, the computational power available and communication costs fluctuate. These issues require careful consideration when developing grid enabled applications. The fact that these resources are distributed, heterogeneous and non-dedicated, make writing parallel grid-aware applications much more challenging [14]. While in theory optimization problems should easily benefit from grid computing, in practice appropriate design, careful tuning and thorough reevaluation of parallel implementations are necessary. Most of all, this requires a thorough understanding of how metaheuristics behave in such environments.

This work aims to investigate the practical benefits that large scale parallel processing can bring to metaheuristics for combinatorial optimization problems. In particular, this paper describes four simple but efficient strategies for the parallelization in grid environments to improve the extended GRASP with ILS heuristic for the mirrored traveling tournament problem proposed in [34]. The sequential strategy substitutes the local search phase of a GRASP heuristic by an ILS procedure, obtaining high-quality solutions that are among the best known in the literature for benchmark instances of this problem [38].

The remainder of the paper is organized as follows. The following section reviews the formulation of the mirrored traveling tournament problem. Section 3 summarizes the extended GRASP with ILS sequential heuristic. In Section 4, some important issues concerning the parallel implementation of metaheuristics are reviewed. Section 5 describes the four parallel implementations for the mirrored traveling tournament problem. Section 6 presents and compares experimental results obtained with the proposed strategies. Results on a computational grid employing 82 resources from sites in three different cities are reported in Section 7. Concluding remarks are made in the last section.

2.   The mirrored traveling tournament problem

We consider a tournament played by n teams, where n is an even umber. In a *simple round-robin* (SRR) tournament, each team plays every other exactly once in $n-1$ prescheduled rounds. In a *double round-robin* (DRR) tournament, each team plays every other twice, once at home and once away. A *mirrored double round-robin* (MDRR) tournament is a simple round-robin tournament in the first $n-1$ rounds, followed by the same tournament with reversed venues in the last $n-1$ rounds. We assume that each team in the tournament has a stadium in its home city and that the distances between the home cities are known. Each team is located at its home city at the beginning of the tournament, to where it returns at the end after playing the last away game. Whenever a team plays two consecutive away games, it goes directly from the city of the first opponent to the other, without returning to its own home city.

The Traveling Tournament Problem (TTP) was first established by Easton et al. [11]. Given n teams and the distances between their home cities, the TTP consists in finding a DRR tournament such that every team does not play more than three consecutive home or away games, no repeaters (i.e., two consecutive games between the same two teams at different venues) occur, and the sum of the distances traveled by the teams is minimized. Benchmark instances are available in [38]. To date, even small benchmark instances of the TTP with $n = 10$ teams cannot be solved exactly. The largest instance for which the optimal solution is known ($n = 8$ teams) took four days of processing time using twenty processors in parallel [10]. We also refer to this problem as the non-mirrored TTP, for which both mirrored and non-mirrored solutions are feasible.

The mirrored Traveling Tournament Problem (mTTP) has an additional constraint: the games played in round k are exactly the same played in round $k + (n - 1)$ for $k = 1, \ldots, n - 1$, with reversed venues. Repeaters do not occur in mirrored schedules. Mirrored tournaments are a common tournament structure in Latin America.

The TTP has raised significant interest in recent years especially after challenge instances were proposed in [38]. Easton, Nemhauser and Trick [11] applied an Integer Linear Programming approach to the TTP. They modified their three-phase approach [27] previously used to schedule the Atlantic Coast Conference (ACC) basketball league, generating new high-quality solutions. Some of the results were later improved upon by Benoist et al. through the combination of Lagrangian relaxation and constraint programming [6]. Anagnostopoulos et al. [4] proposed a simulated annealing algorithm for the TTP to explore a large neighborhood with complex moves. The algorithm included a strategic oscillation technique and applied re-heat to both balance the exploration of feasible and infeasible regions and to escape from local minima at very low temperatures. Lim et al. [22] used a two-stage approach consisting of simulated annealing and hill-climbing techniques were able to improve even further some of the results obtained by [4]. Recently, Rasmussen and Trick [28] also considered Benders decomposition approaches to the TTP. Gaspero and Schaerf [17] explored a composite-neighborhood tabu search approach. It is important to point out that these works provide solutions for the non-mirrored Traveling Tournament Problem, while our work focus on the mirrored version. As far as we know, there is no parallel

metaheuristic solutions for the mTTP (nor TTP), which is the focus of
the work proposed here.

3.   Extended GRASP with ILS heuristic

The GRASP (Greedy Randomized Adaptive Search Procedure) metaheuristic [29] is a multi-start or iterative process, in which each iteration
consists of two phases: construction and local search. The construction phase builds a feasible solution, whose neighborhood is investigated
during the local search phase until a local minimum is found. The best
overall solution is kept as the result.

The construction and local search phases are problem-dependent and
should be customized for each problem. GRASP has experienced continued development and has been applied in a wide range of areas [13].
Resende and Ribeiro [29, 30] described successful implementation techniques and parameter tuning strategies, as well as enhancements, extensions, and hybridizations of the original algorithms.

The ILS (Iterated Local Search) metaheuristic [23] starts from a locally optimal feasible solution. A random perturbation is applied to the
current solution, which is then followed by a local search. If the local
optimum obtained after these steps satisfies some acceptance criterion,
then it is accepted as the new current solution, otherwise the latter does
not change. The best solution is, if necessary, updated and the above
steps are repeated until some stopping criterion is met.

A hybridization of the GRASP and ILS metaheuristics into an effective hybrid heuristic for the mTTP was proposed in [34]. Basically, the
authors substituted the local search phase of GRASP by an ILS procedure. The pseudo-code in Algorithm 16.1 summarizes the main steps
of the GRILS-mTTP heuristic for finding approximate solutions for the
mirrored traveling tournament problem.

The outer **while** loop in Algorithm 16.1 executes a GRASP *construction phase* followed by an ILS *local search phase*, until a stopping criterion is met. Typically, the algorithm continues executing until a solution
is found with a cost that is as good as or better than a given *target* value,
or until a given period of time has elapsed.

During the GRASP phase of each iteration, an initial solution S is
constructed to which a local search algorithm is then applied, returning
a new current solution S. This solution is also used to initialize the best
solution $\underline{S}$ in the current iteration.

The ILS phase of the iteration is the inner **repeat** loop which applies a perturbation to the current solution S obtaining a new solution
S'. A local search algorithm is applied to S', where four neighborhood

---

Procedure GRILS-mTTP();
1.   **while** .NOT.*StoppingCriterion* **do**
2. $S \leftarrow$ BuildGreedyRandomizedSolution();
3. $\underline{S}, S \leftarrow$ LocalSearch(S);
4.   **repeat**
5. $S' \leftarrow$ Perturbation(S);
6. $S' \leftarrow$ LocalSearch(S');
7. $S \leftarrow$ AcceptanceCriterion(S, S');
8. $S^* \leftarrow$ UpdateGlobalBestSolution(S, S^*);
9. $\underline{S} \leftarrow$ UpdateIterationBestSolution($S, \underline{S}$);
10.   **until** *ReinitializationCriterion*;
11. **end**;
12. **return** S^*;

Figure 16.1. Pseudo-code of the GRASP with ILS heuristic for the mTTP.

structures are used. The first three are simple exchanges in which TS (team swap), HAS (home-away swap) and PRS (partial round swap) neighborhoods are explored by local searches. The GR (game rotation) ejection chain neighborhood, explored only as a diversification move, is performed less frequently by the heuristic as a perturbation [34].

A first-improving strategy similar to the VND (Variable Neighborhood Descent) procedure [19] was used to implement the local search algorithm. Once a local optimum with respect to the TS neighborhood is found, a quick local search using the HAS neighborhood is performed. Next, the PRS neighborhood is investigated, followed again by a local search using the HAS neighborhood. This scheme is repeated until a local optimum with respect to these three neighborhoods is found.

In this context, the new solution S' is accepted or not as the new current solution, depending on an acceptance criterion. The best overall solution S^* and the best solution in the current GRASP iteration are updated, if necessary, and a new cycle starts with the perturbation of the current solution, until a re-initialization criterion is met.

A new GRASP iteration starts if 50 consecutive deteriorating moves to neighbor solutions have been accepted since the last time $\underline{S}$ (the best solution found in this GRASP iteration) was updated. Re-initialization occurs if too many perturbations followed by local search are performed without improving the best solution in the current GRASP iteration. It is important to notice that a GRASP iteration is not interrupted if the current solution S is still being improved.

The parallelization of this algorithm does not only aim to reduce the total running time, but also to improve its effectiveness and robustness. The use of several processors concurrently to explore different search trajectories, as described in Section 5, may lead to a more thorough investigation of the neighborhoods.

4.   Parallel implementation of metaheuristics

Programming paradigms commonly used to develop low communication parallel programs on distributed clusters include the *master-slave* (also often referred to as task farming) and the *client-server* models [14]. These approaches are especially attractive, since they can generally be applied to take advantage of all available resources in a grid environment.

Cung et al. [9] reviewed some major issues on parallel implementations of metaheuristics, such as the types of parallelism as well as appropriate parallel programming models and parallelization strategies for this class of heuristics. With respect to parallelization strategies [9, 39], two main approaches are used: single-walk and multiple-walk. Each iteration of a metaheuristic generally starts with the construction of an initial solution, followed by a search to improve the solution. New neighboring solutions are evaluated by making a series of minor alterations to a given solution. The sequence of solutions evaluated is known as a *walk* or *trajectory*. In the case of a single-walk parallelization, one unique search trajectory is traversed in the solution space and the search for the best neighbor at each iteration is performed in parallel. The neighborhood search is performed faster in parallel, but the search trajectory is the same as the one followed in the corresponding sequential implementation. On the other hand, a multiple-walk parallelization strategy is characterized by the investigation in parallel of multiple trajectories, each of them performed by a different processor. A search "thread" is a process running in each processor traversing a walk in the solution space. These processes can be either independent (where no information is exchanged among processes) or cooperative (the information collected along a trajectory is disseminated and used by other processes to improve or to speed up the search).

Cooperative strategies are the most general and promising, but often incur in additional costs in terms of communication and storage. However, if cooperation is well explored and implemented, it can globally lead to better solutions in smaller computation times even if each individual iteration may take longer, see e.g. [32].

Developing and tuning efficient parallel implementations of metaheuristics require a thorough programming effort and keen implementation skills. In the context of grid computing, where communication is at a premium, one of the most difficult aspects to be determined is the nature of the information to be shared, in order to improve the search without taking too much additional memory or time to be collected, as well as the frequency at which this information is exchanged. The information shared by the search threads can be implemented either as global variables stored in a shared memory, or as a pool in the local memory of a dedicated central processor. In the case of the latter, information is exchanged with the other processors via message passing.

5.   Parallel strategies for the extended GRASP with ILS heuristic

This section presents four simple, but efficient, strategies for the parallelization of the best known algorithm (the hybrid heuristic GRILS-mTTP [34] summarized in Section 3) for solving the mTTP. Besides obtaining speedups in execution times, improvements in solution quality are also sought. All four versions are based on the master-worker programming paradigm and adopt a multiple-walk search strategy. This work aims to investigate how degrees of cooperation and increased diversity (in terms of number of trajectories investigated and the amount of information being shared) affect the GRILS-mTTP heuristic.

Initially, the master process generates and distributes distinct seeds to be used by the pseudo-random number generator of each worker process. As the number of workers increases, this will foster greater diversity. In order to reduce the chance that processes search the same neighborhood (i.e., evaluate the same solutions), each process uses a different sequence of pseudo-random numbers. The Mersenne Twister random number generator of Matsumoto and Nishimura [24] was chosen based on the recommendation in [33].

5.1   Parallel strategy with independent processes

This version, denoted by *PAR-I*, is representative of executing the sequential algorithm simultaneously on multiple machines independently of each other (e.g. as a parameter sweep application).

After receiving their seeds, each worker starts a cycle in which it generates a new solution during a GRASP construction phase and then executes an ILS local search phase until the re-initialization criterion is met. This cycle is repeated until a solution with a cost equal to or better than a given target value (used as a stopping criterion) is found. Although no communication takes place between the independent searches, once the stopping criterion has been met, a controller process (master) receives and records the solution found and responds by broadcasting a halt message to each worker to terminate their execution.

5.2   Parallel strategy with one-off cooperation

This version, *PAR-O*, is identical to the previous one, with the exception of the first iteration of the main loop. After each worker executes the GRASP phase, the best initial solution encountered by each of them is sent to the master, which in turn selects and broadcasts back to all the workers the best overall solution. Therefore, all workers will execute the ILS local search phase of the first iteration using the same initial solution. The following iterations are executed independently. This is called one-off cooperation because this exchange only occurs during the first iteration.

5.3   Parallel strategy with one elite solution

One of the possible shortcomings of the previous versions is the lack of continuous cooperation between the workers during their execution, i.e., each worker process does not learn from searches carried out in parallel (or from solutions found) in previous iterations by other workers. In the earlier strategies, the current best solution is not available to all workers. Information gathered from good solutions should be used to implement more effective strategies [12, 31]. Typically, in these history-based parallel cooperative strategies, the master manages the exchange of information collected along the trajectories investigated by each worker.

In this version, PAR-1P, the master keeps the best (or elite) solution received from any worker. Each time the best solution is improved, the master broadcasts the solution's cost to all workers to avoid unnecessary communication from them. The intuition is to use this information not only to converge faster to a target solution, but also to find better solutions than the independent search strategies.

In PAR-1P, there is no one-off cooperation during the first iteration. Instead, each time a worker completes the ILS local search phase, it will compare the cost of the solution found with that of the best solution held by the master. If it is lower, the worker sends its solution to the master, otherwise the solution is discarded. After this, two outcomes are possible. Either, the worker requests the best solution held by the master to repeat the ILS local search phase with this solution, or the worker continues with the next iteration (i.e. re-initialization causes a new initial solution during the GRASP construction phase to be created and proceeds with the next steps of the sequential heuristic) as in the previous versions.

The probability of each outcome is denoted by Q and $1 - Q$, respectively. Q was fixed at 10% in all experiments reported in Section 6. In this way, workers indirectly exchange elite solutions found along their search trajectories. This parallel cooperative strategy promotes a more thorough search of the space around good solutions, characteristic of single-walk parallelization approaches.

5.4 Parallel strategy with a pool of elite solutions

In this cooperative strategy, PAR-MP, the master is dedicated to managing a centralized pool of elite solutions (and their costs), including collecting and distributing them upon request. As in the previous version, workers start their searches from different initial solutions and can exchange and share elite solutions found along their search trajectories.

In PAR-MP, the master will update the elite solution pool with a newly received solution according to given criteria which are based on the quality of the solutions already in the pool (as described below). When a worker completes an iteration, it can either request an elite solution from the pool or construct a new initial solution randomly, again with probabilities of Q and $1 - Q$, respectively. Q was fixed at 10% in all experiments reported in Section 6.

The goal to be achieved by cooperation through the pool of elite solutions is to exchange meaningful information in a timely manner, so as that the parallel search finds better solutions than the simple concatenation of the results of the individual methods. Developments in metaheuristics have proved to be particularly successful when their basic concepts are combined with cooperative methods. These hybrid cooperative approaches maintain a reference set of high quality solutions which are repeatedly used during the search to guarantee a fruitful balance between diversification and intensification [18].

Pool management A very important aspect of this strategy is the management of the pool of elite solutions. However, maintaining such a pool is not trivial. The main issue consists in finding a balance between the attempt to collect a number of high quality solutions, which often share similar properties, and the need to guarantee a certain degree of diversity in the pool. Empirically, previous research (see e.g. [12])

observed that history-based heuristics are less likely to be successful if the recorded solutions are very similar.

The pool consists of a limited number M of positions, which are initialized with null solutions. The pool manager supports two essential operations: the insertion of a new solution into its appropriate position in the pool and the selection of a solution from the pool from which a worker will initiate a new search.

To guarantee the diversity within the pool, the insertion of a new solution depends on the state of the pool and on how the solution was generated. When the new candidate solution has been derived from an elite solution in the pool, the cost of the new solution must be better than the cost of the elite solution from which it was generated. If true, the new solution will obligatorily take the place of that elite solution. On the other hand, if the solution was derived from a solution produced by the GRASP construction phase, the solution can be inserted directly into any vacant position. In the case where the pool is full, the solution is inserted only if it is as good as the worst elite solution already in the pool (thus replacing the latter).

When a worker process requests an elite solution from the master, a solution is selected at random from the pool and sent back to worker process.

6.   Experimental results

The four parallel algorithms PAR-I, PAR-O, PAR-1P, and PAR-MP, described in Section 5, were implemented using C++ and version 7.0.6 of the LAM grid-enabled implementation [21] of the message passing interface standard MPI [25]. For evaluation purposes, the experiments reported in this section were executed in isolation on a dedicated cluster (of 1.7 GHz Pentium 4 processors, each of which with 256 Mbytes of RAM memory, interconnected by a Fast Ethernet network) to avoid external influences in performance. Each processor has a local copy of the executable code and the problem data.

Two sets of benchmark instances have been proposed for the traveling tournament problem[11]. The first is made up of *circle instances*, artificially generated to represent easier instances. The name circ$n$ is used to denote a circle instance with $4 \leq n \leq 20$ teams. Each circle instance is built from a graph, generated as follows. Nodes are placed at equal unit distances along a circumference and labeled $0, 1, \ldots, n - 1$. There are edges only between nodes i and $i + 1 \bmod n$, for $i = 0, 1, \ldots, n - 1$. In the corresponding circle instance, the distance between the home cities of teams i and j (with $i > j$) is given by the length of the shortest path between them in the graph and is equal to the smaller of $i - j$ and $j - i + n$. The second set are realistic instances generated using the distances between the home cities of a subset of teams playing in the National League of the MLB (Major League Baseball) in the United States. These national league instances are denoted by nl$n$, with $4 \leq n \leq 16$. We did not consider the smaller instances with $n = 4$ and $n = 6$, for which optimal solutions have already been found. Furthermore, an additional real-life instance has been created by Ribeiro and Urrutia [34], named br24. This instance is made up of the home cities of the 24 teams playing in the first division of the 2003 edition of the Brazilian soccer championship. All instances and their best known solutions are available from [38].

The experiments aim to investigate how parallel computing can be used to harness cooperation and diversity, improving solution quality and convergence when executing the GRILS-mTTP heuristic in distributed computing environments. The parameter M was set to P, where P is the number of worker processes used in the parallel execution. The probability, Q, of choosing a solution from the pool was fixed at 10%.

Table 16.1 displays, for each instance, the cost of the best known solution at the time of writing obtained by the sequential implementation of the GRILS-mTTP heuristic after five days of processing time [38]. These are compared with the cost of the best solutions found during the following experiments by the four parallel implementations of the GRILS-mTTP heuristic. The execution time required varies with the number of processors used and these details are described in the following experiments. Notice that PAR-I, PAR-O, and PAR-1P found the same cost solutions for each of the benchmark instances, while the PAR-MP implementation was able to improve the best solution found by the three others in the case of three instances. The last column gives the relative improvement obtained by PAR-MP over the cost of the best known sequential solution.

In the experiments reported next, the costs of the best solutions found by the sequential heuristic, as reported in Table 16.1, are referred to as the **easy** targets. The costs of the solutions obtained by the PAR-I, PAR-O, and PAR-1P implementations are referred to as the **medium** targets, for the instances for which the best solution obtained by these versions improved the easy targets (instances circ10, circ16, circ18, nl16, and br24). The PAR-MP implementation further improved the best known solutions for three of these instances, and these best solution costs are referred to as the **hard** targets (for instances circ16, circ18, and nl16).

Table 16.1. Solutions found by the sequential and parallel implementations.

Instance	Sequential	PAR-I/O/1P	PAR-MP	Improvement (%)
circ8	140	140	140	-
circ10	276	272	272	1.45
circ12	456	456	456	-
circ14	714	714	714	-
circ16	1004	984	978	2.59
circ18	1364	1308	1306	4.25
circ20	1882	1882	1882	-
nl8	41928	41928	41928	-
nl10	63832	63832	63832	-
nl12	120655	120655	120655	-
nl14	208086	208086	208086	-
nl16	285614	280174	279618	2.09
br24	506433	503158	503158	0.65

The scalability of the parallel strategies was evaluated to study the benefits of searching an increasing number of multiple trajectories. Executing with more processes offers a greater diversity due to the use of multiple distinct initial seeds and solutions. Table 16.2 (resp. Table 16.3) shows the average execution times over five runs with different seeds for the sequential and the PAR-I and PAR-O (resp. PAR-1P and PAR-MP) parallel versions. These tables present the time in seconds required to find a solution whose cost is at least as good as the corresponding easy target, using one processor for the sequential implementation and eight, 16, and 24 processors for each parallel version.

The parallel versions converged faster than the sequential one for all instances. As the number of processors used increases, all parallel versions were able to find the corresponding easy targets faster, as reflected by the speedups presented in Tables 16.4 and 16.5. The speedups of the PAR-MP parallel version were greater than those of the other implementations. For example, the average speedups of the four algorithms on 24 processors were 12.51 for PAR-I, 12.62 for PAR-O, 13.19 for PAR-1P and 13.41 for PAR-MP.

The following experiment considers the computation times taken by the parallel versions to find solutions at least as good as the medium targets (exclusively for the instances for which the latter are smaller than the corresponding easy targets), addressing the benefits of exchanging information between the workers instead of letting them execute independently. The average processing times in seconds, based on five executions, on 24 processors are reported in Table 16.6. Results show that

Table 16.2.   Average computation times in seconds to find the easy targets on eight, 16, and 24 processors (PAR-I and PAR-O parallel versions).

Instance	Sequential $P = 1$	PAR-I $P = 8$	$P = 16$	$P = 24$	PAR-O $P = 8$	$P = 16$	$P = 24$
circ8	0.87	0.27	0.15	0.14	0.26	0.15	0.14
circ10	197.64	26.32	18.32	15.98	26.31	18.32	15.97
circ12	4.48	1.33	1.11	0.55	1.32	1.11	0.54
circ14	3.46	0.87	0.82	0.73	0.86	0.73	0.71
circ16	413.73	56.54	33.93	24.78	56.58	33.97	24.77
circ18	175.14	43.19	30.06	13.62	43.24	29.99	13.62
circ20	800.94	177.28	82.99	43.47	176.67	83.05	43.47
nl8	0.79	0.23	0.08	0.07	0.22	0.08	0.06
nl10	453.54	113.22	39.03	19.52	113.22	39.03	19.51
nl12	22.13	4.23	1.83	1.34	4.23	1.83	1.33
nl14	33.23	5.32	4.80	4.34	5.32	4.80	4.35
nl16	1433.05	474.12	243.14	73.62	474.26	243.11	73.58
br24	156.32	40.87	33.83	29.08	40.54	33.70	28.97

Table 16.3.   Average computation times in seconds to find the easy targets on eight, 16, and 24 processors (PAR-1P and PAR-MP parallel versions).

Instance	Sequential $P = 1$	PAR-1P $P = 8$	$P = 16$	$P = 24$	PAR-MP $P = 8$	$P = 16$	$P = 24$
circ8	0.87	0.19	0.16	0.12	0.19	0.15	0.11
circ10	197.64	26.31	17.04	15.97	26.31	17.04	15.96
circ12	4.48	1.33	0.39	0.35	1.33	0.38	0.34
circ14	3.46	0.65	0.53	0.52	0.65	0.53	0.51
circ16	413.73	55.30	33.93	24.45	55.31	33.94	24.50
circ18	175.14	24.73	21.96	13.60	24.70	23.96	13.62
circ20	800.94	106.34	70.50	43.47	106.30	70.47	43.46
nl8	0.79	0.17	0.10	0.08	0.16	0.08	0.07
nl10	453.54	113.21	39.01	19.52	125.43	39.05	19.51
nl12	22.13	3.91	1.61	1.33	3.91	1.64	1.33
nl14	33.23	5.85	4.80	4.35	5.84	4.80	4.34
nl16	1433.05	323.33	236.36	73.59	322.27	223.61	73.54
br24	156.32	33.99	25.29	20.23	30.77	23.71	16.55

PAR-MP presents the smallest computation time in most cases. Note that this version makes use of a cooperative strategy based on a pool of M elite solutions ($M = 24$). Although PAR-1P also shares information, it only records one elite solution. Therefore, the degree of diversity is smaller than in PAR-MP, possibly leading the workers to search the same region and, consequently, taking longer to converge to the target.

Table 16.4. Speedups on eight, 16, and 24 processors (PAR-I and PAR-O parallel versions).

Instance	PAR-I			PAR-O		
	$P = 8$	$P = 16$	$P = 24$	$P = 8$	$P = 16$	$P = 24$
circ8	3.23	5.69	5.87	3.28	5.70	6.21
circ10	7.50	10.79	12.37	7.51	10.79	12.38
circ12	3.37	4.02	8.21	3.39	4.03	8.21
circ14	3.99	4.21	4.74	4.00	4.72	4.74
circ16	7.32	12.20	16.70	7.31	12.18	16.71
circ18	4.06	5.83	12.87	4.05	5.84	12.86
circ20	4.52	9.65	18.42	4.53	9.64	18.43
nl8	3.51	9.36	11.15	3.61	9.38	12.34
nl10	4.00	11.62	23.24	4.01	11.62	23.25
nl12	5.23	12.09	16.56	5.23	12.12	16.63
nl14	6.25	6.93	7.65	6.24	6.93	7.64
nl16	3.02	5.89	19.47	3.02	5.89	19.48
br24	3.83	4.62	5.38	3.86	4.64	5.40
average	4.60	7.91	12.51	4.62	7.96	12.62

Table 16.5. Speedups on eight, 16, and 24 processors (PAR-1P and PAR-MP parallel versions).

Instance	PAR-1P			PAR-MP		
	$P = 8$	$P = 16$	$P = 24$	$P = 8$	$P = 16$	$P = 24$
circ8	4.42	5.40	7.23	4.57	5.65	7.51
circ10	7.51	11.60	12.37	7.51	11.60	12.39
circ12	3.36	11.23	12.67	3.35	11.58	13.07
circ14	5.30	6.50	6.69	5.29	6.54	6.78
circ16	7.48	12.19	16.92	7.48	12.19	16.89
circ18	7.08	7.98	12.87	7.09	7.31	12.86
circ20	7.53	11.36	18.42	7.53	11.37	18.43
nl8	4.80	7.24	9.58	4.99	9.90	9.95
nl10	4.01	11.63	23.24	3.62	11.62	23.25
nl12	5.66	13.75	16.59	5.67	13.50	16.64
nl14	5.68	6.92	7.64	5.68	6.92	7.66
nl16	4.43	6.06	19.47	4.45	6.40	19.49
br24	4.60	6.18	7.72	5.08	6.59	9.45
average	5.53	9.08	13.19	5.56	9.32	13.41

Results in Table 16.1 have shown that the PAR-MP implementation found better solutions than those obtained by the three other parallel implementations for three instances. Table 16.7 summarizes the results obtained by PAR-MP and gives the average overall computation times

Table 16.6. Computation times in seconds to find solutions at least as good as the medium targets.

Instance	Target	PAR-I	PAR-O	PAR-1P	PAR-MP
circ10	272	5,768.81	4,650.31	7,209.82	3,725.68
circ16	984	5,366.12	735.73	3,881.09	959.24
circ18	1308	9,323.88	8,565.20	13,972.76	10,620.86
nl16	280174	12,207.18	11,488.44	7,058.52	3,171.40
br24	503158	4,322.45	4,268.41	5,046.40	2,220.69

(based on five executions), in seconds, required to find the new solutions using 24 processors and the relative improvement with respect to the best solutions found by the other parallel implementations (i.e. the medium targets). We notice that the solution obtained by PAR-MP for instance circ18 is also the best known solution for the corresponding instance of the non-mirrored version of the TTP. Nevertheless, PAR-MP still requires just under six hours on average to find the solution.

Table 16.7. New best solutions obtained by the parallel version PAR-MP.

Instance	New best solution cost	Improvement (%)	Time (s)
circ16	978	0.61	4,690.83
circ18	1306	0.15	20,883.81
nl16	279618	0.20	14,586.73

The following experiment addresses the robustness of the parallel implementations from another point of view. Compared to the times needed by version PAR-MP to find the hard targets, we investigate whether PAR-I, PAR-O, and PAR-1P can also manage the same feat. Given the time taken by PAR-MP to find the best known solutions reported in the Table 16.7, the other parallel versions PAR-I, PAR-O, and PAR-1P were allowed to run for approximately twice this time, again using 24 processors. The values of the best solutions found by each version for each instance are presented in Table 16.8. These results show that the other parallel implementations were not able to find solutions as good as those obtained by PAR-MP, even if significantly more processing time is given, illustrating the effectiveness of the cooperation scheme implemented in the latter.

We used *time-to-target solution value* plots [1, 2] for the measured computation times to further evaluate and compare the behavior of the four parallel versions running on different numbers of processors. This

Table 16.8. Solutions found by PAR-I, PAR-O, and PAR-1P when executed for twice the time taken by PAR-*MP*.

Instance	Time (s)	Target	PAR-I	PAR-O	PAR-1P
circ16	10,000	978	984	984	984
circ18	40,000	1306	1308	1308	1308
nl16	30,000	279618	280174	280174	280174

approach is based on plots showing empirical distributions of the random variable *time-to-target solution value.* To plot the empirical distribution, we first fix a given problem instance and a target solution value. Next, each algorithm is executed N times, recording the running time to find the first solution at least as good as the target value. For each algorithm, we associated with the i-th sorted running time t_i a probability $p_i = (i - \frac{1}{2})/N$ and plot the points $z_i = (t_i, p_i)$, for $i = 1, \ldots, N$.

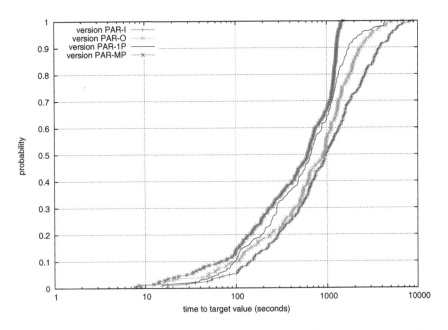

Figure 16.2. Empirical distributions of the random variable time-to-target solution value for the parallel versions PAR-I, PAR-O, PAR-1P, and PAR-*MP* using 8 processors on instance nl16.

Figure 16.2 displays the empirical distributions of the *time-to-target solution value* for the four parallel versions associated with the instance nl16 and the target cost of 284000 (a value between the easy and medium target), obtained from $N = 200$ independent runs of each version on

$P = 8$ processors. Version PAR-MP behaves better than the other versions, finding the target value in less than 1,400 seconds with probability 95% compared to 2,445 seconds with the same probability for the next fastest amongst the others (PAR-1P). PAR-MP took at most approximately 1,500 seconds in the slowest run, while PAR-1P took more than 4,300 seconds in the worst run. The behavior depicted in this plot is common to different instances and target values. The plot of the empirical distribution associated with the PAR-MP strategy is clearly to the left of those of the other versions, illustrating that the former is more robust since it is able to find with higher probabilities the same solutions found by the others in the same computation time.

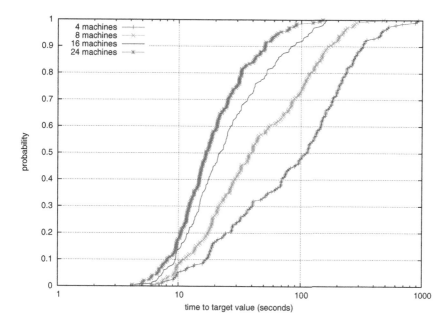

Figure 16.3.   Empirical distributions of the random variable time-to-target solution value for the parallel version PAR-MP using 4, 8, 16, and 24 processors for the instance br24.

The plot in Figure 16.3 further illustrates the scalability of the parallel version PAR-MP, as already shown in Tables 16.2 and 16.3. This plot depicts the empirical distributions for four, eight, 16, and 24 processors, obtained from $N = 200$ runs on instance br24 using the easy target.

7.   Grid implementation and experiments

Numerous scientists and engineers, from diverse scientific and technological fields, are showing interest in exploiting the potential of grid computing. This section briefly highlights a number of observations with respect to executing metaheuristics on a computational grid. For performance evaluation purposes, the experiments presented earlier in Section 6 were necessarily obtained with exclusive access to a dedicated cluster of processors, since grids are inherently shared computing environments. This sharing, together with the fact that grid resources are heterogeneous, means that the computing power available from resources is neither identical nor constant.

As computational environments scale to hundreds of individual computational resources, failures are more likely to occur especially during the execution of long-running applications. Users naturally want their programs to adapt to both faults and changes in available performance in order to continue executing efficiently. While the current implementations of MPI are suitable for use in the static environments like computing clusters, in practice they are not robust enough for computational grids. For example, grid enabled implementations of MPI do not provide support for dynamic rescheduling of processes. Furthermore, a single process failure will cause the whole application to abort.

In an effort to hide the intricacies of grid environments, grid engineers have been developing *grid middleware* (an intermediate layer of software) to provide tools which insulate users from the underlying complexities, and management systems, that automatically and efficiently adapt grid-enabled applications to the dynamically changing characteristics of the grid.

The EasyGrid AMS middleware [7] provides for a robust and efficient execution of programs in grid environments. Parallel MPI applications are transformed *automatically* into *system-aware* versions by incorporating grid middleware into the user's application without modification to the latter. These system-aware applications or *Smart G-Apps* are adaptive, robust to resource failure (fault tolerant), self-scheduling programs capable of reacting to changes which occur in shared, dynamic, unstable distributed environments like computational grids.

This approach aims to relieve programmers and users of the task of enabling (existing) applications to execute efficiently in grid environments. Turning aspects related to the grid transparent to the programmer avoids the need to develop one version of the application for a local cluster computing platform and another for the grid. The methodology is based on application-centric middleware which provides services (e.g. static and dynamic scheduling, and integrated fault tolerance strategies) specifically tuned to the needs of each individual application [26].

Given that the PAR-MP implementation produced the best results for the mTTP when compared to the other parallel strategies, a grid enabled version of PAR-MP based on the EasyGrid AMS was evaluated in a real grid environment. *Grid Sinergia*computational grid is an initiative to create and operate a research oriented, production level computational grid across three states (Rio de Janeiro, São Paulo, and Espírito Santo) in the south east of Brazil. The objectives include providing researchers with a realistic and practical environment for distributed computing research and offering system administrators practical experience in management and operation of grid computing environments. Grid Sinergia currently employs the Globus Toolkit middleware [3] across the participating sites which are interconnected by the Brazilian National Research Network's experimental high speed (10Gbit) optical network *Rede Giga*.

An initial experiment was carried out employing 82 resources from the following three sites of Grid Sinergia, located in three different cities within the state of Rio de Janeiro: (a) two clusters in the city of Rio de Janeiro, one with 30 Linux PCs (Pentium II 400 MHz) and the other with 24 Linux PCs (Pentium IV 1.7GHz), each of them connected by a fast ethernet network; (b) in Niterói (a distance of 40 Km from cluster (a)), a cluster of 26 Linux PCs (Pentium IV 2.6 GHz) interconnected via Gigabit switches; (c) and in Petrópolis (approximate 100 Km from both clusters (a) and (b)) two Linux PCs (Pentium IV 3.2 GHz).

Table 16.9 displays the average processing times (measured over five runs) in seconds required to find a solution whose cost is at least as good as the corresponding medium target, on the shared resources of the computational grid during the day (i.e. normal working hours) and during the night (after working hours, the resources tend to be utilized less). We notice that, although the resources were being shared with other users, the practical benefits obtained with the computational grid were outstanding. For example, in the case of instance circ18, for the medium target, an almost four fold improvement was achieved by the grid with respect to the dedicated 24-processor cluster.

Table 16.9.   Average processing times of the parallel cooperative strategy PAR-MP when executed in 82 resources from three sites of Grid Sinergia.

Instance	Dedicated cluster (24 CPUs)	Time (seconds) Shared grid (82 CPUs, working hours)	Shared grid (82 CPUs, after hours)
circ10	3,725.68	1,077.38	333.48
circ16	959.24	747.94	513.63
circ18	10,620.86	2,376.76	2,313.73
data16	3,171.40	1,044.31	908.95

8.   Concluding remarks

Metaheuristics have found their way into the standard toolkit of combinatorial optimization methods. Parallel implementations of metaheuristics can be applied to hard combinatorial optimization problems, often allowing reductions in computation times. Although independent strategies can already obtain good computational results, parallelizations based on cooperative search strategies lead to more robust implementations.

The computational results reported in this paper show that the sequential heuristic for the mirrored traveling tournament problem benefits from low communication parallel implementations, which are capable of finding better solutions with respect to their sequential counterpart. In particular, the use of a pool of elite solutions offers a diversity of high quality solutions from which workers can restart their searches for better solutions. The pool also provides a mean to implement cooperation and to achieve faster convergence.

Consistent speedups were obtained on experiments performed on a dedicated cluster with up to 24 processors. The cooperative implementation PAR-MP obtained average results systematically better than the others. In particular, it was able to improve the hard targets to several instances and to disclose previously unknown solutions.

This parallel strategy was also implemented and tested using a true grid platform. We reported original results from pioneer computational experiments on a shared computational grid formed by 82 machines distributed over four clusters in three cities, illustrating the potential of the application of computational grids in the fields of metaheuristics and combinatorial optimization.

Given the above favorable results and the enormous potential of computational grids, a broader investigation is underway, exploring the use of a significantly larger number of processors and investigating new programming challenges.

References

[1] R.M. Aiex, M.G.C. Resende, and C.C. Ribeiro. Probability distribution of solution time in GRASP: An experimental investigation. *J. of Heuristics*, 8:343–373, 2002.

[2] R.M. Aiex, M.G.C. Resende, and C.C. Ribeiro. TTTPLOTS: A perl program to create time-to-target plots, 2005, submitted for publication.

[3] Globus Alliance. Globus. online reference at http://www-unix.globus.org/toolkit/, last visited on May 04, 2005.

[4] A. Anagnostopoulos, L. Michel, P. Van Hentenryck, and Y. Vergados. A simulated annealing approach to the traveling tournament problem. In *Proceedings of CPAIOR'03*, 2003.

[5] B.C. Ball and D.B. Webster. Optimal schedules for even-numbered team athletic conferences. *AIIE Transactions*, 9:161–169, 1997.

[6] T. Benoist, F. Laburthe, and B. Rottembourg. Lagrange relaxation and constraint programming collaborative schemes for traveling tournament problems. In *Integration of AI and OR Techniques in Constraint Programming*, pages 15–26, Kent, 2001.

[7] C. Boeres and V.E.F. Rebello. EasyGrid: Towards a framework for the automatic grid enabling of legacy MPI applications. *Concurrency and Computation Practice and Experience*, 17:425–432, 2004.

[8] D. Costa. An evolutionary tabu search algorithm and the NHL scheduling problem. *INFOR*, 33:161–178, 1995.

[9] V.-D. Cung, S.L. Martins, C.C. Ribeiro, and C. Roucairol. Strategies for the parallel implementation of metaheuristics. In C.C. Ribeiro and P. Hansen, editors, *Essays and Surveys in Metaheuristics*, pages 263–308. Kluwer, 2002.

[10] K. Easton, G. Nemhauser, and M. Trick. Solving the travelling tournament problem: a combined integer programming and constraint programming approach. In E. Burke and P. Causmaecker, editors, *Selected Papers from the 4th International Conference on the Practice and Theory of Automated Timetabling*, volume 2740 of *Lecture Notes in Computer Science*, pages 100–109. Springer-Verlag, 2003.

[11] K. Easton, G.L. Nemhauser, and M.A. Trick. The traveling tournament problem: Description and benchmarks. In T. Walsh, editor, *Principles and Practice of Constraint Programming*, volume 2239 of *Lecture Notes in Computer Science*, pages 580–589. Springer-Verlag, 2001.

[12] E.R. Fernandes and C.C. Ribeiro. Using an adaptive memory strategy to improve a multistart heuristic for sequencing by hybridization. In S.E. Nikoletseas, editor, *4th International Workshop on Experimental and Efficient Algorithms*, volume 3503 of *Lecture Notes in Computer Science*, pages 4–15. Springer-Verlag, 2005.

[13] P. Festa and M.G.C. Resende. GRASP: An annotated bibliography. In C.C. Ribeiro and P. Hansen, editors, *Essays and Surveys in Metaheuristics*, pages 325–367. Kluwer, 2002.

[14] I. Foster. *Designing and Building Parallel Programs.* Addison-Wesley, 1995.

[15] I. Foster and C. Kesselman, editors. *The GRID: Blueprint for a New Computing Infrastructure.* 2nd edition. Morgan Kaufmann, 2004.

[16] I. Foster, C. Kesselman, and S. Tuecke. The anatomy of the grid: Enabling scalable virtual organizations. *International Journal of Supercomputer Applications,* 15:200–222, 2001.

[17] L. Gaspero and A. Schaerf. A composite-neighborhood tabu search approach to the traveling tournament problem. *Journal of Heuristics,* to appear.

[18] P. Greistorfer and Stefan Voss. Controlled pool maintenance for metaheuristics. In C. Rego and B. Alidaee, editors, *Metaheuristic Optimization via Memory and Evolution: Tabu Search and Scatter Search,* pages 387–421. Springer, 2004.

[19] P. Hansen and N. Mladenovic. Developments of variable neighborhood search. In C.C. Ribeiro and P. Hansen, editors, *Essays and Surveys in Metaheuristics,* pages 415–439. Kluwer, 2002.

[20] M. Henz. Scheduling a major college basketball conference revisited. *Operations Research,* 49:163–168, 2001.

[21] LAM/MPI parallel computing. Online document at http://www.lam-mpi.org/, last visited on July 25, 2005.

[22] A. Lim, B. Rodrigues, and X. Zhang. A simulated annealing and hill-climbing algorithm for the traveling tournament problem. *European Journal of Operations Research,* 2005 (to appear).

[23] H.R. Lourenço, O. Martins, and T. Stutzle. Iterated local search. In F. Glover and G. Kochenberger, editors, *Handbook of Metaheuristics,* pages 321–353. Kluwer, 2002.

[24] M. Matsumoto and T. Nishimura. Mersenne twister: A 623-dimensionally equidistributed uniform pseudo-random number generator. *ACM Transactions on Modeling and Computer Simulation,* 8:3–30, 1998.

[25] Message Passing Forum. MPI: A message passing interface. Technical report, University of Tennessee, 1995.

[26] A.P. Nascimento, A.C. Sena, J.A. da Silva, D.Q.C. Vianna, C. Boeres, and V.E.F. Rebello. Managing the execution of large scale MPI applications on computational grids. In C. Amorim, G. Silva, V. Rebello, and J. Dongarra, editors, *Proceedings of the 17th International Symposium on Computer Architecture and High*

Performance Computing, pages 69–76, Rio de Janeiro, 2005. IEEE Computer Society Press.

[27] G.L. Nemhauser and M.A. Trick. Scheduling a major college basketball conference. *Operations Research*, 46:1–8, 1998.

[28] R. Rasmussen and M. Trick. A Benders approach to the constrained minimum break problem. *European Journal of Operational Research*, to appear.

[29] M.G.C. Resende and C.C. Ribeiro. Greedy randomized adaptive search procedures. In F. Glover and G. Kochenberger, editors, *Handbook of Metaheuristics*, pages 219–249. Kluwer, 2003.

[30] M.G.C. Resende and C.C. Ribeiro. GRASP with path-relinking: Recent advances and applications. In T. Ibaraki, K. Nonobe, and M. Yagiura, editors, *Metaheuristics: Progress as Real Problem Solvers*, pages 29–63. Kluwer, 2005.

[31] C.C. Ribeiro and I. Rosseti. A parallel GRASP for the 2-path network design problem. In B. Monien and R. Feldman, editors, *Parallel Processing: 8th International Euro-Par Conference*, volume 2400 of *Lecture Notes in Computer Science*, pages 922–926. Springer-Verlag, 2002.

[32] C.C. Ribeiro and I. Rosseti. Efficient parallel cooperative implementations of GRASP heuristics, 2005, submitted for publication.

[33] C.C. Ribeiro, R.C. Souza, and C.E.C. Vieira. A comparative computational study of random number generators. *Pacific Journal of Optimization*, 1:565–578, 2005.

[34] C.C. Ribeiro and S. Urrutia. Heuristics for the mirrored traveling tournament problem. *European Journal of Operational Research*, to appear.

[35] R.A. Russell and J.M. Leung. Devising a cost-effective schedule for a baseball league. *Operations Research*, 42:614–625, 1994.

[36] J.A.M. Schreuder. Combinatorial aspects of construction of competition Dutch professional football leagues. *Discrete Applied Mathematics*, 35:301–312, 1992.

[37] J.M. Thompson. Kicking timetabling problems into touch. *OR Insight*, 12:7–15, 1999.

[38] M.A. Trick. Challenge traveling tournament instances. Online document at `http://mat.gsia.cmu.edu/TOURN/`, last visited on May 29, 2005.

[39] M.G.A. Verhoeven and E.H.L. Aarts. Parallel local search. *Journal of Heuristics*, 1:43–65, 1995.

[40] M.B. Wright. Scheduling English cricket umpires. *Journal of the Operational Research Society*, 42:447–452, 1991.

[41] J.T. Yang, H.D. Huang, and J.T. Horng. Devising a cost effective basketball scheduling by evolutionary algorithms. In *Proceedings of the 2002 Congress on Evolutionary Computation*, pages 1660–1665, Honolulu, 2002.

Part VII

Algorithm Tuning, Algorithm Design and
Software Tools

Chapter 17

USING EXPERIMENTAL DESIGN TO ANALYZE STOCHASTIC LOCAL SEARCH ALGORITHMS FOR MULTIOBJECTIVE PROBLEMS

Luís Paquete,[1] Thomas Stützle[2] and Manuel López-Ibáñez[3]

[1] *Faculdade de Economia &*
CSI – Centro de Sistemas Inteligentes
Universidade do Algarve
Campus de Gambelas, 8005-139 Faro
Portugal
lpaquete@ualg.pt

[2] *IRIDIA, CP 194/6*
Université Libre de Bruxelles
Avenue F. Roosevelt 50, 1050 Brussels
Belgium
stuetzle@ulb.ac.be

[3] *School of the Built Environment*
Napier University
Merchiston Campus, 10 Colinton Road, EH10 5DT Edinburgh
UK
m.lopez-ibanez@napier.ac.uk

Abstract   Stochastic Local Search (SLS) algorithms can be seen as being composed of several algorithmic components, each playing some specific role with respect to overall performance. This article explores the application of experimental design techniques to analyze the effect of components of SLS algorithms for Multiobjective Combinatorial Optimization problems, in particular for the Biobjective Quadratic Assignment Problem. The analysis shows that there exists a strong dependence between the choices for these components and various instance features, such as the structure of the input data and the correlation between the objectives.

Keywords: Approximation methods and heuristics, combinatorial optimization, multiple objective and goal programming, design of experiments

1. Introduction

Stochastic Local Search (SLS) algorithms [14] for Multiobjective Combinatorial Optimization Problems (MCOPs) typically involve the selection and parameterization of many *algorithmic components* whose role with respect to the overall performance and relation to certain instance features is often not clear. This problem is even further increased by the recent trend towards *hybrid* approaches [7].

In this article, we take a *modular* perspective to the design and analysis of SLS algorithms for MCOPs that are solved with respect to the notion of Pareto optimality. An SLS algorithm is understood as a combination of SLS *components* that can be *coupled* and that can result in different behaviors. The effect of these components on the overall performance can then be analyzed by means of experimental design techniques. Roughly speaking, each component is considered a *factor*, that is, an abstract characteristic of an SLS algorithm that can affect random variables such as solution quality and computation time; each factor has associated *levels* that are possible instantiations of the SLS component.

Given the stochasticity of the algorithms under study, we apply a sound empirical methodology for evaluating the solution quality reached by the algorithms. We base our analysis on the *better relation* [13, 31] and *attainment functions* [11]; the latter allows the application of statistical hypothesis tests on the performance of two or more SLS algorithms [21]. Finally, an exploratory data analysis tool is used to illustrate the effects of the SLS components and their parameter settings in the objective space in order to derive more fine-grained conclusions.

This analysis shows that different choices for these components can affect the SLS algorithm in several different ways, and that the same choices can lead to different behaviors in dependence of certain instance features. We illustrate these results for SLS algorithms applied to the Biobjective Quadratic Assignment Problem (BQAP). The BQAP is defined as follows: given n facilities and n locations, one $n \times n$ matrix $\mathbf{A}$ where a_{ij} is the distance between locations i and j, and two $n \times n$ matrices $\mathbf{B}^1$ and $\mathbf{B}^2$ where b_{rs}^1 is the first flow and b_{rs}^2 is the second flow between facilities r and s, the objective function in the BQAP is stated as:

$$\min_{\phi \in \Phi} \begin{cases} \sum_{i=1}^{n} \sum_{j=1}^{n} a_{\phi_i \phi_j} b_{ij}^1 \\ \\ \sum_{i=1}^{n} \sum_{j=1}^{n} a_{\phi_i \phi_j} b_{ij}^2 \end{cases}, \tag{17.1}$$

where Φ is the set of all permutations of $\{1, 2, \ldots, n\}$, ϕ_i gives the location assigned to facility i in the solution $\phi \in \Phi$, and "min" refers to the notion of Pareto optimality. This analysis is based on *straightforward* extensions of simple SLS algorithms for the single-objective QAP, which are applied to solve several scalarizations of the objective function vector for instances of different size, structure of the input data, and correlations between the flow matrices.

Algorithm 1 Algorithm framework

---

for all weight vectors $\vec{\lambda}$ **do**
  s = Choose solution
  s' = SLS$(s, \vec{\lambda})$
  Add s' to Archive
Filter Archive
return Archive

---

2.   SLS Algorithms and SLS Components

One large class of SLS approaches for MCOPs is based on solving several scalarizations of the objective function vector [23]. In a biobjective problem such as the BQAP, the objective function vector is *scalarized* according to a weight vector $\vec{\lambda} = (\lambda_1, \lambda_2)$ with non-negative components. The type of scalarization chosen is the well-known weighted sum formulation given by

$$f(s) = \lambda_1 f_1(s) + \lambda_2 f_2(s) .$$

It is known that the efficacy of approaches based on weighted sum depends strongly on the number of efficient solutions that are non-supported, that is, those solutions that do not lie in the convex-hull of the efficient set in the objective space. Non-supported solutions are not optimal for any weight vector and, therefore, their large number can affect negatively the performance of such SLS algorithms. Despite this pitfall, the usage of scalarizations has been shown to be a very successful strategy [3][8][13][16][22][30]. Moreover, a strong advantage of this approach is that for tackling multi-objective problems, the SLS algorithms developed for the single-objective counterpart are directly applicable. The necessary adaptation for handling MCOP s mainly lies on the strategy for changing the weight vectors at run-time in order to return a set of solutions that is well spread in the objective space.

Algorithm 1 presents the pseudo-code of the algorithmic framework, where the procedure SLS corresponds to the SLS algorithm for solving each scalarization and Archive is a data structure that maintains the solutions that are found at the end of each scalarization. The choice for the underlying SLS algorithm is obviously crucial for tackling an MCOP; typically, it will be advisable to apply the state-of-the-art algorithms for the single-objective case, if available. The procedure Filter removes dominated solutions from the Archive. Note that in this framework the weight vectors are given *a priori* and the generation of the initial solution is left open. The simplicity of this framework allows an easy identification and parameterization of different SLS components. In particular, the following four components were considered:

Dispersion policy.   An usual requirement on the set of solutions returned by SLS algorithms for MCOPs is that they are *spread* in the objective space. For this reason, we applied SLS algorithms that use *maximally*

Algorithm 2 `Restart` search strategy

---

   for all weight vectors $\vec{\lambda}$ **do**
      s is a randomly generated solution
      $s' = \mathsf{SLS}(s, \vec{\lambda})$
      Add s' to Archive
   Filter Archive
   return Archive

---

dispersed weight vectors [25]: given Q objectives and $\binom{z+Q-1}{Q-1}$ distinct weight vectors, each $\vec{\lambda} = (\lambda_1, \ldots, \lambda_Q)$ is normalized such that $\sum_{q=1}^{Q} \lambda_q = 1$ and its components can take values from $\{0, \frac{1}{z}, \frac{2}{z}, \ldots, 1\}$. In the biobjective case, z corresponds to the number of scalarizations. The main parameter for the dispersion policy is the number of scalarizations. In principle, it can be expected that for an increased number of scalarizations also an increased number of solutions is returned. However, the growth rate of the number of solutions with the number of scalarizations is unknown in advance and it possibly depends on the number of non-supported solutions.

Search strategy. We consider two search strategies: `Restart` and `2phase`, shown in Algorithms 2 and 3, respectively. The `Restart` strategy starts the underlying SLS algorithm for each scalarization from a randomly generated solution. The `2phase` strategy consists of the following two phases. Firstly, it obtains a high quality solution for one objective. Next, in the second phase, a sequence of scalarized problems is solved. The starting solution for each scalarization is the best solution found for the previous scalarization, except for the first iteration, where the starting point is the solution returned in the first phase. The change of the weights in the second phase is performed as follows [22]: given two objectives, the weight vector in the first and last position in the sequence of weight vectors are $(1, 0)$ and $(0, 1)$, respectively; then, two successive weight vectors differ by only $\pm\frac{1}{z}$ in any two components. This approach for generating the weight vectors can easily be extended to an arbitrary number of objectives using an algorithm to generate all compositions of z in Q parts in a combinatorial Gray code order [17]. Note that different strategies for changing the weights, for example, allowing steps larger than size $\pm\frac{1}{z}$, may result in different behavior. However, such a study is beyond the scope of this article.

Intensification mechanism. Besides spread, also the quality of the individual solutions is important, since it affects the quality of the returned set. Improved solution quality for each scalarization can be obtained, for example, by simply increasing the number of iterations of the underlying SLS algorithm. This can be seen as an *intensification* of the search process, whereas the dispersion policy introduces *diversification*.

Algorithm 3 2phase search strategy

---

s is a randomly generated solution

$s' = \mathsf{SLS}_1(s)$ /* First phase */

for all weight vectors $\vec{\lambda}$ **do**

  $s = s'$

  $s' = \mathsf{SLS}_2(s, \vec{\lambda})$ /* Second phase */

  Add s' to Archive

Filter Archive

return Archive

Component-wise step.   Since the number of solutions is bounded by the number of scalarizations, a further step is to accept non-dominated solutions in the neighborhood of each solution returned by each scalarization. This is done by a procedure we call *component-wise step*. Similar ideas to this component-wise step have been proposed earlier [1][9][12][22][29].

As said, our goal is to analyze the contribution of each of these four SLS components on overall performance. This type of experimental analysis is studied under the topic of Experimental Design, which consists of a set of techniques that allows the planning of experiments with the purpose of comparing the effects of different *factors* at different *treatment levels* on a given number of *experimental units* [4]. A factor is any feature of the experimental conditions that might affect the output of an experiment, and, thus, each of the four SLS components presented above is considered as a factor.

3.   BQAP Instances

One goal of this experimental analysis is to investigate the influence of instance features on SLS performance. Since certain instance features may affect algorithm performance in different ways, three features of this problem are taken into account: structure of the input data, correlation between flow matrices, and instance size. The instances were generated by two instance generators proposed by Knowles and Corne [18] that are available at http://dbkweb.ch.umist.ac.uk/knowles/mQAP. These generators allow to obtain BQAP instances for different combinations of parameters for the features mentioned above. The features and parameters for the two generators are as follows.

Structure of input data.   One generator yields *unstructured* instances, where each entry of the distance matrix and the flow matrices is randomly generated according to a uniform distribution within a given range. The second instance generator gives instances with some underlying *structure*, where the distribution of the flows is clearly non-uniform and a significant part of the flows is zero. The parameter values for generating the instances were defined as for the single-objective unstructured instances in the class `Taixxa` and the structured ones in the class `Taixxb` [28].

Correlation between flow matrices.   Little is known on how the correlation between objectives could affect the performance of SLS algorithms. Therefore, we also considered different degrees of correlation between objectives by generating different correlations between the flow matrices. Given a value from a flow matrix and a parameter ρ, the corresponding matrix entry of the second or further flow matrices is generated by influencing the probability of accepting a value taken from a random distribution. For the purpose of this analysis, the values considered were $\{-0.75, 0.0, 0.75\}$.

Instance size.   Another aspect investigated is the understanding of how the performance of the SLS algorithms scales with instance size. In the BQAP case, the number of locations corresponds to the instance size n. We generated instances of sizes 25, 50 and 75, which are named here as *small*, *medium*, and *large*, respectively. Note that a QAP instance with more than 30 locations cannot, in practice, be solved to optimality by exact algorithms.

All instances are *symmetric*, that is, $a_{ij} = a_{ji}$ and $b_{kl} = b_{lk}$ and only two objectives were considered. For each combination of the instance parameters 3 instances were generated, resulting in a total of 54 instances, which are available at `w3.ualg.pt/~lpaquete/qap/bQAP.tar.gz`.

4.   Performance Assessment Methodology

The performance assessment of algorithms for MCOPs is more complex than for the single-objective case and fundamental criticisms have been raised against the use of unary quality indicators [31]. We avoid these known drawbacks by using methodologies to which these criticisms do not apply; these are the *better relation* [13, 31], which provides the most basic assertion of performance, and *attainment functions* [11], which allow to test hypotheses with respect to the distribution of the solution quality over multiple runs and to detect large differences of performance in the objective space.

Better relation.   Following Zitzler et al. [31], we consider a set of points A to be *better* than a set of points B if, and only if, $A \neq B$ and every point in B is weakly dominated by, at least, a point in A. This relation is also one of the *outperformance relations* introduced by Hansen and Jaszkiewicz [13]. Thus, as a first step, we count how many times each outcome associated to each level of a component is *better* than the outcomes from another level of the same component is counted. However, we restrict the comparison of outcomes to those that were produced *within* the same levels of other components, in order to reduce variability. This allows to detect if some level is clearly responsible for a good or bad performance. On the other hand, if no clear-cut answer is obtained from this first analysis, one can intuitively conclude that the outcomes are mostly *incomparable*. However, it is not known to which degree they really differ and further analysis must be carried out by the attainment functions.

Attainment functions. Fonseca and Fleming [5] associate the performance of an SLS algorithm for multiobjective problems to the probability of attaining *an arbitrary point* in the objective space in one single run. This function is called *attainment function* [11] and it can be seen as a generalization of the distribution function of the solution cost [14] to the multiobjective case. From the outcomes obtained in several runs of an SLS algorithm, these probabilities can be empirically estimated to construct the *empirical attainment function* (EAF). Hence, using the outcomes of several algorithms for a certain problem instance, allows hypothesis testing on the equality of the corresponding EAFs. The applied statistical tests are based on the maximum absolute distance between two EAFs for the two-sample case and the maximum absolute distance between k EAFs, for the k-sample case [2]. In this latter case, if the hypothesis of equality is rejected, pairwise comparisons between the k EAFs must be performed and the returned p-values must be corrected by Holm's procedure [15]. Due to the nature of these tests, permutation tests [10] are required and the permutation procedure has to be specified according to the chosen experimental design [24, 21].

Location of differences. If the null hypothesis on the equality of EAFs should be rejected, a visual inspection of the largest differences of performance can be obtained by plotting the points in the objective space where the absolute difference of the corresponding EAFs is higher than a certain value. We assume that differences below 20% are negligible. Finally, the sign of the differences indicates which level or configuration presented better performance in which points in the objective space.

Figure 17.1 gives an example. The two plots in the top, (a) and (b), show the EAFs associated to two algorithms, where the darker the color, the higher is the probability of attaining these points. At each plot in the top, the bottom-left dark line corresponds to the attainment surface associated to the solutions found by the algorithm with the minimum probability. The white line in the top right corresponds to the attainment surface associated to the solutions dominated with probability one by the corresponding algorithm. The two plots in the bottom, (c) and (d), indicate the location of the positive differences in favor of the two algorithms, where the darker the color, the higher is the difference. (In the plots given in Section 5, we occasionally suppress one of the two plots, if no differences in favor of one of the two algorithms were found.) The line in the bottom-left corresponds now to the attainment surface associated to the solutions found by *both* algorithms with minimum probability, while the line in the top-right corresponds to the attainment surface associated to the solutions found by *both* algorithms with probability one. Hence, any difference would be located within these two lines. This visual inspection allows to see that the algorithm associated to the left plots has a much higher probability of attaining

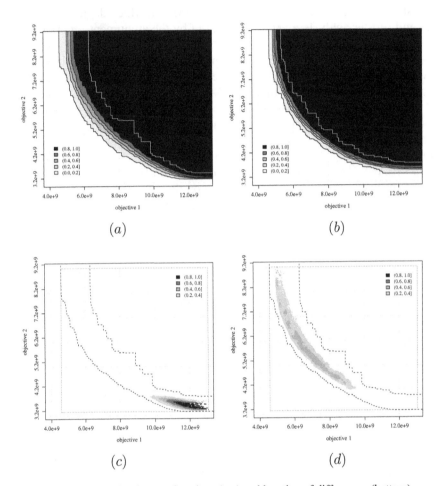

Figure 17.1. Attainment functions (top) and location of differences (bottom).

the region towards the minimization of the second objective, whereas the other algorithm is able to attain a wider region of the trade-off curve towards the center and towards the minimization of the first objective.

4.1   CPU time

In the experimental set-up, also the CPU time taken by each algorithm was measured as a response variable. Hence, in dependence of the particular parameters of components, the CPU time may vary significantly. Note that, differently from solution quality, the CPU time taken from several runs is described by a univariate distribution and it can be analyzed by parametric statistics, if the assumptions of independence of error terms, equal variance and normality are met [4].

5. Experimental Setup and Results

As the underlying SLS algorithm, the Robust Tabu Search (RoTS) proposed by Taillard [27]. RoTS is a rather standard Tabu Search algorithm based on the 2-swap neighborhood and it is one of the best known and best performing SLS algorithms for the QAP. RoTS starts from a random solution and at each iteration the best non-tabu neighboring solution is chosen (even if it is worse than the current one). A neighboring solution that assigns facilities i and j to locations r and s, respectively, is tabu if in the last tl iterations this same assignment occurred. The variable tl is a parameter of the algorithm and is called *tabu list length* or *tabu tenure*. The tabu status of an assignment can be overridden by the *aspiration criterion*, that is, if the tabu neighboring solution is the best solution found so far. Taillard considered a further rule: if a facility i has not been assigned to a location r during the last u iterations, any neighboring solution that does not consider that assignment is forbidden. Here, the same parameters as provided by Taillard [27, 28] were used, that is, the tabu list length is chosen randomly every $2.2n$ iterations from the interval $[0.9n, 1.1n]$, while $u = 3n^2$. Our implementation also applies the fast evaluation of neighboring solutions described by Taillard [27].

For the experimental analysis, we considered a full factorial design, that is, we ran all algorithms that resulted from all possible combinations of parameters of the four SLS components described in Section 2. The two search strategies we tested were Restart and 2phase; the second phase of 2phase starts from the solution returned by the first phase, whose length was fixed to $10n^2$ tabu iterations. The number of scalarizations were defined in dependence of the instance size as n, $5n$ and $10n$. Similarly, for the intensification mechanism, $50n$ and $100n$ tabu iterations of the underlying RoTS algorithm were considered, as well as an iterative improvement algorithm based on the 2-swap neighborhood (II). Finally, the effect of the component-wise step was also considered in the experimental setup. This resulted in a total of 36 configurations that were ran 5 times on each instance on an Intel(R) Xeon(TM) 2.4 GHz CPU with 512 KB of cache under Debian GNU/Linux.

In the following, we describe and discuss the experimental results of this analysis with respect to the solution quality associated to each of the four SLS components under study using the methodology outlined in the previous sections. The results for the better relation were computed as percentage of all pairwise comparisons, averaged over the three instances of the same type, size and correlation. Each permutation test for the hypothesis test with respect to the equality of the EAFs consisted of 10 000 permutations and the significance level was set to 0.05.

5.1 Search Strategy

Better relation.   Only for medium and large unstructured instances with positively correlated flow matrices, we observed some evidence that the

2phase strategy performs better than the Restart strategy. All the remaining results were inconclusive.

Attainment functions.   The null hypothesis of the equality of the attainment functions was always rejected. Hence, the 2phase and Restart search strategies produce statistically different outcomes on all instances tested.

Location of differences.   Figure 17.2 shows the location of the differences above 20% on few unstructured and structured instances. Differences higher than 20% associated to both search strategies occur, which means that the Restart and 2phase strategies perform better in different regions of the objective space. Common to all plots with differences in favor of the 2phase strategy is the fact that these differences occur towards the improvement of the objective where the second phase of the 2phase strategy starts (objective 2). In addition, the correlation seems to have a strong influence on the shape of the approximation to the efficient set on unstructured instances. Furthermore, on these instances, we observed that differences in favor of the Restart strategy cover a wider region for instances with negatively correlated flow matrices than for positively correlated ones. Finally, large differences between the search strategies were observed on the structured instances; the differences in favor of the Restart strategy cover a wider region than the differences in favor of the 2phase strategy.

5.2   Number of Scalarizations

Better relation.   The increase on the number of scalarizations does not generally correspond to a noticeable better performance with respect to the better relation. Only for unstructured instances with positively correlated flow matrices some exceptions are found.

Attainment functions.   The null hypothesis of equal performance is mostly rejected, except when comparing $5n$ against $10n$ scalarizations on large, structured instances; this observation seems to be independent of the correlation between the flow matrices. This result indicates the existence of some limiting behavior above $5n$ scalarizations.

Location of differences.   Figure 17.3 gives the location of the differences above 20% for an unstructured and a structured instance, respectively. Given are the differences between n and $5n$ scalarizations in favor of the latter and between $5n$ and $10n$ scalarizations in favor of the latter (differences in favor of n scalarizations, in the first case, and $10n$ scalarizations, in the second case, were below 20%). The differences between n and $10n$ scalarizations are not shown since they are slightly larger than the two cases above. It can be seen that the increase of the number of scalarizations corresponds to a slight improvement of the solution quality.

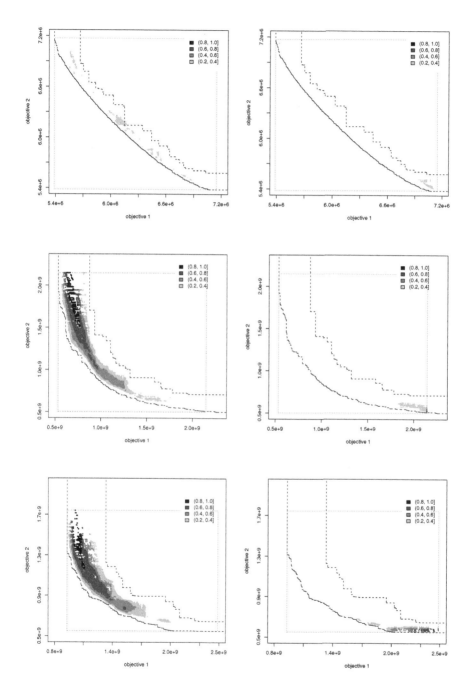

Figure 17.2. Location of differences between the search strategies; examples are given for an unstructured instance with $\rho = -0.75$ in favor of `Restart` (top-left) and `2phase` (top-right) and for two structured instances with $\rho = 0.75$ in favor of `Restart` (center-left) and `2phase` (center-right) and with $\rho = -0.75$ in favor of `Restart` (bottom-left) and `2phase` (bottom-right); all instances have size 50.

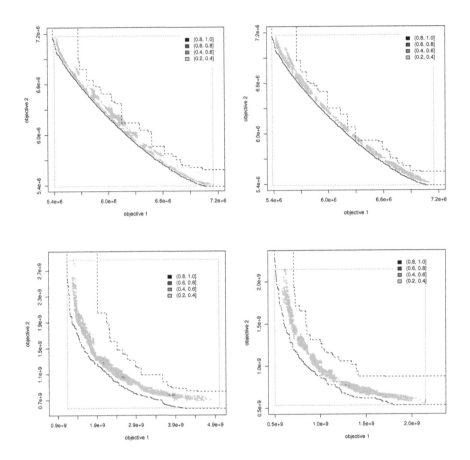

Figure 17.3. Location of differences between the different number of scalarizations (plots, where no points are in favor of one parameter setting, are not given); examples are given for an unstructured instance with $\rho = -0.75$ between n and $5n$ scalarizations in favor of the latter (top-left) and between $5n$ and $10n$ scalarizations in favor of the latter (top-right); below are the results for a structured instance with $\rho = 0.0$ between n and $5n$ scalarizations in favor of the latter (bottom-left) and another structured instance with $\rho = -0.75$ between n and $5n$ scalarizations in favor of the latter (bottom-right). All instances have size 50.

5.3 Tabu iterations

Better relation. For this SLS component, the analysis of the better relation provides the clear result that the RoTS approach (any of $50n$ and $100n$ tabu iterations) is preferable to using only an iterative improvement algorithm (II) for all unstructured instances, while only little differences could be found between $50n$ and $100n$ tabu iterations. The latter observation indicates some limiting performance after $50n$ tabu iterations. However, for structured instances no strong conclusion can be taken since almost every pair of outcomes is incomparable. In addition, size and correlation seems not to interfere with these observations in both, structured and unstructured instances.

Attainment functions.   Most statistical tests suggested a rejection of the null hypothesis on unstructured instances; the few exceptions occur when comparing $50n$ with $100n$ tabu iterations. The null hypothesis was also frequently rejected when comparing the outcomes on structured instances of size 25, while this was not the case for the larger instances. Different correlations of the flow matrices do not influence these observations.

Location of differences.   Figure 17.4 gives the location of large differences in terms of EAFs for an unstructured and a small structured instance, respectively. They show the differences between II and $50n$ tabu iterations in favor of the latter (left plots), and between $50n$ and $100n$ iterations in favor of the latter (right plots). The differences in favor of II, in the first case, and $50n$, in the second case, were below 20%. The search intensification has a different effect for unstructured and structured instances; while the largest differences for the former are found between II and $50n$ tabu iterations, for the structured instances these occur between $50n$ and $100n$ tabu iterations. In addition, the difference between II and RoTS seems to be higher for unstructured instances (see magnitude of the differences in the top-left plot of Figure 17.4).

5.4   Component-wise Step

Better relation.   The comparison between the use or not of the component-wise step with respect to the better relation was inconclusive, since every pairwise comparison resulted in an incomparable case.

Attainment functions.   In most cases, the null hypothesis was rejected. The null hypothesis was not rejected only for two unstructured instances of size 25 and for all unstructured instances of size 50 and 75 with positively correlated flow matrices.

Location of differences.   Figure 17.5 gives the location of the differences above 20% in unstructured (top plots) and structured (bottom plots) instances. Only the differences in favor of the use of this step are shown, since all the differences above 20% were in its favor. The top-left plot of Figure 17.5 shows their location for an unstructured instance with size 25 and with positively correlated flow matrices. These differences are almost imperceptible and are confined to a very small part of the trade-off. However, as the correlation of the flow matrices decreases in unstructured instances, the benefit of using this component-wise step becomes more relevant. The same result applies for structured instances, although the relation between the correlation of flow matrices and the performance of this component is less evident as for the unstructured instances.

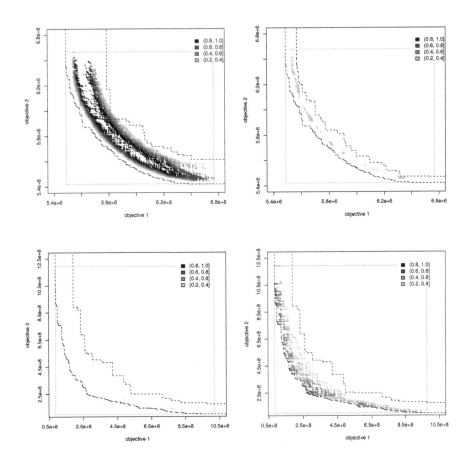

Figure 17.4. Location of differences between II and $50n$ tabu iterations in favor of the latter (left) and between $50n$ and $100n$ tabu iterations in favor of the latter (right) for an unstructured instance with $n = 50$ (top) and a structured instance with size 25 (bottom), each with $\rho = 0.0$. (Plots, where no points are in favor of one parameter setting, are not given.)

5.5   Summary

The strength of the underlying SLS algorithm for the scalarized problem, here represented by the number of tabu iterations, has a strong effect on the overall performance in unstructured instances. On the other hand, this fact seems to be irrelevant on medium size and large structured instances. The addition of the component-wise step and the increase from n to $5n$ scalarizations corresponds, in general, to an increase of the solution quality; the only exception occurs in unstructured instances with positively correlated flow matrices, where no significant improvement was found. Finally, the increase from $5n$ to $10n$ scalarizations did not correspond to a clear increase of performance, indicating some limiting behavior above these values.

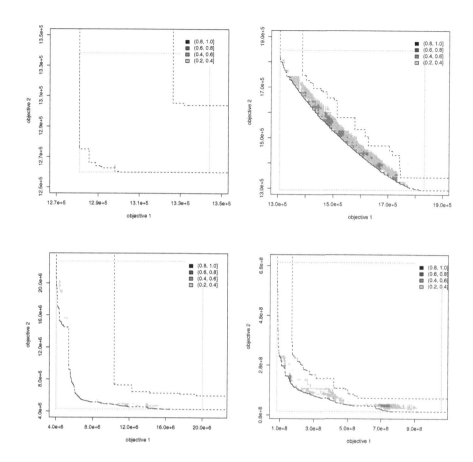

Figure 17.5. Location of differences for the use of the component-wise step for unstructured (top) and structured (bottom) instances with size 25 and $\rho = 0.75$ (left) and $\rho = -0.75$ (right); all differences are in favor of the use of component-wise step (plots in favor of not using the component-wise step are suppressed).

The statistical difference observed between search strategies on all the instances indicates that the Restart and 2phase strategies are far from producing similar outcomes. A noticeable difference is found on structured instances, which indicates a better performance of the Restart strategy. This behavior is actually very different from the one found for the TSP [20], where 2phase outperforms very clearly Restart. However, the 2phase strategy performed reasonably well for unstructured instances with positive correlation, although this is mainly due to the longer first phase of this search strategy.

Finally, the instance structure and the correlation between flow matrices have a strong effect on the performance of the SLS algorithms tested here. The differences observed between search strategies seem to be strongly related to the underlying structure of the input data. In addition, different correlations between flow matrices on unstructured instances are also influential on the choice for the component-wise step. Interestingly, the instance size has little effect on performance. A summary of the analysis is also given in Table 17.1, which gives the main findings classified according to correlation and instance class (structured vs. unstructured).

Table 17.1. Summary of the results of the experimental analysis.

	Unstructured	*Structured*
0.75	• 2phase is better than `Restart` on larger instances; • addition of component-wise step is not significant; • high no. of tabu iterations is better but few differences above $5n$; • high no. of scalarizations is better.	• `Restart` covers wider region than 2phase; • addition of component-wise step is better; • high no. of tabu iterations is better on small instances; • tabu iterations not significant on medium and large instances; • high no. of scalarizations is better on small instances; • no. of scalarizations not significant on medium and large instances.
0.0	• addition of component-wise step is better; • high no. of tabu iterations is better but few differences above $5n$; • high no. of scalarizations is better.	• as above.
−0.75	• `Restart` covers wider region than 2phase; • addition of component-wise step is better; • high no. of tabu iterations is better but few differences above 5n; • high no. of scalarizations is better.	• as above.

5.6   Computation time

Our experimental design uses different settings for the levels of several algorithm components and the computation time is a variable whose value depends on the particular configuration. There is also one more specific reason of leaving the computation time variable: when fixing computation time without an exact cost model for the dependence between computation time and parameter settings, it is very difficult to choose the algorithm components and parameter settings in such a way as to guarantee the termination of the algorithm within the given time limits and to still have a balanced experimental design.

When analyzing the computation time, it is obvious that, all other parameters being equal, the computation time clearly increases, for example, with an increasing number of tabu search iterations. While some of these trade-offs are obvious, a more fine-grained examination is of interest in this article; examples are the analysis of the interactions between the four components with respect to computation time or more particular questions like whether the component-wise step incurs a significant overhead in computation time.

These questions are answered by performing an ANOVA with respect to computation time. In order to meet the required assumptions for the ANOVA analysis, the computation times recorded were divided according the presence or lack of structure in the instance. Furthermore, instance size and the correlation of the flow matrices were defined as *crossed blocks* [4].

The analysis indicated the presence of the following two interactions in both structured and unstructured instances: *tabu iterations × scalarizations × size* and *tabu iterations × scalarizations × search strategy*. The first interaction means that, since the instance size affects the size of the neighborhood, this is reflected in the computation time resulting by changes in the number of tabu iterations or the number of scalarizations. The second interaction means that search strategies behave differently with the change of tabu iterations and the number of scalarizations. Rather obviously, as also indicated by ANOVA, the number of iterations and scalarizations have the largest effect on the computation times, while interestingly the component-wise step has very little effect.

Two more observations are noteworthy. The first is that `Restart` is faster than `2phase` under these experimental conditions. The reason is mainly due to the long first phase of `2phase`. The second is that there are no statistically significant differences between the `2phase` search strategies with an increasing number of scalarizations when using `II` as the underlying algorithm, due to the long time taken by the first phase. (More in detail, this is true between n and $5n$ on unstructured instances, and among all scalarizations tested here on structured instances).

6.   Conclusions

This article presented an extensive and sound experimental analysis of SLS algorithms for the BQAP and tested the effect of different choices for algorithmic components of these SLS algorithms. This analysis was based on the use of the better relation and attainment functions as well as an exploratory data analysis technique that provided a clear location of the largest differences between outcomes in the objective space. This performance assessment methodology avoids the known pitfalls of most of the measures of performance of SLS algorithms for multiobjective problems [31]. A further extension of this analysis is to use second-order attainment functions [6] to analyze the pairwise relationship between solutions.

One interesting outcome of this analysis was to verify that the use of the component-wise step often results in a significantly improved performance and that this step requires only a very small additional computation time. Similar results hold for the biobjective TSP [22] and it is therefore to be expected that the component-wise step is also effective for many other MCOPs. From a wider perspective, the analysis also emphasized the strong dependency between SLS algorithm performance and instance features like the structure and the correlation of the input data. Hence, these dependencies need to be taken into account when designing and implementing SLS algorithms for MCOPs.

Finally, some of the algorithms tested in this experimental analysis presented already, under the same experimental conditions, a comparable performance to a much more complex but high-performing SLS algorithm for unstructured instances of the BQAP [19]. For structured instances, taking into account the known better performance of Iterated Local Search algorithms over RoTS [26], much improved performance was obtained by some straightforward modifications of the 2phase search strategy presented here [20]. Therefore, besides clarifying the effect of several components of an SLS algorithm on performance, a careful experimental analysis like in this study can be used for designing high-performance, yet conceptually simple SLS algorithms.

References

[1] K. Andersen, K. Jörnsten, and M. Lind. On bicriterion minimal spanning trees: An approximation. *Computers & Operations Research*, 23(12):1171–1182, 1996.

[2] J. Conover. *Pratical Nonparametric Statistics*. John Wiley & Sons, New York, NY, 1980.

[3] P. Czyzak and A. Jaszkiewicz. Pareto simulated annealing - a metaheuristic technique for multiple objective combinatorial optimization. *Journal of Multi-Criteria Decision Analysis*, 7:34–47, 1998.

[4] A. Dean and D. Voss. *Design and Analysis of Experiments*. Springer Verlag, New York, NY, 1999.

[5] C. M. Fonseca and P. Fleming. On the performance assessment and comparison of stochastic multiobjective optimizers. In H. M. Voigt et al., editors, *Proceedings of PPSN-IV, Fourth International Conference on Parallel Problem Solving from Nature*, volume 1141 of *LNCS*, pages 584–593. Springer Verlag, Berlin, Germany, 1996.

[6] C. M. Fonseca, V. Grunert da Fonseca, and L. Paquete. Exploring the performance of stochastic multiobjective optimisers with the second-order attainment function. In C. C. Coello, A. H. Aguirre, and E. Zitzler, editors, *Evolutionary Multi-criterion Optimization (EMO 2005)*, volume 3410 of *LNCS*, pages 250–264. Springer Verlag, Berlin, Germany, 2005.

[7] X. Gandibleux and M. Ehrgott. 20 years of multiobjective metaheuristics. But what about the solution of combinatorial problems with multiple objectives? In C. C. Coello, A. H. Aguirre, and E. Zitzler, editors, *Evolutionary Multi-criterion Optimization (EMO 2005)*, volume 3410 of *LNCS*, pages 33–46. Springer Verlag, Berlin, Germany, 2005.

[8] X. Gandibleux, N. Mezdaoui, and A. Fréville. A tabu search procedure to solve multiobjective combinatorial optimization problems. In R. Caballero, F. Ruiz, and R. Steuer, editors, *Advances in Multiple Objective and Goal Programming*, volume 455 of *LNEMS*, pages 291–300. Springer Verlag, 1997.

[9] X. Gandibleux, H. Morita, and N. Katoh. Use of a genetic heritage for solving the assignment problem. In C. M. Fonseca, P. Fleming, E. Zitzler, K. Deb, and L. Thiele, editors, *Evolutionary Multi-criterion Optimization (EMO 2003)*, volume 2632 of *LNCS*, pages 43–57. Springer Verlag, Berlin, Germany, 2003.

[10] P. I. Good. *Permutation Tests: A pratical guide to resampling methods for testing hypothesis*. Springer Verlag, New York, USA, 2nd edition, 2000.

[11] V. Grunert da Fonseca, C. M. Fonseca, and A. Hall. Inferential performance assessment of stochastic optimizers and the attainment function. In E. Zitzler, K. Deb, L. Thiele, C. C. Coello, and D. Corne, editors, *Evolutionary Multi-criterion Optimization (EMO 2001)*, volume 1993 of *LNCS*, pages 213–225. Springer Verlag, Berlin, Germany, 2001.

[12] H. V. Hamacher, M. Labbé, and S. Nickel. Multicriteria network location problems with sum objectives. *Networks*, 33:79–92, 1999.

[13] M. P. Hansen and A. Jaszkiewicz. Evaluating the quality of approximations to the non-dominated set. Technical Report IMM-REP-1998-7, Institute of Mathematical Modelling, Technical University of Denmark, Lyngby, Denmark, 1998.

[14] H. Hoos and T. Stützle. *Stochastic Local Search – Foundations and Applications*. Morgan Kaufmann Publishers, San Francisco, CA, 2004.

[15] J. Hsu. *Multiple Comparisons - Theory and Methods*. Chapman & Hall/CRC, 1996.

[16] A. Jaszkiewicz. Genetic local search for multiple objective combinatorial optimization. Technical Report RA-014/98, Institute of Computing Science, Poznań University of Technology, Poznań, Poland, 1998.

[17] P. Klingsberg. A gray code for compositions. *Journal of Algorithms*, 3:41–44, 1982.

[18] J. Knowles and D. Corne. Instance generators and test suites for the multiobjective quadratic assignment problem. In C. M. Fonseca et al., editors, *Evolutionary Multi-criterion Optimization (2003)*, volume 2632 of *LNCS*, pages 295–310. Springer Verlag, Berlin, Germany, 2003.

[19] Manuel López-Ibáñez, Luis Paquete, and Thomas Stützle. Hybrid population-based algorithms for the bi-objective quadratic assignment

problem. *Journal of Mathematical Modelling and Algorithms*, 5(1):111–137, 2006.

[20] L. Paquete. *Stochastic Local Search Algorithms for Multiobjective Combinatorial Optimization: Methodology and Analysis*. PhD thesis, Fachbereich Informatik, Technische Universität Darmstadt, 2005.

[21] L. Paquete and C. Fonseca. A study of examination timetabling with multiobjective evolutionary algorithms. In *Proceedings of the 4th Metaheuristics International Conference (MIC 2001)*, pages 149–154, Porto, Portugal, 2001.

[22] L. Paquete and T. Stützle. A two-phase local search for the biobjective traveling salesman problem. In C. M. Fonseca, P. Fleming, E. Zitzler, K. Deb, and L. Thiele, editors, *Proceedings of the Evolutionary Multicriterion Optimization (EMO 2003)*, volume 2632 of *LNCS*, pages 479–493. Springer Verlag, Berlin, Germany, 2003.

[23] L. Paquete and T. Stützle. Stochatic local search for multiobjective optimization problems. In T. F. Gonzalez, editor, *Approximation Algorithms and Metaheuristics*. Chapman & Hall / CRC, In press.

[24] K. J. Shaw, C. M. Fonseca, A. L. Nortcliffe, M. Thompson, J. Love, and P. J. Fleming. Assessing the performance of multiobjective genetic algorithms for optimization of a batch process scheduling problem. In *Proceedings of the 1999 Congress on Evolutionary Computation (CEC'99)*, volume 1, pages 34–75, 1999.

[25] R. E. Steuer. *Multiple Criteria Optimization: Theory, Computation and Application*. Wiley Series in Probability and Mathematical Statistics. John Wiley & Sons, New York, NY, 1986.

[26] T. Stützle. Iterated local search for the quadratic assignment problem. *European Journal of Operational Research*, 2006. In press.

[27] É. D. Taillard. Robust taboo search for the quadratic assignment problem. *Parallel Computing*, 17:443–455, 1991.

[28] É. D. Taillard. Comparison of iterative searches for the quadratic assignment problem. *Location Science*, 3:87–105, 1995.

[29] E. G. Talbi. A hybrid evolutionary approach for multicriteria optimization problems: Application to the flow shop. In E. Zitzler, K. Deb, L. Thiele, C. C. Coello, and D. Corne, editors, *Evolutionary Multi-criterion Optimization (EMO 2001)*, volume 1993 of *LNCS*, pages 416–428. Springer Verlag, Berlin, Germany, 2001.

[30] E. L. Ulungu. *Optimisation combinatoire multicritére: Détermination de l'ensemble des solutions efficaces et méthodes interactives*. PhD thesis, Université de Mons-Hainaut, Mons, Belgium, 1993.

[31] E. Zitzler, L. Thiele, M. Laumanns, C. M. Fonseca, and V. Grunert da Fonseca. Performance assessment of multiobjective optimizers: An analysis and review. *IEEE Transactions on Evolutionary Computation*, 7(2):117–132, 2003.

Chapter 18

DISTANCE MEASURES AND FITNESS-DISTANCE ANALYSIS FOR THE CAPACITATED VEHICLE ROUTING PROBLEM

Marek Kubiak[1]

[1]*Institute of Computing Science*
Poznan University of Technology
Piotrowo 2, 60-965 Poznań
Poland
Marek.Kubiak@cs.put.poznan.pl

Abstract   The way a metaheuristic algorithm is adapted to a given problem is a central issue, as it may considerably influence the efficiency of the resulting algorithm. Certain schemes of such adaptation rely on statistical analyses of the fitness landscape of instances of the problem, e.g. on the fitness-distance analysis. This kind of analysis requires that distance measures for solutions of the problem are defined.

   The paper presents the fitness-distance analysis of the Capacitated Vehicle Routing Problem (CVRP). Certain distance metrics are defined, experiments with these metrics and other measures on well-known instances of the CVRP described, and results of the analysis provided. These results reveal traces of some structure ('big valley') in fitness landscapes of more than half of the considered instances, which may provide plausible explanation for efficiency of a well-known metaheuristic algorithm for the problem. They also confirm that fitness-distance analysis could become a tool used by designers of metaheuristics to guide and justify their design choices.

Keywords: fitness-distance analysis, fitness landscape, distance measures, design of metaheuristics, capacitated vehicle routing problem

1.   Introduction

Metaheuristics, usually inspired by nature, define schemes of algorithms which have to be further adapted to a given problem in order to be useful (e.g. neighbourhood operators have to be chosen for local search, crossover operators have to be chosen or designed for an evolutionary algorithm).

One of the conclusions which may be derived from the No Free Lunch theorem [16] is that there is no algorithm which performs better than any other on all possible optimisation problems. Thus, an algorithm may be efficient only on a certain subclass of all problems, while in other cases it will perform worse than others, even random search or complete enumeration of solutions. This conclusion applies to metaheuristics as well. In this context the process of adaptation of a metaheuristic may be viewed as a means of moving the 'efficient' (for this algorithm) subclass of problems in the space of all possible problems.

This process of adaptation, the choice or design of components of a metaheuristic, may profoundly influence the efficiency of an algorithm. For a class of metaheuristics, namely evolutionary algorithms, this phenomenon was even observed experimentally [9]. Thus, the adaptation of a metaheuristic to a problem is a crucial operation which should be well justified and carefully performed.

In many cases, however, this adaptation is done by a designer based on his or her intuition and experience. It may result, of course, in an efficient algorithm for the problem, which is quite often the case, but the design process may also result in a poorly performing algorithm, which happens not so rarely. What is more important, this way of adaptation provides neither justification for the choices made by the designer, nor gives knowledge which might be useful for others in the future.

A different way of adaptation of a metaheuristic algorithm to a given problem has received some attention recently. It is based on statistical analyses of the search space (fitness landscape) of instances of the problem. Such analyses, e.g. fitness-distance analysis [2, 3, 6, 8], random-walk correlation measurement [8], estimation of the size of atractor of local optima [8], provide objective information about certain properties of the fitness landscape and may justify the design or choice of components for a metaheuristic algorithm. One example of such design technique is the construction of so-called distance preserving crossover operators (DPX) based on results of fitness-distance analysis [3, 6, 8]; these operators create an offspring which is not further from its parents than these parents are from each other with respect to a distance measure. Such operators are deemed useful in evolutionary optimisation if the 'big valley' structure exists in the fitness landscape of the considered problem, i.e. when good solutions of the problem are close (similar) to each other [3, 6, 8].

Therefore, the first goal of this paper is to present results of statistical analysis, namely the fitness-distance analysis, of some instances of the Capacitated Vehicle Routing Problem (CVRP) taken from literature. As Boese stated in his paper [2] about the analysis of a fitness landscape: 'understanding this cost [(fitness)] surface can help both to explain the success of previous heuristics (e.g. simulated annealing) and to motivate new, more effective heuristics'.

The second goal of this work is to present some new distance metrics for solutions of the CVRP and compare their properties with other measures developed earlier [13] or at the same time [14] in order to provide some guidelines on their proper use.

Additionally, in the author's opinion many publications about algorithms for combinatorial optimisation problems describe fitness landscapes using the notion of distance without actually providing its definition: moves in the landscape are told to be near of far jumps, search strategies are said to intensify the search in a promising region of the search space or to diversify the search by identifying new regions with good solutions. However, accurate, objective, quantitative information about what is near or what is a region is rarely provided, with few exceptions [11, 13, 14, 17]. Thus, this paper demonstrates that also in the case of combinatorial structures distance may be defined strictly and become a useful tool for verification of research intuition.

2. The Capacitated Vehicle Routing Problem

The CVRP [1, 12, 15] is a very basic formulation of a problem which a transportation company may face in its everyday operations. The goal is to find the shortest-possible set of routes for the company's vehicles in order to satisfy demands of customers for certain goods. Each of identical vehicles starts and finishes its route at the company's depot, and must not carry more goods than its capacity specifies. All customers have to be serviced, each exactly once by one vehicle. Distances between the depot and customers are given.

The version of CVRP considered here does not fix the number of vehicles (it is a decision variable); also the distance to be travelled by a vehicle is not constrained. Names of instances used in this study are provided in table 18.1, in section 5.0.

In order to describe distance metrics properly in the next section, some basic definitions related to solutions of the CVRP have to be given. A solution s of the CVRP is a set of $T(s)$ routes:

$$s = \{t_1, t_2, \ldots, t_{T(s)}\}.$$

A route is a sequence of customers (nodes) starting with the depot, v_0, and has the form:

$$t_i = (v_0, v_{i,1}, v_{i,2}, \ldots, v_{i,n(t_i)}) \qquad \text{for } i = 1, \ldots, T(s),$$

where $n(t_i)$ is the number of customers on route t_i.

Some additional specific constraints should also be satisfied by a feasible solution, but they are not provided here. Refer to [1, 12, 15] for more information about the CVRP.

3. New distance metrics for solutions of the CVRP

The distance metrics presented in this section correspond to certain structural properties of solutions of the CVRP which are deemed important for their quality: existence of certain edges (or even paths) or specific ways of partitioning of the set of customers into routes (clusters). Although these metrics might seem simple at first sight, their strength lies in the fact that they are linked directly to the mentioned properties of CVRP solutions, not to any specific solution representation.

Distance in terms of edges: d_e

The idea of this metric is based on a very similar concept formulated for the travelling salesman problem (TSP): the number of common edges in TSP tours [2]. In the cited research it was found that good solutions of the TSP have many common edges and are, thus, very close to each other. Due to similarity between solutions of the TSP (one tour) and the CVRP (a set of disjoint tours/routes) the idea of common edges may be easily adapted to the latter.

In order to introduce the distance metric some definitions are required:

$$E(t_i) = \Big\{\{v_0, v_{i,1}\}\Big\} \cup \Big(\bigcup_{j=1}^{n(t_i)-1} \big\{\{v_{i,j}, v_{i,j+1}\}\big\} \Big) \cup \Big\{\{v_{i,n(t_i)}, v_0\}\Big\}$$

$$E(s) = \bigcup_{t_i \in s} E(t_i)$$

$E(t_i)$ is a multiset of undirected edges appearing in route t_i. $E(s)$ is a multiset of edges in solution s. The notion of a multiset is required here, because routes in some solutions of the CVRP may include certain edges twice.

Using the general concept of distance between subsets of the same universal set, as defined by Marczewski and Steinhaus [7] (cited after [5]), the distance d_e between two solutions s_1, s_2 of the same CVRP instance may be defined as:

$$d_e(s_1, s_2) = \frac{|E(s_1) \cup E(s_2)| - |E(s_1) \cap E(s_2)|}{|E(s_1) \cup E(s_2)|}$$

Due to the fact that d_e is only a special case of the Marczewski-Steinhaus distance, it inherits all its properties of a metric; its values are also normalized to the interval [0,1].

This distance metric perceives solutions of the CVRP as multisets of edges: solutions close to each other will have many common edges; distant solutions will have few common ones. However, closer investigation of the metric reveals that it is not intuitively 'linear' (although it is 'monotonic'), e.g. $d_e = 0.5$ does not mean that exactly half of each $E(s_i)$ is common; 50% of common edges implies $d_e \approx 2/3$.

Distance in terms of partitions of customers: d_{pc}

The concept behind the second distance metric is based on the 'cluster first – route second' heuristic approach to the CVRP [1, 15]: first find a good partition of customers into clusters and then try to find routes (solve TSPs) within these clusters, separately. According to this idea the distance metric should identify dissimilarities between solutions perceived as partitions of the set of customers.

An example of a distance metric for partitions of a given set may be found in [5] (it is even more generally defined there, for hypergraphs or binary trees). This example is easily adaptable to solutions of the CVRP. Let us define:

$$C(s) = \{c_1(s), c_2(s), \ldots, c_{T(s)}(s)\}$$

$$c_i(s) = \{v_{i,1}, v_{i,2}, \ldots, v_{i,n(t_i)}\}$$

$$\sigma(c_i(s_1), c_j(s_2)) = \frac{|c_i(s_1) \cup c_j(s_2)| - |c_i(s_1) \cap c_j(s_2)|}{|c_i(s_1) \cup c_j(s_2)|}$$

$C(s)$ is a partition of the set of customers into clusters; one cluster, $c_i(s)$, holds customers from route t_i of s; $\sigma(\cdot)$ is the distance between two clusters.

According to [5], the distance between solutions may be defined as:

$$d_{pc}(s_1, s_2) = 1/2\left\{ \max_{i=1}^{T(s_1)} \min_{j=1}^{T(s_2)} \sigma(c_i(s_1), c_j(s_2)) + \max_{i=1}^{T(s_2)} \min_{j=1}^{T(s_1)} \sigma(c_i(s_1), c_j(s_2)) \right\}$$

This function is a distance metric for partitions; it is also normalized. It is not exactly a metric for solutions of the CVRP, because $d_{pc}(s_1, s_2) = 0$ does not imply $s_1 = s_2$ (the number of solutions which are not discriminated by d_{pc} may be exponentially large).

The formula for d_{pc} has the following sense: firstly, the best-possible assignment (matching) of clusters from $C(s_1)$ to clusters from $C(s_2)$ is made (the one which minimizes $\sigma(\cdot)$), and vice versa; that is the idea behind internal min operators. Secondly, two worst assignments are chosen among those pairs (the max operators), and distances in those two assignments are averaged to form the overall distance between partitions. Thus, it may be concluded that d_{pc} is somehow 'pessimistic' in the choice of 'optimistic' matches of clusters.

This mixture of max and min operators in d_{pc} makes interpretation of its values difficult. Certainly, values near to 0 indicate great similarity of solutions. However, larger values do not necessarily indicate very dissimilar partitions; it is sufficient that there are 'outliers' in partitions, which can hardly be well assigned to clusters in the other solution, and the max operator will result in large values, implying distant solutions.

Distance in terms of pairs of nodes: d_{pn}

The third distance metric, d_{pn}, is based on the same idea as d_{pc}: distance between solutions viewed as partitions of the set of customers. However, this idea has a different, more straightforward mathematical formulation in d_{pn}. Here, the Marczewski-Steinhaus [7] concept of distance is applied to sets of *pairs of nodes* (customers).

Lets define:

$$PN(t_i) = \bigcup_{j=1}^{n(t_i)-1} \bigcup_{k=j+1}^{n(t_i)} \left\{ \{v_{i,j}, v_{i,k}\} \right\}$$

$$PN(s) = \bigcup_{t_i \in s} PN(t_i)$$

$PN(t_i)$ is the set of undirected pairs of nodes (customers) which are assigned to the same route t_i (it is a complete graph defined over the set of customers in route t_i). $PN(s)$ is the set of all such pairs in solution s.

The distance d_{pn} between solutions is defined as:

$$d_{pn}(s_1, s_2) = \frac{|PN(s_1) \cup PN(s_2)| - |PN(s_1) \cap PN(s_2)|}{|PN(s_1) \cup PN(s_2)|}$$

Similarly to d_{pc}, this function is not exactly a metric for solutions of the CVRP, but for partitions implied by those solutions.

The formula for d_{pn} has more straighforward sense than the one for d_{pc}; here, the value of distance roughly indicates how large are parts of sets of pairs which are not shared by two compared solutions. If $d_{pn} = 0$ then two solutions imply identical partitions; $d_{pn} = 1$ implies completely different partitions (not even one pair of nodes is assigned to a route in the same way in s_1 and s_2).

4.   Distance measures defined in the literature

In recent years some more distance measures and metrics for solutions of the CVRP have been described in the literature. These are:

- the edit distance [13],

- the add-remove edit distance for Prins' split representation [14],

- the stop-centric and route-centric distance measures [17].

The author managed to analyse and implement the two first measures, so they are described in the sections below. He did not manage, however, to implement the measures developed at the same time by Woodruff and Lokketangen [17], so these distances are not considered here. Nevertheless, it is important to remember that such measures exist and should also be, in the near future, compared to those described in this work.

It is not the purpose of this paper to provide detailed definitions of all existing distance measures for solutions of the CVRP. Therefore, in the sections below the measures are only shortly described and their properties discussed. For the detailed definitions, implementation issues and examples the interested reader is referred to the cited publications.

Edit distance for CVRP solutions: d_{eu}

Sorensen [13] defined a distance measure for solutions of the CVRP based on the concept of the edit distance between strings. An edit operation on a string is a modification of one its character by means of an elementary operation: insertion, deletion or substitution. Sorensen describes how to define an edit distance on permutations. Further, he extends this distance to the case of permutations with reversal independence (or undirected permutations, like single routes in the CVRP) and to the case of sets of such permutations (like solutions of the CVRP). The sets of permutations (routes) of two CVRP solutions are matched in this process (in an optimal way) by solving the minimal-cost assignment problem. Therefore, it is possible to determine which routes in one solution correspond to which routes in the other solution. This distance measure will be called d_{eu} in this paper (the edit distance for undirected routes).

Although the edit distance for strings and undirected permutations is a metric, it is not clear whether d_{eu} is a metric for solutions of the CVRP; this matter is not clarified in [13] (although it is not the most important property of a measure and is not required here). This measure is not normalized, as well.

The value of this distance is the minimal number of elementary edit operations required to transform one set of permutations (a CVRP solution) into another set. Thus, $d_{eu} = 0$ implies that compared solutions are identical (there is no edit operation required).

This measure focuses on the same order of customers in the matched routes; if this order is disturbed somehow, then some edit operations are required to perform the necessary transformation. In this sense, the function d_{eu} is similar to d_e, which also stresses the aspect of order (by inspecting edges and paths). For this edit distance, however, it is also important that long identical subpaths are in the same places of (absolute positions in) routes in two solutions. Even if such long subpaths exist in matched routes, a difference in their positions in these routes may incur some additional edit cost. In consequence, this property of d_{eu} makes it different from d_e, which disregards positions of customers in routes and only takes edges into account.

Since the order of customers in routes is important for d_{eu}, it means that the same suites of vertices in routes of two solutions (the same clusters) are not enough for this measure to make these solutions close. This fact should make it different from metrics which concentrate on clusters only: d_{pc} and d_{pn}.

It is also worth noting that the distance d_{eu} is inflated when numbers of routes in two compared solutions differ. This is due to the fact that the assignment problem involved in the distance computation has to match some routes of one solution to artificially added empty routes in the other one; it implies performing additional insertions or deletions.

Add-remove edit distance for split representation: d_{ear}

In their work on a path relinking procedure for the CVRP, Sorensen et al. [14] proposed another kind of distance measure for CVRP solutions, which is also based on the concept of edit operations. This measure, however, compares solutions encoded in the split representation (proposed earlier by Prins in [10]), which consists of only one permutation of customers and is decoded into a CVRP solution by an optimisation algorithm. The distance between such permutations defined in [14] is the edit distance wihtout the operation of substitution; only insertions and deletions are considered. The authors call it the 'add-remove edit distance', so it will be denoted d_{ear} hereafter. The cost of one such edit operation is set to $1/2$ in [14], but here it is assumed to be equal to 1.

This measure is not normalized, but this might be amended easily by introducing in its formula a factor being the reciprocal of twice the number of customers. This measure is a metric for permutations, but is not exactly a metric for CVRP solutions, because not every solution of the problem may be encoded in the split representation and decoded back without changes. Nevertheless, it seems not to be a great disadvantage of this distance function if one imagines an algorithm working only on solutions encoded in this representation. However, in case any two CVRP solutions had to be compared by a distance measure, this distance would not be directly useful, unless all solutions were encoded as permutations (perhaps with some loss of information on the actual routes). In order to have the possibility of comparison of measures, this approach is applied in this paper.

It is harder to provide interpretation of values of this measure than in the previous case. This value is, of course, the minimal number of edit operations required to transform one permutation into another one, but it is not clear how an edit operation influences the underlying CVRP solution. Due to the nature of the split representation it is unknown which edges actually exist in a solution, and where each route starts and finishes. Thus, an edit operation on a permutation may imply in the decoded solutions additional modifications of vertices which are not directly involved in the edit operation itself. This phenomenon is visible in the example provided by the authors in [14] (page 844, 3 last move operations). It seems that this property of d_{ear} might decrease its utility.

5. Fitness-distance analysis of the CVRP

Random solutions vs. local optima

The first stage of the fitness-distance analysis focuses on possible differences between sets of local optima and random solutions of a given instance in terms of distance in these sets.

In order to check these differences in case of the CVRP, large random samples of 2000 different solutions of each type were generated:

- a random solution: first, a random sequence of customers was drawn (each permutation having the same probability of being chosen); then, this permutation was split into a number of routes, with this number drawn from the binomial distribution; each infeasible solution was abandoned and a new one created;

- a local optimum: a random solution was subject to a greedy local search with a neighbourhood of 3 joined operators: 2-opt, exchange of 2 customers, joining of 2 routes [6].

Finally, statistics on values of distance in these samples were computed, as shown in table 18.1 and figure 18.1. It may be seen in the table and the figure that average values of each type of distance in samples of local optima are usually much lower than the corresponding averages for random solutions. The highest difference is visible, quite surprisingly, for d_{ear}. The next one is d_{pn}, and d_{eu} and d_e follow. The smallest differences appear for distance d_{pc}, perhaps due to its 'pessimistic' nature mentioned earlier.

During inspection of plots of fitness versus distance d_e for the sets of random solutions an interesting observation was made: there were visible trends in these sets, indicating that *better* random solutions actually tend to be *further* from each other than worse ones. The author's guess is that this phenomenon was observed because worse random solutions usually have more routes than better ones. In consequence, solutions with low quality usually shared many edges which start at the depot node, so the average values of d_e between random solutions were artificially decreased. Thus, these average values given in table 18.1 are biased.

From this experiment it may be concluded that local optima are clustered together in some parts of the fitness landscapes rather than scattered all over it, like solutions generated at random. This also means that they usually share many common properties: the same assignments of customers to routes or the same edges/subpaths.

Table 18.1. Average values of distance in sets of random solutions and local optima for each instance.

Instance name	Random solutions					Local optima				
	d_e	d_{pc}	d_{pm}	d_{eu}	d_{ear}	d_e	d_{pc}	d_{pm}	d_{eu}	d_{ear}
c100	0.7250	0.8606	0.9909	82.5	166.4	0.6729	0.7661	0.7311	74.9	109.6
c100b	0.7250	0.8596	0.9908	82.6	166.6	0.5397	0.7033	0.5891	55.0	75.2
c120	0.7271	0.8661	0.9918	98.8	203.0	0.7136	0.7941	0.7459	92.0	138.4
c150	0.7279	0.8714	0.9936	123.0	258.1	0.6785	0.8104	0.7592	112.6	168.6
c199	0.7305	0.8771	0.9951	162.6	349.0	0.7010	0.8263	0.7913	153.8	234.6
c50	0.7147	0.8336	0.9820	41.2	77.5	0.6053	0.7433	0.6586	32.5	45.1
c75	0.7110	0.8381	0.9887	60.9	121.7	0.6883	0.7989	0.7517	52.1	75.8
f134	0.7298	0.8702	0.9929	110.5	228.4	0.6326	0.7817	0.7095	104.9	150.4
f71	0.7170	0.8472	0.9865	58.6	114.4	0.4770	0.6460	0.3384	37.5	47.4
tail100a	0.7119	0.8518	0.9910	80.7	166.4	0.5947	0.7849	0.6904	64.8	90.9
tail100b	0.7121	0.8522	0.9911	81.0	166.4	0.6105	0.8061	0.6977	67.6	94.9
tail100c	0.7149	0.8525	0.9910	81.2	166.6	0.6255	0.8086	0.7070	66.7	96.9
tail100d	0.7116	0.8529	0.9905	80.9	166.6	0.6088	0.8241	0.6766	71.1	99.7
tail150a	0.7183	0.8651	0.9937	121.3	258.1	0.6272	0.8191	0.7308	106.9	153.8
tail150b	0.7229	0.8682	0.9938	122.1	257.9	0.6627	0.8258	0.7306	108.0	158.7
tail150c	0.7202	0.8667	0.9936	121.8	257.9	0.6290	0.8118	0.7357	106.0	152.7
tail150d	0.7185	0.8654	0.9939	121.3	258.2	0.6356	0.8289	0.7496	113.7	161.6
tai385	0.7128	0.8798	0.9976	303.8	699.9	0.7451	0.8784	0.8606	312.9	476.9
tai75a	0.6994	0.8391	0.9887	60.2	121.6	0.6212	0.7952	0.7151	49.9	69.0
tai75b	0.7022	0.8389	0.9885	60.2	121.5	0.6176	0.8079	0.7324	51.5	73.4
tai75c	0.7005	0.8410	0.9884	60.0	121.6	0.5813	0.7883	0.6822	48.5	67.7
tai75d	0.7090	0.8427	0.9886	60.9	121.7	0.5193	0.7626	0.5488	42.7	56.4

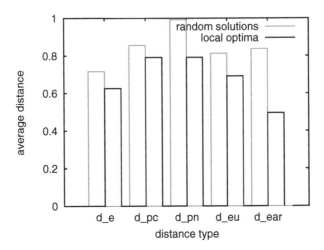

Figure 18.1.   Average values of distance in sets of random solutions and local optima.

Trends in sets of local optima

The second stage of the fitness-distance analysis is an attempt to find trends in the sets of local optima themselves and verify the 'big valley' hypothesis: if better solutions tend to be closer (more similar) to each other [2, 3, 6, 8, 11]. Such trends are usually revealed by means of the fitness-distance correlation coefficient (FDC). In case of this study positive values of correlation would indicate a 'big valley' structure: solutions with lower cost would be closer to each other in search space.

In order to verify this hypothesis and compute the values of FDC for instances of the problem the following experiment was conducted:

1 In all sets of local optima (the same as in the previous experiment) and for each pair of solutions two values of the objective function were computed $(f(s_1), f(s_2))$; also values of all distance measures were evaluated $(d(s_1, s_2))$. Consequently, one pair of solutions constituted a single, 3 dimensional observation.

2 In each set of local optima two values of correlation were computed for each distance: $r(f(s_1), d(s_1, s_2))$ and $r(f(s_2), d(s_1, s_2))$; the value of the linear determination coefficient was computed as:

$$r^2 = r^2(f(s_1), d(s_1, s_2)) + r^2(f(s_2), d(s_1, s_2)).$$

3 Plots of fitness versus distance were generated and inspected for existence of the described trends.

The computed values of the determination coefficient should not be biased because no correlation between $f(s_1)$ and $f(s_2)$ was observed. Therefore, values of $r^2(f(s_1), d(s_1, s_2))$ and $r^2(f(s_2), d(s_1, s_2))$ could be added. These values are presented in table 18.2.

A comment on this FDA method should be given. This way of computing FDC (its square, actually) is different from those proposed earlier, e.g. in [2–4, 6, 8, 11] and others. It is based on a 3 dimensional model of a fitness landscape, where *pairs* of solutions (s_1, s_2) are subject to measurement of values: $f(s_1), f(s_2), d(s_1, s_2)$. This way of analysis does not require the knowledge about global optima, as it is in the case of the approaches cited above, so it may be used for problems for which such solutions remain unknown.

Table 18.2. Values of the linear determination coefficient r^2 between fitness and each distance measure for all considered instances of the CVRP.

Instance name	Linear determination coefficient				
	r_e^2	r_{pn}^2	r_{pc}^2	r_{eu}^2	r_{ear}^2
c100	**0.1971**	0.1255	0.0622	0.1156	0.0037
c100b	**0.3469**	**0.4684**	**0.2110**	**0.4140**	0.0266
c120	0.0200	*0.1524*	0.0905	0.0654	0.0140
c150	**0.2959**	**0.2320**	0.0150	**0.2073**	0.0246
c199	**0.2371**	**0.2415**	0.0136	**0.2044**	0.0077
c50	*0.1665*	*0.1729*	0.0687	0.1496	0.0158
c75	0.1123	0.1119	0.0107	0.0792	0.0011
f134	*0.1597*	0.0447	0.0993	0.0833	0.0099
f71	**0.2369**	**0.3782**	0.0457	**0.3263**	0.0099
tai100a	0.0789	0.1099	0.0342	0.1144	0.0196
tai100b	**0.2382**	**0.2333**	0.0321	**0.2120**	0.0084
tai100c	**0.2458**	**0.3797**	0.0363	**0.2897**	0.0312
tai100d	*0.1779*	**0.2717**	0.0820	**0.1982**	0.0172
tai150a	0.0086	0.0003	0.0033	0.0019	0.0013
tai150b	0.0716	**0.2234**	0.0350	0.1264	0.0210
tai150c	**0.1896**	**0.2174**	0.0462	**0.2158**	0.0376
tai150d	0.0420	0.0573	0.0111	0.0438	0.0096
tai385	0.1225	*0.1577*	0.0473	0.1059	0.0013
tai75a	*0.1654*	*0.1561*	0.0207	*0.1634*	0.0172
tai75b	0.0246	0.0750	0.0262	0.0703	0.0077
tai75c	**0.1874**	**0.2576**	0.0338	**0.2002**	0.0164
tai75d	**0.2440**	**0.2775**	0.0587	**0.1993**	0.0057
Avg.	0.1622	0.1975	0.0493	0.1630	0.0140
Std. dev.	0.0931	0.1161	0.0448	0.0975	0.0099

The values of r^2 emphasized in boldface in table 18.2 are those greater than 0.18. One such value corresponds to two independent values of correlations $r^2(f(s_1), d(s_1, s_2))$ and $r^2(f(s_2), d(s_1, s_2))$ being at least 0.3. Although these values are not large, the author thinks they are significant as indicators of fitness-distance correlation. According to one of the first pieces of work on FDC [4], even single values of correlation as large as 0.15 might be indicators of a 'big valley'.

All cases with $r^2 \in [0.15, 0.18)$ are typeset in italic. These are values which are deemed 'borderline cases'; perhaps there exists a 'big valley', but there is more doubt about it.

First general observation based on the values in table 18.2 is that d_{ear} is not correlated with fitness at all. Thus, it seems that this type of distance does not reveal any 'big valley' in the CVRP. A very similar conclusion might be derived from values of r^2_{pc}: d_{pc} does not correlate with fitness except for one instance, c100b.

Conclusions are different in case of the three other measures. Firstly, d_e reveals fitness-distance correlation for 10–14 instances out of 22. Significant values of FDC indicate that in these cases better solutions tend to contain more common edges.

Secondly, when d_{pn} is taken into account, it appears as though there are 'big valleys' in 11–15 cases (mostly the same as for d_e). It means that for these instances good local optima usually contain similar clusters (assignments of customers to routes).

Finally, d_{eu} reveals fitness-distance correlation in 10-11 cases (again, usually the same as for d_e). This result suggests that good solutions of the CVRP are closely related in terms of edit operations on routes.

Looking at table 18.2 from the point of view of instances, one can see that 9 instances out of 22 reveal 'big valleys' for 3 distance measures: d_e, d_{pn}, and d_{eu}. This result suggests that in good solutions of these instances it is important to preserve contents of routes (clusters) and the order of customers in routes (edges/paths/sequences). Consequently, these instances should be easy for algorithms which preserve these properties of solutions. The easiest instance should be c100b, which reveals the largest values of r^2.

There are also 5 instances which do not reveal fitness-distance correlation with respect to any distance measure used here: c75, tai100a, tai150a, tai150d, tai75b. Values of FDC for each of them are very small. Therefore, these instances should be hard for optimisation by means of algorithms preserving clusters or edges.

The other instances listed in table 18.2 are intermediate cases: there is some indication of 'big valley' with respect to one measure, but not when others are taken into account.

The conclusions derived from values of FDC may be further verified through inspection of fitness-distance plots. In this study, 2 dimensional FD plots are constructed based on the mentioned 3 dimensional observations: pairs of solutions with approximately the same fitness (max. 5% difference) are selected and plotted against distance. In figures 18.2 and 18.3 one can see samples of such plots for all distance measures. Inspection of these plots confirms the conclusions derived from the values of FDC. As expected, the strongest trends (not very strong, though) are revealed for d_e, d_{pn} and d_{eu}, while for d_{pc} and d_{ear} solutions are more evenly distributed with respect to distance, or form a curious shape with several horizontal levels in case of instance f71.

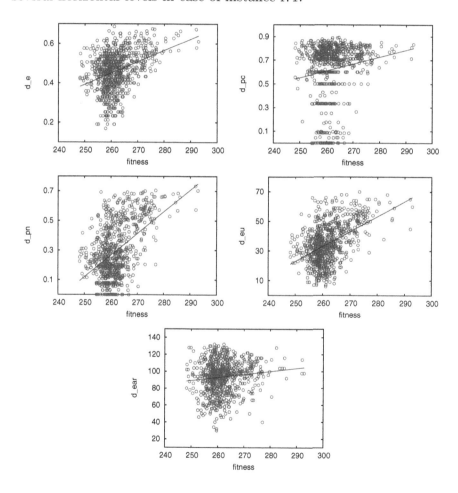

Figure 18.2. Fitness-distance plots with local optima for instance f71 and all types of distance, together with lines of regression.

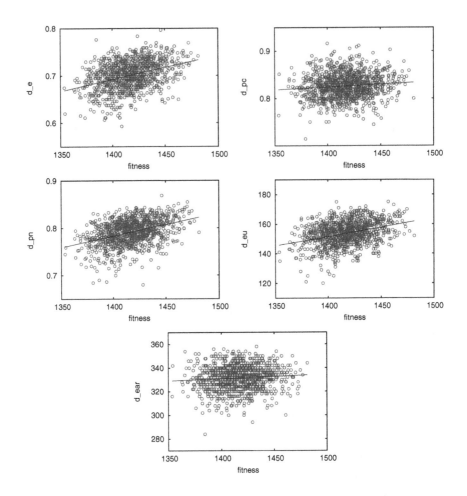

Figure 18.3. Fitness-distance plots with local optima for instance c199 and all types of distance, together with lines of regression.

6.   Relationships between distance measures

The relationships between distance measures are another interesting issue. Because it is difficult to track such relationships analytically, the author attempted to reveal them through a sampling experiment. In order to achieve this goal, values of the correlation coefficient for each pair of distances were computed for each problem instance, using the sets of local optima generated earlier. These values, averaged over all instances, are shown in table 18.3.

This table reveals that measures d_{pn} and d_{eu} are highly correlated ($r \approx 0.8$). This value suggests that the edit distance is focused on

Table 18.3.   Values of correlations between distance measures.

	d_e	d_{pn}	d_{pc}	d_{eu}	d_{ear}
d_e	1.0000	0.5686	0.2954	0.6663	0.1483
d_{pn}	—	1.0000	0.3845	0.8046	0.2051
d_{pc}	—	—	1.0000	0.3696	0.0745
d_{eu}	—	—	—	1.0000	0.2040
d_{ear}	—	—	—	—	1.0000

existence of the same clusters (assignments of customers to routes). On the other hand, the correlation between d_e and d_{eu} is weaker ($r \approx 0.67$) implying that common edges are not as important for the edit distance as clusters are. Given the observation made earlier about properties of d_{eu} this is not surprising: many common edges are not enough to make solutions close to each other in the sense of edit distance; they have to come in sequences in order to be useful for edit operations.

Distances d_{pn} and d_e are correlated on the intermediate level ($r \approx 0.57$), which is also not surprising: if a common edge exists in two compared solutions, then the end vertices of this edge form a common pair of nodes. Therefore, low values of d_e imply low values of d_{pn} (but the opposite is not true).

Much lower correlation exists between d_{pc} and other measures. The highest, $r \approx 0.38$, is between d_{pc} and d_{pn}, which is understandable when their common background is accounted for. Even lower correlations exist between d_{ear} and other measures. In the author's opinion, these observations suggest that d_{pc} and d_{ear} are measuring distance between CVRP solutions in some manner which is difficult to explain in terms of edges or clusters of customers.

7.   Conclusions

This study revealed that local optima of the considered instances of the CVRP are closer to each other than random solutions. This fact may explain why efficient algorithms for this problem should include local search components; it simply reduces the size of space to be searched for good solutions.

Moreover, the experiments described here unveiled traces of the 'big valley' structure in more than half of the instances and with respect to 3 distance measures. This discovery indicates that metaheuristics which preserve the properties of solutions correlated with fitness (clusters, edges or subpaths) could be very efficient for the CVRP, at least in case of these instances.

The correlations of fitness and distance might also explain why the intensification technique presented in [12], which has been extremely conservative in changing 'good' edges or routes, has been so successful.

Given these results it may be concluded that fitness-distance analysis could become a tool widely used by designers of metaheuristic algorithms, because once the notion of distance is clearly and unambiguously defined, the analysis provides objective information about properties of solutions which are important to the overall fitness. Research intuition might be, therefore, confirmed or rejected by this kind of analysis [3, 6, 8].

Additionally, this paper examined relationships between distance measures themselves. The author deems that the most sensible measures among those described here are d_{pn}, d_e and d_{eu}: each of them is correlated with fitness to some extent; the first two have also very straightforward interpretations. The other measures, d_{pc} and d_{ear}, are less valuable, being uncorrelated with fitness and rather difficult to understand.

Continuation of this work should firstly focus on the examination of efficiency of the optimisation techniques which rely on fitness-distance correlation (e.g. distance preserving crossovers [3, 6, 8]). Such work should verify the impact of FDC on the process of optimisation.

Another important direction of research would be a comparative study of distance measures described here and by Woodruff and Lokketangen [17]. Such a study could ultimately answer the question which measures are sensible and useful in the case of the CVRP.

Yet another interesting matter seems to be the issue of convergence of metaheuristic algorithms. Since there are distance measures and metrics for solutions of the CVRP available, such an analysis could focus on convergence in the space of *solutions* rather than in the space of fitness values.

These lines of research, which unfold with this and similar studies, may, hopefully, lead to development of practical guidelines for designers of future metaheuristic algorithms.

Acknowledgements

This research has been supported by the Polish State Committee for Scientific Research (KBN) through grant no. 3T11F00426.

References

[1] Aronson, L. D. (1996): "Algorithms for Vehicle Routing - a Survey". Technical Report [DUT-TWI-96-21]. Delft University of Technology, The Netherlands.

[2] Boese, K. D. (1995): "Cost Versus Distance in the Traveling Salesman Problem". Technical Report [TR-950018]. UCLA CS Department, California.

[3] Jaszkiewicz, A., and Kominek, P. (2003): "Genetic Local Search with Distance Preserving Recombination Operator for a Vehicle Routing Problem". In: *European Journal of Operational Research* **151**, 352–364.

[4] Jones, T., and Forrest, S. (1995): "Fitness Distance Correlation as a Measure of Problem Difficulty for Genetic Algorithms". In: L. J. Eshelman (ed.), *Proceedings of the 6th Int. Conference on Genetic Algorithms*, Kaufmann, 1995, 184–192.

[5] Karonski, M., and Palka, Z. (1977): "On Marczewski-Steinhaus Type Distance between Hypergraphs". In: *Applicationes Mathematicae* **16** (1), 47–57.

[6] Kubiak, M. (2004): "Systematic Construction of Recombination Operators for the Vehicle Routing Problem". In: *Foundations of Computing and Decision Sciences* **29** (3), 205–226.

[7] Marczewski, E., and Steinhaus, H. (1958): "On a Certain Distance of Sets and the Corresponding Distance of Functions". In: *Coll. Math.* **6**.

[8] Merz, P. (2004): "Advanced Fitness Landscape Analysis and the Performance of Memetic Algorithms". In: *Evolutionary Computation* **12** (3), 303–325.

[9] Michalewicz, Z. (1993): "A Hierarchy of Evolution Programs: An Experimental Study". In: *Evolutionary Computation* **1** (1), 51–76.

[10] Prins, C. (2001): "A Simple and Effective Evolutionary Algorithm for the Vehicle Routing Problem". Presented at *The 4th Metaheuristics International Conference, MIC2001*, Porto, Portugal, 143–147.

[11] Reeves, C. R. (1999): "Landscapes, Operators and Heuristic Search". In: *Annals of Operations Research* **86**, 473–490.

[12] Rochat, Y., and Taillard, É. D. (1995): "Probabilistic Diversification and Intensification in Local Search for Vehicle Routing". In: *Journal of Heuristics* **1**, 147–167.

[13] Sorensen, K. (2003): "Distance Measures Based on the Edit Distance for Permutation-type Representations". Presented at *The Adoro Workshop, GECCO 2003*, Chicago.

[14] Sorensen, K., Reimann, M., and Prins, C. (2005): "Path Relinking for the Vehicle Routing Problem Using the Edit Distance". Presented at *The 6th Metaheuristics International Conference, MIC2005*, Vienna, Austria, 839–846.

[15] Toth, P., and Vigo, D. (eds.) (2002): "The Vehicle Routing Problem". Society for Industrial and Applied Mathematics, Philadelphia, 2002.

[16] Wolpert, D. H., and Macready, W. G. (1997): "No Free Lunch Theorem for Optimization". In: *IEEE Transactions on Evolutionary Computation* 1 (1), 67–82.

[17] Woodruff, D. L., and Lokketangen, A. (2005): "Similarity and Distance Functions to Support VRP Metaheuristics". Presented at *The 6th Metaheuristics International Conference, MIC2005*, Vienna, Austria, 929–933.

Chapter 19

TUNING TABU SEARCH STRATEGIES VIA VISUAL DIAGNOSIS

Steven HALIM[1] and Hoong Chuin LAU[2]

[1] *stevenha@comp.nus.edu.sg, School of Computing, National University of Singapore;*
[2] *hclau@smu.edu.sg, School of Information Systems, Singapore Management University*

Abstract: While designing working metaheuristics can be straightforward, tuning them to solve the underlying combinatorial optimization problem well can be tricky. Several tuning methods have been proposed but they do not address the new aspect of our proposed classification of the metaheuristic tuning problem: tuning search strategies. We propose a tuning methodology based on *Visual Diagnosis* and a generic tool called *Visualizer for Metaheuristics Development Framework* (V-MDF) to address specifically the problem of tuning search (particularly Tabu Search) strategies. Under V-MDF, we propose the use of a Distance Radar visualizer where the human and computer can collaborate to diagnose the occurrence of negative incidents along the search trajectory on a set of training instances, and to perform remedial actions on the fly. Through capturing and observing the outcomes of actions in a Rule-Base, the user can then decide how to tune the search strategy effectively for subsequent use.

Key words: Metaheuristics, Software Framework, Tuning Problem, Visualization

1.   INTRODUCTION

Metaheuristics have been used extensively to solve hard combinatorial optimization problems, often with significant success. Given that metaheuristics do not guarantee optimality in general, the challenge is not so much to design a *working* algorithm but to tune it so as to obtain the *best* possible result. One way to measure the goodness of a metaheuristic algorithm is by checking its result against a set of *benchmark* problem instances.

Since different problems or even instances of the same problem may require the metaheuristic algorithm to be configured with different search parameters, components and/or strategies in order to work optimally, some resort to trial-and-error tuning through extensive experiments. Others use their past knowledge or experiences to tune the algorithm. From the industry standpoint, this process is unproductive especially against a backdrop of tight development schedules.

Alternatively, human [1] intelligence and machines can collaborate to shorten development time through the use of a well-designed visualization and interaction tool. The human-plus-computer collaboration has obtained considerable success in solving complex tasks, e.g. CAD/CAM. With the help of a well-designed visual diagnostic tool, an algorithm designer is able to examine search trajectories more systematically, steer the search, and readily see the impact of his action. We argue that this significantly reduces the time to design good search strategies which in turn speed up the overall development time.

Using visualization to assist optimization has been proposed in the seminal work of (Jones, 1996). In this paper, we propose a visualization scheme that determines quickly a set of rules that are helpful to the underlying metaheuristic algorithm. Unlike works due to (Klau *et al.*, 2002) and others which focused on *problem-specific* visualization, we emphasize the design of a *generic* problem-independent tool called *Visualizer for Metaheuristics Development Framework* (*V-MDF*). This work is an extension of *MDF* proposed in (Lau *et al.*, 2004b, 2006).

Instead of relying on specific problem domain information, V-MDF seeks to capture a pictorial view of the search trajectories and reports any anomalies to the human user. By visual inspection of these anomalies, the user can determine with higher accuracy the problems encountered during search, and apply remedial actions (such as tuning the parameters, adjusting the components of metaheuristics, or deriving better adaptive search strategies). With V-MDF, the algorithm designer begins with a metaheuristic on some defined search strategies, observes the search run-time dynamics, and dynamically improves the search strategies.

V-MDF differs from existing approaches for tuning metaheuristic which focused on the design of an efficient method for automatically choosing the best parameter values and/or metaheuristic components in black-box fashion (Adenso-Diaz and Laguna, 2006; Birattari, 2004). Instead, we extend the idea of visualizing the search process by (Kadluczka *et al.*, 2004) and that of analyzing the search landscape by (Fonlupt *et al.*, 1999; Merz, 2000; Hoos and Stuetzle, 2005), to help the users in designing better metaheuristics. This feature makes V-MDF especially useful for designing metaheuristics for new combinatorial optimization problems where search strategies have not been well-defined.

---

[1] The terms *human, user, or algorithm designer* are used interchangeably to refer to those who specialize in the development of metaheuristic algorithms.

This paper proceeds as follows: In Section 2, we discuss metaheuristic tuning problem in a broader sense. In Section 3, we review several tuning methods in the literature and classify them appropriately. In Section 4, a Visual Diagnosis Tuning methodology is proposed, followed by a discussion of V-MDF, the tool to support this methodology, in Section 5. A case study of the usage of V-MDF is given in Section 6. Section 7 gives the conclusions and future directions.

2. THE METAHEURISTICS TUNING PROBLEM

Recently, there is a growing interest in addressing the metaheuristics tuning problem. There are on-going discussions in the literature about the proper definition and scope of the tuning problem, e.g. (Birattari, 2004), as well as various proposals of tuning methods. However, several aspects are often overlooked. In this section, we propose a new classification to put into perspective our broader view of the metaheuristic tuning problem, especially in tuning search strategies.

2.1 Different Types of Tuning Problem

The term 'tuning' is often too broad. In the context of metaheuristics, we classify tuning problem into three types.

2.1.1 Type-1: Calibrating Parameter Values

In this 'easiest' type of tuning problem, the metaheuristic algorithm has been completely defined; and all the designer needs to do is to set the appropriate parameter values, e.g. setting the tabu tenure, setting the size of candidate list, etc. Different parameter values may influence the overall metaheuristic performance. Seemingly easy as it sounds, the challenge is that varying the value of one parameter may affect the optimal setting of the other parameter values, since the parameters are often correlated. Furthermore, in many practical situations, the range of parameter values is too large for the algorithm designer to determine their values through trial-and-error.

2.1.2 Type-2: Choosing Best Components

In this type of tuning problem, the algorithm designer needs to choose several components that will be used in a particular metaheuristic algorithm, e.g. choosing neighborhood (2/3/k-Opt), tabu list (tabu move/attribute), etc. Typically, each choice of metaheuristic component has its own strengths and weaknesses. (Charon and Hudry, 1995) show that different components have different effects to the performance of metaheuristics. Finding the optimal mix of components of the metaheuristic is often a challenging task as one needs to try a large number of combinations. This type of tuning problem is considered to be more complex than type-1, because once a good configuration is found, one may still need to properly set the parameter values of the components in the chosen configuration.

2.1.3   Type-3: Tuning Search Strategies

In this type of tuning problem, the algorithm designer needs to design good search strategies to optimize the *run-time dynamics* of the algorithm. In general, a good search behavior has the following characteristics: intensify the search on a good region to yield better solutions and diversify when the region is depleted of its potential, e.g. the adaptive/reactive strategies of Reactive Tabu Search (Battiti and Tecchiolli, 1994).

Unfortunately, search strategies are often problem-specific and deriving them is tricky. The effectiveness of search strategies is strongly dependent on the correct timings in which they are applied, which in turn introduce more parameters and rules. Recently, more complex intensifying and diversifying strategies have been proposed in the form of *hybridization*, in which one metaheuristic is hybridized with other metaheuristics and/or with other techniques such as linear programming and branch & bound. While such hybridization can further exploit the beneficial effect of intensification or diversification, it also adds another dimension of complexity in tuning the strategies.

Finding the effective search strategies to enhance the performance of the search algorithm is challenging because the number of possible search strategies can only be limited by one's own imagination. Moreover, due to several circumstances, a search strategy does not always perform what it is intended to perform as illustrated in various 'failure modes' (Watson, 2005). Thus, the need to explore a lot of strategies and then verify its correctness and effectiveness has made this type of tuning problem a tedious task.

2.2   The Need for a Good Solution

Tuning problem is a serious issue. Comments from experts highlight both the importance and the difficulty of addressing this tuning problem:

- "The selection of parameter values that drive heuristics is itself a scientific endeavor, and deserves more attention than it has received in the Operations Research literature...", (Barr *et al.*, 1995).

- "The design of a good metaheuristic remains an art...", (Osman and Kelly, 1996).
- "For obtaining a fully functioning algorithm, a metaheuristic needs to be configured: typically some modules need to be instantiated (Type-2) and some parameters (Type-1) need to be tuned.", (Birattari, 2004).
- "There is anecdotal evidence that about 10% of the total time dedicated to designing and testing of a new heuristic or metaheuristic is spent on development, and the remaining 90% is consumed (by) fine-tuning (its) parameters", (Adenso-Diaz and Laguna, 2006).
- "...Optimization of an Iterated Local Search may require more than the optimization of the individual components..." and "...There is no a priori single best size for the perturbation. This motivates the possibility of modifying the perturbation strength and adapting it during the run.", Stuetzle in (Glover and Kochenberger, 2003).

Any metaheuristic algorithm designer will face this tuning problem and they must find the solution: metaheuristic that is optimally configured to solve the underlying combinatorial optimization problem under its current context.

Formerly, due to the difficulty of tuning problem, algorithm designers choose to deal with the type-1 and type-2 problems only. Unfortunately, this effort does not guarantee good performance. One may have a good set of components of the metaheuristic algorithm and has all its parameters properly set. But, if the metaheuristic does not exploit adaptive memory and conducting intelligent exploration of the search space, it will often be outperformed by a dynamic, adaptive, self-correcting, and more intelligent counterpart. A simple example has been shown by Reactive-Tabu Search (Battiti and Tecchiolli, 1994), where a good search strategy which is able to adaptively adjusts the tabu tenure can outperform the performance of the original, static Tabu Search, on the set of unknown future instances --- even if the tabu tenure setting of the *static* Tabu Search is the *best* over the set of training instances.

The new classification that we propose put this type-3 tuning problem to be equally important with the other types. Ideally, we believe that to obtain the best solution for the metaheuristic tuning problem, all types of tuning problem must be addressed properly.

In this paper, we propose a new approach for addressing the tuning problem, especially in tuning search strategies.

3.   LITERATURE REVIEW

There are several proposals to address the metaheuristic tuning problem. We classify these tuning methods into two major types: Black-Box versus White-Box tuning methods. The details of the classification are shown in the Table 19-1.

Table 19-1. Black-Box versus White-Box Tuning Methods

	Black-Box Tuning Methods	**White-Box Tuning Methods**
Definition	• Treat metaheuristics as 'black-box'. Usually in form of automated tools that systematically search for the best parameter values or combination of metaheuristic components.	• Open up the 'box' to allow the algorithm designer to inspect the inner-working of the algorithm and to assist in designing a better algorithm. • Require collaboration with human.
Strengths	• Can relieve the burden of addressing type-1 and type-2 tuning problem from human.	• Can address type-1, type-2, and especially type-3 tuning problem. • Allow for possible human creativity, innovation or invention.
Weaknesses	• Do not allow for human creativity, innovation, or invention. • Have difficulty in handling type-3 tuning problem. • Often try too many configurations, thus in tight development time, they can only run/test each configuration in a relatively short period of time.	• Do not relieve the burden of tuning from human. • The human must understand the behavior of the metaheuristic. • Tuning results are inconsistent as different users do tuning differently. • The time required to conduct tuning is also a variable as it depends on the expertise of the user.

3.1   Black-Box Tuning Methods

CALIBRA.   (Adenso-Diaz and Laguna, 2006) proposed a tool to automatically calibrate parameter values when given pre-defined ranges. It works by iteratively calling the target algorithm with various set of parameter values and then uses the objective value feedbacks to determine which set of parameter values should be used in the next iteration. CALIBRA uses Taguchi's fractional factorial design to keep the number of parameter values being tried to be within acceptable limit. Iteratively, CALIBRA can narrow down the range of the algorithm parameters until the values converge. After the maximum number of iterations elapsed, CALIBRA will return the best set of parameters found so far. This way, it manages to solve the type-1 tuning problem quite effectively.

CALIBRA has limitations. The current version can only tune up to 5 parameters, the other parameters must be fixed to 'appropriate values'. The need to supply initial range is also problematic when one does not know a good starting range for certain parameters. Furthermore, CALIBRA is not designed to address type-2 and type-3 tuning problem.

F-RACE. (Birattari, 2004) proposed the racing algorithm, a method that was previously known in the machine learning community. The racing algorithm (F-Race), paraphrasing from his work, can be summarized as follows: First, feed F-Race with a (possibly large) set of candidate configurations. F-Race will estimate the expected performance of the candidate configurations in an incremental way and discard the worst ones as soon as sufficient statistical evidence is gathered against them. This allows a better allocation of computing power because rather than wasting time in the evaluation of low-performance configurations; the algorithm focuses on the assessment of the better ones. As a result, more data is gathered concerning the configurations that are deemed yielding better results, and eventually a more informed and sharper selection is performed among them. Finally, the last configuration is declared as the winner (best) configuration. This process is very much analogous with the real life racing.

The number of possible configurations to test can be very large, thus by not trying every possible configuration blindly, F-Race is much better than systematic brute force try-all approach. F-Race is classified as type-1 and type-2 tuning problem solver as it can be used to find good parameter values and proper combination of components of the algorithm simultaneously.

However, F-Race also has several inherent limitations, which arise from the fact that it is a black-box tuning method. Similar to CALIBRA, F-Race is unable to help the algorithm designer to give solution beyond the best configuration found in the set of possible configurations initially supplied to the tuning algorithm. One should also be aware of the 'combinatorial explosion' of the number of configurations to be tried, as it will require enormous computation time that may possibly exceeding the maximum allowed development time. Hence, to keep the size of initial set of configurations small, the algorithm designer must intelligently decide which should be included in the set, a process that preferably not be done *blindly*.

3.2   White-Box Tuning Methods

STATISTICAL ANALYSIS.   The search space (a.k.a. fitness landscape) of combinatorial optimization problems can be enormously large. Even if one is unlikely to explore the entire search space, one may gather crucial statistical properties of the search space, such as the structure of the search space, the distribution of local optima, the existence of 'big valleys', etc. The result of such analysis, if interpreted and reasoned correctly, may yield interesting discoveries that can be exploited to improve the design of the metaheuristic algorithms.

Some of the widely used methods for statistical analysis are Fitness Distance Correlation (FDC) and Run Time Distribution (RTD) analysis. The works by (Fonlupt *et al.*, 1999; Merz, 2000; Hoos and Stuetzle, 2005), are typical works that utilize statistical methods in metaheuristics design.

Statistical methods for metaheuristics design can be used to address all types of tuning problem. However, this process is not straightforward. Knowing the statistical information about the fitness landscape of a combinatorial optimization problem is a necessary but not sufficient condition to design a good metaheuristic for that problem. A significant amount of human effort is still required to reason on the facts found using statistical analysis before a good solution for tuning problem can be produced. In the context of tuning problem, this lengthy process is undesirable due to tight development time. We argue that without a proper computer-aided tool, it is difficult if not impossible to generate the required solution within tight development time, with merely statistical data.

HUMAN-GUIDED SEARCH.   As shown in many experiments, human is known to have advantages in visual perception and intelligence over today's computer. Human-guided search tries to utilize these advantages by providing the user with a good visualization and interaction tool to view the problem-specific visualization of the *current solution* (e.g. TSP tours, etc) and to control the search, respectively. In general, human know the ingredients of good solutions of the combinatorial optimization problem, thus human guidance may be able to assist the algorithm to obtain good results quicker.

Research on interactive man-machine optimization can be found in as early as (Michie *et al.*, 1968) and (Krolak *et al.*, 1971). Recently, this line of work is re-surfaced in (Klau *et al.*, 2002).

Human-guided search is an indirect form of white-box tuning method, where one can add the strategies that were adopted when manually guiding the search into the underlying search algorithm. However, guiding the search for a prolonged period of time is tedious in practice as its effectiveness will be limited by the stamina and patience of the human user.

VISUALIZATION OF SEARCH ALGORITHM BEHAVIOR.   Rather than visualizing problem-specific information as in human-guided search above, the generic attributes of the search algorithm can also be visualized. By monitoring them, one can gauge the algorithm's performance and can use the information to tune the search algorithm accordingly.

(Kadluczka *et al.*, 2004) proposed a generic visualizer to visualize the search coverage. The authors proposed a mapping for N-dimensional objects to 2-D space which can be displayed on the screen. By plotting the positions of the N-dimensional solutions in 2-D space, one can approximately identify which search space has/has not been explored by the metaheuristic search algorithm. This information can be used as a guidance to tune the algorithm. The limitation of this approach is that the huge gap in size between of the exponential search space and the polynomial screen space renders such visualization inappropriate for larger values of N. Furthermore, the static visualization adopted in this work does not convey the dynamic run-time behavior of metaheuristic search well.

3.3   Remarks

Each tuning method has its own strengths and weaknesses. However, their effectiveness can only be compared relatively --- not only since they are customized to address different types of tuning problems, but also many subjective issues are involved, especially the tuning methods with human intervention. In Table 19-2, we provide our subjective view of the differences of each tuning methods with respect to V-MDF as the basis of comparison. In general, most tuning methods have difficulty in handling the type-3 tuning problem. We also observe that most of the works (other than statistical methods) have yet to release their tools for public use (CALIBRA is available on the web[2]).

There are few other works around tuning problem, e.g. agent based approach +CARPS (Monett-Diaz, 2004); self adaptive algorithms (Battiti and Tecchiolli, 1994); metaheuristic to tune other metaheuristic: meta-evolution (Pilat and White, 2002); visualization of 2-variables problem (Syrjakow and Szczerbicka, 1999). All of them belong to either black-box or white-box tuning method depending on whether these methods treat the metaheuristic algorithm being tuned as black-box or not.

Table 19-2. Comparison of several factors between existing tuning methods:

Tuning Methods	A	B	C	D	E	F
Type of Method	Black	Black	White	White	White	White
Can address type-1?	Easy	Easy	Hard	Hard	Hard	Average
Can address type-2?	N/A	Easy	Hard	Hard	Hard	Average
Can address type-3?	N/A	N/A	Hard	Average	Average	Easy
Ease of Usage	Easy	Easy	Hard	Average	Average	Average

Legends: **A** (CALIBRA), **B** (F-Race), **C** (Statistical Methods), **D** (Human Guided Search), **E** (Visualization of Search), **F** (V-MDF)

---

[2] The current version is available at http://coruxa.epsig.uniovi.es/~adenso/file_d.html

4.   VISUAL DIAGNOSIS TUNING METHODOLOGY

4.1   Background

The goal of visual diagnosis tuning is to enable the user to address the tuning problem, especially in finding good search strategies quickly, through visual interaction with the search process.

In the past, visualization has been applied for understanding information, e.g. data can be visualized via graphical charts. Good visualization, see (Tufte, 1983, 1990, 1997), conveys information about underlying data or processes and it plays a crucial enabling role in our ability to comprehend large and complex data, to be aware of the situation (*Situation Awareness* theory (Endsley, 2000)). Via such visualization, human can gain insights of the data and possibly, create innovations --- something that is hard to be done by today's computer.

An interaction tool channels the human's idea back to the machine. This cycle of {... – visualization – interaction – visualization – ...} forms the interactive optimization concept as discussed in (Jones, 1996; Scott *et al.*, 2002), and in human guided search that was discussed previously (Michie *et al.*, 1968; Krolak *et al.*, 1971; Klau *et al.*, 2002).

In the context of tuning metaheuristics, if the user is given the proper visualization of the inner-workings of a metaheuristic algorithm, the user may be able to discover interesting properties that are hard for the machine to identify automatically. Such visualization can be used to help answering the 'why' question on the run-time dynamics (i.e. type-3 tuning) of the metaheuristic algorithm, which is the first necessary step to create effective search strategies. As an illustration, suppose that the current results produced by a Tabu Search algorithm are very poor. If presented with the visualization of the inner-workings of Tabu Search plus prior knowledge of the desired Tabu Search behavior, the algorithm designer may become aware of the situation that may be the source of the problem (e.g. the search is trapped in solution cycling) and subsequently, the algorithm designer may find possible treatments to rectify the improper behavior (e.g. increase the tabu tenure).

Visual diagnosis tuning is tied to the metaheuristic algorithm being used and *not* to the combinatorial optimization problem itself, and thus it inherits the generic characteristic of metaheuristic. Hence, such visual diagnosis tuning can be applied to virtually any combinatorial optimization problem as long as a metaheuristic algorithm exists to solve it. For illustrative purpose, our focus in this paper will be for Tabu Search (TS) only.

4.2 {Cause-Action-Outcome} Rules

A *search trajectory* is stated as the path taken by the algorithm from the start until the end of the search. Along this trajectory, the search may encounter *basic events* (e.g. arrive in local optima, an uphill/downhill move, etc).

We define a more generic term *incidents* as the occurrence of a basic event or a sequence/combination of basic events. These incidents can be diagnosed visually to portray the current state of the search trajectory. We define *positive incidents* as incidents that shows the search is along a good trajectory (e.g. new best solution found) and *negative incidents* as incidents that shows the search is along a bad trajectory (e.g. solution cycling).

In response to negative incidents (or *cause*), the user might decide to perform a remedial *action* – such as adjusting search parameter(s), changing component(s) of the algorithm, or applying intensification/diversification strateg(ies). The hope is that this action will result in a positive, *user-defined* desired incident (or *outcome*) within a reasonable time. This {cause-action-outcome} sequence is defined and captured as a *rule*.

To measure the effectiveness of a rule, we compute its *success score*. The range of success score is (0..1]. In this paper, it is measured by the exponential function: $e^{-\Delta iteration/C}$ where $\Delta iteration$ is the number of iterations between the first execution of the action until the observation of the desired outcome. C is a constant and is used to adjust the rate of diminishing success score. We set C to be 30 in this paper. Intuitively, this function dictates that the success score diminishes slowly over time. Observe that the success score = 1 when $\Delta iteration = 0$ (the desired outcome is immediately observed) and it tends to 0 when it takes a very long time (or perhaps never) before the desired outcome is observed.

Once a rule is performed, its *total execution* counter is incremented by 1, its success score is updated, and the search state is monitored for the next application of the rule. Typically, some action needs several iterations or even re-applied several times before the desired outcome is observable. Thus to avoid excessive re-applications of the action of the same rule, the next check of the search state is done using probability *(1-success score)*, that is, the action of an effective rule is less likely to be repeated.

The success scores of each rule throughout the search run are then normalized with the total execution to obtain the *normalized success score (NSS)*, see Figure 19-1. This process is repeated over *several* training instances[3] to avoid the danger of 'over-fitting'. If the *averaged-normalized success score (ANSS)* of a rule over several training instances is high, the rule is regarded as *successful* in bringing the search trajectory into a better one. Otherwise, the rule is regarded as *less successful* and the user might

---

[3] Training instances should have different characteristics, e.g. different problem size.

decide to further adjust his search strategy (action) or to refine the definition of his desired outcome.

By visually diagnosing the transformation from the cause incident to the outcome incident and monitoring the averaged-normalized success score, the user can determine the effectiveness of his search strategy.

For example, the high averaged-normalized success score of:

{Non_Improving – Greedy_Random_Restart – At_Good_Region}

signifies the potential effectiveness of this *greedy* random restart strategy to steer the search from bad region to area with good quality solutions; whereas the almost zero averaged-normalized success score of:

{Solution_Cycling – Decrease_Tabu_Tenure – No_Solution_Cycling}

shows the ineffectiveness of decreasing the tabu tenure during solution cycling, and:

{Solution_Cycling – 'Magic_Strategy' – Reach_Optimal_Solution}

illustrates an almost impossible scenario where only a 'magic strategy' can achieve the overly optimistic desired outcome.

We argue in this work that high averaged-normalized success score is a strong measure of the effectiveness of a remedial action, as high score implies that the action frequently steers the search trajectory from negative incidents to desired (positive) outcomes, at least over the several training instances. This argument carries weight if we assume further that future instances have similar characteristics with training instances.

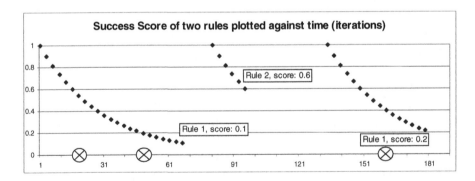

Figure 19-1. This is an example of the success scores of the execution of two rules. The ⊗ sign along the X-axis marks re-applications of the action of the rule before the search manages to arrive at the desired outcome. Here, Rule 1 is executed twice. The scores of 0.1 and 0.2, which is normalized over two executions to obtain NSS of 0.15, imply that either the strategy or the formulation of desired outcome adopted in Rule 1 has problem; whereas the high NSS of 0.6 of Rule 2 implies the potential effectiveness of the strategy used in Rule 2. These rules will then be applied to other training instances to obtain ANSS.

5.   VISUALIZER FOR MDF (V-MDF)

V-MDF is a white-box tuning tool that utilizes the visual diagnosis tuning methodology to address the type-3 tuning problem. In this section, we present the two main components of V-MDF: *Distance Radar* and *Rule-Base*, followed by a discussion on how V-MDF is used for visual diagnosis tuning.

5.1   Distance Radar

The Distance Radar is the underlying graphical user interface for visualizing incidents in the search trajectory. The function of the Distance Radar in this paper is to display incidents that occur along a Tabu Search trajectory. These incidents either indicate the necessity for a remedial action or to display the outcome of an applied strategy. From these incidents, the user can derive rules in form of {cause-action-outcome} discussed above.

Essentially, Distance Radar graphically plots generic properties of *distance*[4], *fitness (objective value), and recency* information of the elite solutions with respect to the *current solution*. In the trajectory based search, the current solution can be seen as the *'current position'* in the search space and the elite solutions, which were found and recorded along the search trajectory traversed so far, can be seen as the signposts or *anchor points* in the search space. By measuring the distance of current solution (current position) to these 'anchor points', coupled with the other two generic properties: fitness and recency information, one can gain information of the relative movement of the search along its trajectory with respect to these 'anchor points'. This new search trajectory tracking concept enables the user to visualize the previously infeasible search trajectory visualization. This is because the size of the set of recorded elite solutions/anchor points is fixed and much smaller than the exponential size of the search space.

The Distance Radar consists of dual 2D graphs: **Radar A** (with **Recency graph**) and **Radar B** (with **Fitness graph**). Each of the radar is used to exhibit distance information from different perspective. In both radars, the X-axes represent the anchor points and Y-axes show the distance between current solution with each of the anchor point. The Y-axes is drawn in logarithmic scale to emphasize the importance of anchor points within short distances with respect to the current solution. Points in the radars are connected with lines to help the user in diagnosing the trend.

---

[4] Discussions about various distance functions can be found in (Sevaux and Soerensen, 2005; Ronald, 1997,1998; Fonlupt *et al.*, 1997), etc. For example, the user can use 'bond distance' and 'hamming distance' to measure the distance of two Traveling Salesman Problem (TSP) and two Military Transport Planning (MTP) solutions, respectively.

Radar A displays the sorted anchor points by their fitness values. The recency values of these anchor points are plotted in the complementary **Recency graph**. Radar A displays only a visually manageable number of anchor points (a small but adjustable fraction with respect to the problem size) and any better elite solution found will replace the poorest recorded anchor point. The effect of Radar A is to approximate the 'goodness' of the region currently being searched. Generally, if Radar A shows a trend that is gradually moving upward (distance to current solution increases) from some anchor points, it indicates that the search is diversifying from the region of these anchor points. On the other hand, if the trend is moving downward, the search is intensifying onto region near these anchor points.

Radar B displays the sorted anchor points by their recency. The fitness of these anchor points is plotted in the complementary **Fitness graph**. Typically the number of recent solutions being recorded is set to be the same as the tabu tenure. Radar B can be seen as a long-term memory mechanism that complements the tabu list (short term memory). As cycling usually occurs around these recently visited solutions (especially local optima), Radar B can detect cycling in them quickly.

All the graphs: Radar A, Recency graph, Radar B, and Fitness graph complements each other to help a user in detecting various incidents. Figure 19-2 illustrates some incidents observeable using these graphs, e.g. *solution cycling, plateau effect, non-improving*, etc.

Figure 19-3 illustrates an example of how the observation of the incidents via Distance Radar can assist the selection of a remedial action. In this example, three elite solutions/anchor points have been found along the search trajectory of a minimizing problem and recorded as Local Optima 1, 2, 3. Now, suppose from the 3rd local optima to current solution, the search experienced a series of non-improving solutions (drawn as dotted lines from Local Optima 3 to current solution). The situation (*cause*) triggered the need for a remedial action. At this point, the algorithm designer may attempt to improve the search by applying a search strategy Z (*action*). Let one of either Solution X or Y be the solutions (*outcome*) reached after applying the strategy Z.

For Solution X, Radar A shows that the search is heading towards the current best local optima solution and Radar B shows that the nearest local optima solution is the one that is 2nd most recently found. Both radars have shown the algorithm designer that after applying strategy Z, the search is heading towards good recently found local optima. Hence, if strategy Z is intended to perform intensification, the observation from the Radar plots shows that it is indeed on the right track; otherwise it is considered as ineffective (as moving towards Solution X is not its intended purpose).

For Solution Y, Radar A and B shows an upward moving horizontal line. This indicates that the current solution is moving away from all known local optima solutions, which is the 'correct' outcome if the purpose of strategy Z is to conduct diversification.

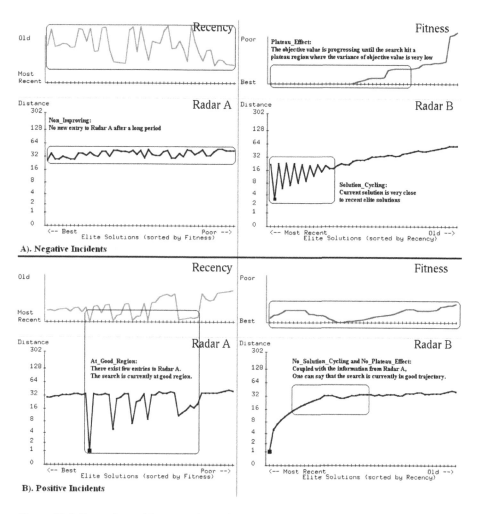

Figure 19-2. Examples and interpretations of several incidents: negative (above) and positive (below).

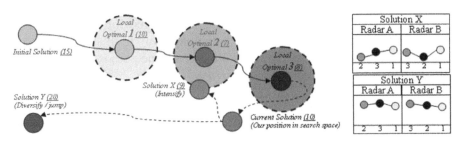

Figure 19-3. This is a visualization of the search trajectory of a minimizing problem. Without the aid of Distance Radar, it is hard to see the search behavior. On the other hand, one can understand the run-time dynamics of the Tabu Search algorithm by observing the plots shown in Distance Radar (moving to X is intensification to region around solution 2 and moving to Y is diversification).

5.2   Rule-Base

The {cause-action-outcome} rules that are derived while observing the incidents using Distance Radar are stored in a repository called the **Rule-Base (RB)**, which maintains the normalized success scores of the rules.

Upon completing visual diagnosis tuning, the user may examine the RB to decide whether to discard statistically inferior rules. The rules that survive eventually will form the basis for the solution of the tuning problem, in the sense that these rules can be either left as search strategies (triggered as needed/type-3) or merged into the metaheuristic algorithm (by modifying the parameters/type-1 or components of the algorithm/type-2).

The {cause-action-outcome} rules are implemented by the event-driven mechanism of MDF (Lau *et al.*, 2004b, 2006), as follows.

First, the user implements a V-MDF's EVENT class that describes an incident and links it with the desired HANDLER class. The user can refine the implementation of these EVENT classes to adjust the accuracy of the sensing of those incidents. The examples of the pseudo-code of V-MDF EVENT classes that describe two negative incidents (which were shown previously in Figure 19-2) are listed in Table 19-3a.

Next, the user needs to define a remedial action: the necessary steps required to alter the search trajectory, in form of the HANDLER class. When V-MDF senses an EVENT (cause) for the first time, it will trigger the associated HANDLER (action), register the ID of the desired outcome in the V-MDF desired_event table, and increment the total execution of the associated rule. However, subsequent re-applications of the action of the same rule will be done using (1-success score) probability. The example of the pseudo-code of V-MDF HANDLER classes is shown in Table 19-3b.

Finally, V-MDF will automatically check the search state after the execution of the HANDLERs with another EVENT (desired outcome). These events have no associated HANDLERs. If this EVENT is expected to occur (by checking the IDs listed in desired_event table), then the success score of the associated rule is computed using the formula explained in Section 4.2. The desired outcomes that occur after a long time will obtain a very low score. The example of the pseudo-code of V-MDF EVENT classes for checking desired outcome is shown in Table 19-3c.

Table 19-3. Examples of cause (EVENT), action (HANDLER), & outcome (EVENT) in *pseudo-code*.

Non_Improving : Event {
  return true if there is no new entry to
  Radar A after a long period, return false
  otherwise;
}
    Set Handler: 'Greedy_Random_Restart'
    Add total execution of this rule by 1.
	Solution_Cycling : Event {
	  return true if the distance to one or more
	  recent elite solutions in Radar B is short,
	  return false otherwise;
	}
	    Set Handler: 'Increase_Tabu_Tenure'
	    Add total execution of this rule by 1.
		A. CAUSE

Greedy _Random_Restart : Handler {
  pick TS current best solution, perturb it in
  greedy fashion, and set TS to resume from
  the newly created solution;
}
    Add 'At_Good_Region'
    in desired_event table.
	Increase_Tabu_Tenure : Handler {
	  get TS current tabu tenure,
	  increase it a bit,
	  set current tabu tenure to the new value;
	}
	    Add 'No_Solution_Cycling'
	    in desired_event table
		B. ACTION

At_Good_Region : Event {
  return true if the fitness difference
  between the current and best found local
  optima is low; return false otherwise;
}
    Add the success score of the
    associated rule if this event's ID
    is found in desired_event table.
	No_Solution_Cycling : Event {
	  return true if the distance to all recent
	  elite solutions in Radar B are far enough,
	  return false otherwise;
	}
	    Add the success score of the
	    associated rule if this event's ID
	    is found in desired_event table.
		C. OUTCOME

Table 19-4. Overall Workflow of V-MDF

A. Implementation Phase
- Implement the metaheuristic algorithm in **MDF framework** (Lau *et al.*, 2004b, 2006)

B. Visual Diagnosis Tuning Phase
- Using V-MDF's **Distance Radar**, diagnose the run-time dynamics of the metaheuristic algorithm when applied to several representative **training instances**.
- For each negative incident that requires an action,
  Write the appropriate {cause (EVENT), action (HANDLER), outcome (EVENT)} rule.
  The success score and total execution of rules will be monitored by **Rule-Base**.
- Human can further diagnose (visually), add new rules, modify or delete existing rules.

C. Rules Selection Phase
- Turn off V-MDF's Distance Radar.
- User can discard rules with low **ANSS** success score. (e.g. instance-specific rules).
- Surviving rules in Rule-Base form the elements of the final metaheuristic algorithm.
  1. Leave the rules as search strategies, or
  2. Merge the rules into the algorithm (i.e. the rules become native to the algorithm).

D. Testing Phase

•   Test the metaheuristic algorithm with **good rules** to the **whole test instances**.

5.3   Putting It All Together

The workflow for implementing and tuning a metaheuristic to solve a combinatorial optimization problem using V-MDF is outlined in Table 19-4.

6.   EXPERIMENTAL RESULTS

In this section, we report the experimental results. The real-life and artificially generated test instances, plus several executables of V-MDF are available at http://www.comp.nus.edu.sg/~stevenha/v-mdf.

6.1   Test Problem: Military Transport Planning (MTP)

We applied V-MDF to tune a Tabu Search implementation for solving an NP-hard combinatorial optimization problem: Military Transport Planning (MTP) which was defined in (Lau *et al.*, 2004a):

> Given service level **q** and a set of **n** requests from military units in tuple: {number_of_vehicle_required, start_time, end_time}, choose **q** out of **n** requests such that the total number of vehicles required to serve all **q** requests is minimized. number_of_vehicle_required ≥ 1 and [start_time .. end_time] lies within the range of a predetermined planning horizon.

Besides experimenting with several real life instances of this problem, we artificially created larger test instances with known optimal values as follows. First, create x random requests and then compute in polynomial time, the minimum number of vehicles z that is required to satisfy *all* $|x|$ requests. Finally, insert y pairs of dummy requests such that $q = |x| + |y|$ and $n = |x| + 2 * |y|$. In this pair of dummy requests (y,y'), every attempt to include y will not increase z while for y', it will always increase the number of vehicles required. The optimal solution is only one: first x requests plus all y requests, ignoring the entire y' requests. The value z will be the optimal value for this artificial test instance.

6.2   Experimental Methodology

The purpose of our experiment is to demonstrate the capability of V-MDF in dealing with the tuning problem that arises during the implementation of a Tabu Search algorithm for MTP. All experiments are conducted using an

Athlon XP 2500+ machine with the following specifications: 1.8 GHz, 512 MB RAM, Windows XP. All codes are developed using VC++ .NET 2003. The experimental methodology is as follows:

1. Prepare a set of real-life and artificially generated test instances.
2. Start with a quick-and-dirty implementation (see Table 19-5).
3. Record the results for all test instances. Tabu Search runs for 1000 iterations for each test instance (see Table 19-7, column 'Before').
4. Tune Tabu Search algorithm with V-MDF using two training instances (T4 and T7). The tuning time taken for the first author to conduct the tuning for the first attempt is approximately 10 man hours.
5. Verify rules in Rule-Base (see Table 19-6) in terms of their effectiveness.
6. Record the results of the tuned algorithm for all test instances again, using the same 1000 iterations limit (see Table 19-7, column 'After').
7. Compare the results.

6.3 Initial Results

Without proper insights on what happens within the search itself, one can only guess which part of the algorithm that needs to be tuned. The only observable fact without using the tool like V-MDF is the trend that the performance of this Tabu Search implementation deteriorates when problem size gets larger (See Table 19-7, column 'Before'). With V-MDF and its Distance Radar, one can detect the possible problems and tune the Tabu Search accordingly.

Table 19-5. Quick-and-dirty TS implementation for MTP using MDF software framework

Component	Remark
Solution	The solution representation is simply a bit string b of size n. $b[i] = 0$ when request i is not satisfied and 1 otherwise.
Initial Solution	Randomly select q requests (the seed is fixed for all the experiments)
Local Move and Neighborhood	Bit-flip move that will transform solution b to b' with 1 bit changed. $(d_{hamming\_distance}(b,b') = 1)$. Thus we have a maximum of O(n) possible neighbors per iteration. Infeasible neighbors are penalized by adding a constant penalty of 1000.
Objective Function	For each satisfied request, add its vehicle requirement to the histogram. The objective value is the maximum value in the resulting histogram.
Tabu List	Same bit flip move can't be applied for the next *tabu_tenure* iterations.
Tabu Tenure	*tabu_tenure* is initially set as 0.1 * n.
Search Strategies	None.

6.4 Tuning Phase

The first problem visually observed is the so-called 'Plateau_Effect'. This phenomenon can be easily explained: The objective values of MTP solutions are discrete and their range is small, thus logically, there will be many MTP solutions that have similar objective value. 'Plateau_Effect' can severely reduce the effectiveness of neighborhood based search. We try several methods and arrive with a penalty function that penalizes *very infeasible* solutions more than *slightly infeasible* solutions and also penalizes solutions that are too far from good solutions found so far. These two modifications help reducing the plateau effect.

The second visual observation reveals 'Solution_Cycling' incidents. We apply V-MDF to observe the behavior of the algorithm while we adjust the tabu tenure, emulating Reactive-TS strategy. We then add a greedy random restart strategy where we perturb the best found solution by randomly pick requests with small number_of_vehicle_required. This acts as a diversifier to enhance the search when it encounters 'Non_Improving' incidents.

All the rules found during the tuning process and their success rates against the training instance are listed in Rule-Base (see Table 19-6). Based on the statistical data, we discard the ineffective rules; merge some of the effective rules into the final algorithm; while the remaining rules are left as search strategies. The results of the algorithm are recorded in Table 19-7.

6.5 Results after Tuning

We observe in Table 19-7 that the result improves substantially compared to the initial results after a relatively short tuning phase. We like to point out that the result per se does not matter much, but rather it is the manner that V-MDF has helped the algorithm designer to identify negative incidents in a timely fashion that is essentially helpful to the tuning of the algorithm. This simple experiment has shown that by understanding the problems encountered by the search algorithm on-the-fly, albeit imperfectly, one can provide better remedies for such problems much faster, compared to blind trial-and-error.

7. CONCLUSIONS AND FUTURE WORKS

In this paper, we studied the issue of tuning metaheuristics through visualization. An extensive review of the existing tuning methods reveals that works in the literature are scarce in handling the type-3 tuning problem. We proposed a new visual diagnosis tuning methodology to address this tuning problem. We presented a generic visualizer tool V-MDF to support this methodology. V-MDF is currently designed for tuning Tabu Search strategies.

Table 19-6. This is the content of the Rule-Base after conducting visual diagnosis tuning using T4 and T7. Observe the column ANSS of a rule over *multiple* (two) training instances. The closer the value to 1.0, the better that rule is. Statistically inferior rules are discarded; good rules are either merged into the final metaheuristic algorithm or left as search strategies.

Cause	Action	Desired Outcome	NSS		ANSS
			Over T4	Over T7	
Effective rules, merged into the original algorithm					
Plateau_Effect	Apply_Penalty_Function	No_Plateau_Effect	-	-	-
Effective rules, left as search strategies					
Solution_Cycling	Increase_Tabu_Tenure	No_Solution_Cycling	4.4/5: 0.87	8.1/10: 0.81	**0.84**
Passive_Search	Decrease_Tabu_Tenure	Aggressive_Search	2.6/4: 0.66	3.9/6: 0.66	**0.66**
Non_Improving	Greedy_Random_Restart	At_Good_Region	2.9/4: 0.71	5.9/7: 0.84	**0.77**
Discarded rules (purposely listed here as illustration)					
Solution_Cycling	Decrease_Tabu_Tenure	No_Solution_Cycling	0.76/2: 0.38	6.6/11: 0.60	**0.49**

Table 19-7. Table of experimental results: before and after tuning. Test instances are divided into two categories and ordered by problem size. T4 and T7 are used as the training instance (shaded) and should not be considered for the evaluation of the final algorithm performance. Observe the improvement of the tuned over the non-tuned algorithm as well as the gap to optimal (for artificially generated test instances).

MTP Test Instances			Vehicles Required		Gap to Optimal		
			Before	After	Optimal	Before	After
Real-life test instances							
T1	n: 39	q: 31	(80%)	6	5	-	-	-
T2	n: 249	q: 186	(75%)	61	35	-	-	-
T3	n: 283	q: 240	(85%)	84	84	-	-	-
T4	n: 302	q: 250	(83%)	277	140	-	-	-
Randomly generated test instances with known optimal							
T5	n: 50	q: 40	(80%)	33	18	16	17 (106%)	2 (13%)
T6	n: 100	q: 85	(85%)	37	37	35	2 (06%)	2 (06%)
T7	n: 200	q: 180	(90%)	54	33	31	23 (74%)	2 (07%)
T8	n: 300	q: 250	(83%)	45	32	24	21 (88%)	8 (33%)
T9	n: 400	q: 300	(75%)	147	87	75	72 (96%)	12 (16%)

Our experience shows that V-MDF is effective in helping the user discover and rectifying negative incidents through proper remedial actions. We believe it is possible to develop a better way for visualizing Tabu Search or other metaheuristic search strategies via statistical methods such as fitness distance correlation plots. We hope to enhance V-MDF by providing decision support for the user to detect negative incidents, to choose better remedial actions, and to measure the performance of the rules. Collaboration between V-MDF and automated methods is also another possible future work. Finally, we see the prospect of using V-MDF as a research tool to invent search strategies not yet known at present.

The progress in metaheuristics research is rapid, but the end-users still require down-to-earth, ready-to-use tools for tuning their metaheuristic algorithms. Currently, research involving metaheuristics tuning problem is still preliminary and there are not many good tools available for public usage. However, we anticipate that several of the tuning methods that are theoretical concepts today will become widely used tools for metaheuristic algorithm design in the near future.

POSTSCRIPT

We have since expanded Distance Radar into a more generic, off-line visualization tool called *Viz* (see Halim *et al.*, 2006).

ACKNOWLEDGEMENTS

The authors acknowledge inputs from Wan Wee Chong for the earlier version of this work, Roland Yap, Mauro Birattari (IRIDIA, Belgium) especially for the discussion of tuning problem, and the anonymous referees.

REFERENCES

Adenso-Diaz, B., and Laguna, M., 2006, Fine-tuning of Algorithms Using Fractional Experimental Designs and Local Search, *Operations Research* **54**(1): 99-114.
Barr, R.S., Golden, B.L., Kelly, J.P., Resende, M.G., and Stewart, W.R., 1995, Designing and Reporting on Computational Experiments with Heuristic Methods, *Journal of Heuristics* **1**:9-32.
Battiti, R., and Tecchiolli, G., 1994, The Reactive Tabu Search, *ORSA Journal on Computing* **6**(2): 126-140.
Birattari, M., 2004, The Problem of Tuning Metaheuristics as seen from a machine learning perspective, PhD Thesis. University Libre de Bruxelles.
Charon, I., and Hudry, O., 1995, Mixing Different Components of Metaheuristics, In *Meta-Heuristics: Theory and Applications*, Osman, I.H. and Kelly, J.P., ed.: Kluwer Academic Press: 589-603.
Endsley, M.R., 2000, Theoretical Underpinnings of Situation Awareness: A Critical Review, in: *Situation Awareness Analysis and Measurement*, Endsley and Garland, ed: Lawrence Erlbaum Associates, Mahwah, NJ.
Fonlupt, C., Robilliard, D., Preux, P., and Talbi, E., 1999, Fitness Landscapes and Performance of Meta-heuristics, in: *Meta-Heuristics - Advances and Trends in Local Search Paradigms for Optimization*, Voss, S., Martello, S., Osman, I.H., Roucairol, C., ed.: Kluwer Academic Press, 18: 255-266.

Glover, F. and Kochenberger, G., 2003, *Handbook of Metaheuristics*, Kluwer Academic Publishers.

Halim, S., Yap, R., and Lau, H.C., 2006, Viz: A Visual Analysis Suite for Explaining Local Search Behavior, To appear in 19th Annual ACM Symposium on User Interface Software and Technology (UIST'06).

Hoos, H.H. and Stuetzle, T., 2005, *Stochastic Local Search: Foundations and Applications*. Morgan Kaufmann.

Jones, T., and Forrest, S., 1995, Fitness Distance Correlation as a Measure of Problem Difficulty for Genetic Algorithms, In Proceedings of 6th International Conference on Genetic Algorithms (ICGA'95): 184-192.

Jones, C.V., 1996, *Visualization and Optimization*, Kluwer Academic Publishers.

Kadluczka, M., Nelson, P.C., and Tirpak, T.M., 2004, N-to-2-Space Mapping for Visualization of Search Algorithm Performance, In Proceedings of 16th IEEE International Conference on Tools with Artificial Intelligence (ICTAI'04): 508-513.

Klau, G.W., Lesh, N., Marks, J., and Mitzenmacher, M., 2002, Human-Guided Tabu Search, In Proceedings of 18th National Conference on Artificial Intelligence (AAAI'02): 41-47.

Krolak, P., Felts, W., and Marble, G., 1971, A Man-Machine Approach Toward Solving The Traveling Salesman Problem, Communications of the ACM **14**(5): 327-334.

Lau, H.C., Ng, K.M., and Wu, X., 2004a, Transport Logistics Planning with Service-Level Constraints, In Proceedings of 19th National Conference on Artificial Intelligence (AAAI'04): 519-524.

Lau, H.C., Wan, W.C., Lim, M.K., and Halim, S., 2004b, A Development Framework for Rapid Meta-Heuristics Hybridization, In Proceedings of International Computer Software and Applications Conference (COMPSAC'04): 362-367.

Lau, H.C., Wan, W.C., Halim, S., and Toh, K. 2006, A Software Framework for Fast Proto-typing of Meta-heuristics Hybridization, To appear in Special Issue of *International Transactions in Operational Research* (ITOR).

Merz, P., 2000, Memetic Algorithms for Combinatorial Optimization Problems: Fitness Landscapes and Effective Search Strategies, PhD Thesis. University of Siegen, Germany.

Michie, D., Fleming, J.G., and Oldfield, J.V., 1968, A Comparison or Heuristic, Interactive, and Unaided Methods of Solving a Shortest-Route Problem. In *Machine Intelligence*, Michie, D., ed: American Elsevier Publishing Co., New York: 245-255.

Monett-Diaz, D., 2004, Agent-Based Configuration of Metaheuristic Algorithms, PhD Thesis. Humboldt University of Berlin.

Osman, I.H. and Kelly, J.P., 1996, *Meta-heuristics – The Theory and Applications*, Kluwer Academic Publishers.

Pilat, M.L. and White, T., 2002, Using Genetic Algorithms to optimize ACS-TSP. In Proceedings of the 3rd International Workshop on Ant Algorithms (ANTS 2002):282-287.

Ronald, S., 1997, Distance functions for order-based encodings, In Proceedings of 1997 IEEE International Conference on Evolutionary Computation (ICEC'97): 43-48.

Ronald, S., 1998, More distance functions for order-based encodings, In Proceedings of the 1998 IEEE International Conference on Evolutionary Computation (ICEC'98): 558-563.

Scott, S.D., Lesh, N., and Klau, G.W., 2002, Investigating Human-Computer Optimization, In Proceedings of Conference on Human Factors in Computing Systems (CHI'02): 155-162.

Sevaux, M., and Soerensen, K., 2005, Permutation distance measures for memetic algorithms with population management, In Proceedings of 6th Metaheuristics International Conference (MIC'05).

Syrjakow, M. and Szczerbicka, H., 1999, Java-based animation of probabilistic search algorithms, In Proceedings of International Conference on Web-based Modeling and Simulation: 182-187

Tufte, E., 1983, *The Visual Display of Quantitative Information*, Graphic Press.

Tufte, E., 1990, *Envisioning Information*, Graphic Press.

Tufte, E., 1997, *Visual Explanations*, Graphic Press.

Watson, J.P., 2005, On Metaheuristics "Failure Modes": A Case Study in Tabu Search for Job-Shop Scheduling, In Proceedings of 6th Metaheuristics International Conference (MIC'05).

Chapter 20

SOLVING VEHICLE ROUTING USING IOPT

Raphael Dorne[1], Patrick Mills[2] and Chris Voudouris[1]

[1]{raphael.dorne, chris.voudouris}@bt.com, Intelligent Enterprise Technologies Group, Intelligent Systems Lab, BT Group CTO Adastral Park, MLB1/PP12, Martlesham Heath, Ipswich, IP5 3RE, United Kingdom; [2]Patrick.Mills@optimalai.com

Abstract: The objective of this paper is mainly to answer one question: "Why use a toolkit such as iOpt to solve a combinatorial optimization problem rather than hard-coding a solution from scratch?" To answer this question, we consider a well studied problem: the Vehicle Routing Problem. We explain in details how to make use of the modeling and solving facilities available in iOpt to tackle this problem. At each step of this building process, we discuss the benefits of using iOpt rather than starting building a solution from scratch. Then we exhibit some experiments comparing the results obtained using the best algorithm built using iOpt and the best known in the literature. The overall conclusion of this work is our toolkit allows the user to maximize reuse of his code, significantly reduce his development time, focus his attention on the design rather than the coding, and exchange problem models or algorithms in a very easy and simple way using XML files within his community. At last, algorithms built using iOpt appear to be very competitive compared to the best hard-wired algorithms found in the literature.

Keywords: iOpt, iSchedule, Metaheuristics, Artificial Intelligence, VRP, Software library

1.   INTRODUCTION

The *Intelligent Optimization Toolkit* (iOpt) can be seen as an advanced software library with additional tools for rapidly designing, building and deploying solutions to combinatorial optimization problems. iOpt's problem modeling facility is based on a technology known as invariants (these act like "one-way" constraints allowing efficient updating of the overall problem model[1]). The solving facility of iOpt is mainly based on heuristic search techniques (Local Search, Genetic Algorithms, Hybrid Algorithms, etc.) where we make use of the fact that these methods share common points allowing breaking any of them down into algorithmic parts namely "search components". In iOpt, any heuristic search algorithm simply becomes a composition of a subset of those components allowing rapid development of existing and new algorithms. A set of visual tools is also provided in the toolkit to view the problem model, algorithm structure and optimization progress along with a 4th generation tool, the Heuristic Search Builder allowing the user to build complete algorithms by dragging-and-dropping algorithmic components on screen [3]. For more information on iOpt, the reader can refer to the following papers [4, 15][2].

iOpt has been successfully used to develop solutions for several real world applications within British Telecom (BT). These successes were mainly due to the high flexibility of iOpt that allows users to easily handle the complexity of customers' requirements (different types of constraints, very specific objective function, ...) and rapidly build prototype algorithms adapted to the customer's needs. For these algorithms, we could not evaluate their efficiency as no other algorithm outside iOpt was available to compare them against. Therefore, in order to make a more informed comparison of iOpt's problem solving power, it was important to compare our solutions to the best-in-class algorithms for a given optimization problem. The choice of vehicle routing (VRP) was justified because many academic benchmarks are available where very specific and efficient hard-wired algorithms have been designed which can evaluate thousands of potential solutions per second. Therefore VRP became relevant to us to find out whether iOpt can compete with these "benchmark-specific" hard-wired solutions or not. Moreover if we can demonstrate that using iOpt is a good approach for solving VRPs, then we can consider that a similar approach is likely to be relevant to tackle other well-known scheduling problems.

2.   ISCHEDULE

iOpt is a sound and complete framework that can be easily extended to address a specific application domain. The iSchedule framework is an

---

[1] The idea of using one-way constraints to model a problem to be used by heuristic search techniques was initially proposed by Michel and Van Hentenryck (2000) for their system *Localizer* [10].

[2] Other toolkits or libraries dedicated to solving combinatorial optimisation problems using metaheuristics are available in the literature such as Templar, HotFrame, Easylocal++, OptQuest, Local Search Toolkit of Ilog Solver, etc. [18].

extension that has been developed to assist non-expert users in addressing scheduling applications. iSchedule implements the basic concepts and business rules that are commonly encountered in scheduling problems. It has all the facilities for representing scheduling problems with unary capacity resources, i.e. resources can only perform one activity at a time (e.g. Vehicle Routing, Job Shop Scheduling, Workforce Scheduling …). The classes of the framework hide the complexity of the invariant-based problem model from the user who has to deal only with entities from his application domain (e.g. Activity, Resource and Break). With regard to visualization, iSchedule provides a set of customizable views of various aspects of a schedule (Gantt chart, tasks table, resources table, resource's timetable, etc.).

Problem model

The modeling facility of iSchedule known as the *Scheduling Modeling Framework* (SMF) contains all the core components available to the user to state his scheduling problem. Each of these components can be extended and specialized to meet the particular requirements of a problem. Most classes can be sub-classed to include additional information. Further, the computational model itself, which is based on invariants, can be modified through the Problem Modeling Framework of iOpt. The main concepts found in SMF are Resources, Activities, Constraints on/between each activity and/or resource, and an objective or cost Function.

Problem solving

Per se, SMF does not make any scheduling decision. This is the role of external processes such as heuristic search algorithms. The modeling framework simply provides the necessary interface to support schedule modifications. This interface consists of two operations: an insertion operation which inserts a task before/after a specified activity in the schedule, and a swap operation which swaps a task with another specified task in the schedule. These two operations suffice to implement any complex move operator/neighborhood.

Of course, search algorithms require more than simple schedule modification operations: e.g. checking consistency, costing schedules, providing access to feasibility and cost information, saving the best solution encountered, resetting/swapping solutions, etc. These requirements are not specific to the scheduling domain but arise in all optimization problems, which is why they are provided at the level of the Problem Modeling Framework and the Heuristic Search Framework of iOpt. The Scheduling Framework like any other domain framework in iOpt uses the Problem Modeling Framework internally to store and manage the computational model associated with schedule objects. Thus, the PMF automatically updates the computational model whenever any modification is made to the problem or the current schedule through the Scheduling Framework. This process is completely transparent for the user reducing the need to develop tedious code each time a new (customer) requirement (constraint, cost, etc.) must be added to the problem model.

3.   VRP MODEL WITH ISCHEDULE

In common with the academic version of a vehicle routing problem, we need to allocate a number of customer visits to a number of vehicles/drivers, working from one common depot. Each driver has start and end times within which they can carry out customer visits and their vehicle a maximum capacity which must not be exceeded. Each visit has a time window (an earliest and latest start times) specifying when the visit can start, a duration, a location and a demand (amount of a vehicle's capacity which will be used up). In the VRP instances considered in this paper, the objective is threefold given by order of importance: 1) allocate a maximum number of visits whilst satisfying the time window and capacity constraints 2) minimize the number of vehicles used 3) minimize the total travel incurred for all vehicles.

This is straightforward to model using iOpt/iSchedule. The Schedule class is extended to represent a VRP problem. We ask iSchedule to automatically generate a decision model for Unallocated Task Cost (to reduce the number of unallocated tasks; cost value set to 1,000,000 per unallocated task), Resource Allocation Cost (to reduce the number of vehicles used; cost value set to 10,000 per vehicle) and Setup Variable Cost (to reduce travel time; cost value equals to travel time in minutes). As you can see these values are carefully chosen to ensure the priority of each sub-objective: the cost of one unallocated task corresponds to the utilization of 100 vehicles, and the cost of using one vehicle corresponds to a drive time greater than 150 hours. Finally, we define the schedule horizon and a capacity timeline to specify the capacity constraint of the vehicles (cf. source code next page).

```
/** Creates a new VRP problem.
 *   @param depot: the depot for the vehicles
 *   @param capacity: the capacity of the vehicles */
public VRP(Depot depot, int capacity) {
    super(Schedule.SETUP_VAR_CST|Schedule.TASK_UNALLOC_CST|Schedule.RES_A
    LLOC_CST);
    this.depot = depot;
    this.capacity = capacity;
    new CapacityTimeline(this, 0, capacity);
    setSetupModel(new VRPSetupModel());
    setServiceModel(new VRPServiceModel());
    beginValueChanges();
    setSetupVariableCostsWeight(1.0);
    setFixedCostsWeight(1.0);
    this.setUnallocatedCostsWeight(1.0);
    this.setHorizonStart(depot.getReadyTime());
    this.setHorizonEnd(depot.getDueDate());
    endValueChanges();
}
```

The *Resource* class of iSchedule is extended to represent a *Vehicle* where start/end locations are specified along with a specific cost for using a resource.

```
/** Creates a resource Vehicle for the specified vehicle routing problem.
 * @param vrp: a vehicle routing problem */
public Vehicle(VRP vrp) {
    super(vrp);
    new VehicleStart(this, vrp.getDepot());
    new VehicleEnd(this, vrp.getDepot());
    setFixedCost(10000.0);    // fixed cost to discourage use of vehicles
    setSetupUnitCost(1.0); // travel related costs.
    setName("Vehicle-" + (getIndex() + 1));
}
```

To represent a customer visit, we extend the *Task* class where the location, capacity demand, time window and visit duration are set.

```
/** Creates a new visit and adds it to the problem.
 * @param vrp: the VRP problem
 * @param id: the ID of the visit
 * @param location: the location for the visit
 * @param demand: the capacity demand for the visit
 * @param readyTime: the time from when the visit can be applied /started
```

```
* @param dueDate: the time when the visit has to be completed
* @param serviceTime: the time required by the resource to complete the visit */
public Visit(VRP vrp, int id, FieldLocation location, int demand, int readyTime, int dueDate,
int serviceTime) {
    super(vrp);
    this.id = id;
    this.location = location;
    this.demand = demand;
    this.readyTime = readyTime;
    this.dueDate = dueDate;
    this.serviceTime = serviceTime;
    this.startsBetween((double) readyTime, (double) dueDate);
    this.setQuantity(vrp.getCapacityTimeline(0), (double) demand);
    this.setName("Visit-" + id);
    this.setUnallocatedCost(1000000.0); // to discourage unallocated activities
}
```

The overall code for modeling a VRP using iSchedule requires only 8 classes and around 100 lines of code. For more information on how to model a problem using iSchedule the reader can refer to the following paper [2].

4.   VRP SOLVING IN ISCHEDULE

The *Heuristic Search Framework* (HSF) of iOpt is composed of a library of algorithmic parts. Each category of components represents a well-known concept that can be encountered in the literature for single solution based algorithm (Simulated Annealing, Tabu Search, Guided Local Search, etc.) and population-based algorithms (Evolutionary computation, Genetic Algorithms, Hybrid algorithms). As an example, HSF holds the following non-exhaustive list of categories: starting point or initial solution generation, local search, neighborhood search, neighborhood, tabu mechanism, mutation (any local search technique can be added as a mutation), crossover, selection, etc.). In addition, HSF being framework-oriented, it automatically provides the interaction between those components making it totally transparent for the user. Thus a user willing to implement a new algorithm using HSF will look at the available algorithmic parts in the library and if any is missing or very specific to his problem, he can focus all his attention only on this component and easily add it to HSF by extending one of the existing categories. For example, one may want to create a new tabu search for his specific problem; most of the components of such an algorithm are generic and therefore already provided by HSF except maybe a specific tabu mechanism to decide the potential neighbors to set tabu each time a move is performed. In such a case, the user adds this component to the library by following the guidelines provided by the framework for each category (similar to the Java Swing library: what are the functions to be extended and when and how they will be called) then plugs it into a meta-tree of search components for a tabu search. An immediate result is this new component is now available to all the other users of iOpt.

With regard to VRP and more generally to any well-studied optimization problem, we can easily make use of HSF and build a set of advanced algorithmic parts trying to be representative of all the most efficient techniques available in the literature which allows the user to re-create these techniques and even to build new ones based on similar principles.

During 2004, we have implemented over 15 new search components for Vehicle Routing Problems. We list most of them in the following paragraphs, sorted by category along with some experiments when available.

4.1 Neighborhood components

In this section, we describe the new neighborhoods added to iOpt, mainly aimed at improving the solution quality produced by iSchedule for Vehicle Routing, but also applicable to other scheduling problems in some cases. For a survey of local search methods and metaheuristics for Vehicle Routing, the reader can refer to both papers from Bräysy O. and Gendreau M. [5].

Table 20-1. Time complexities of searching whole neighborhoods (n = number of sequence elements, m = number of sequences)

Neighborhood	Name in the literature	Complexity
SwapMoveNeighborhood	Exchange	$O(n^2)$
InsertMoveNeighborhood	Relocate	$O(n^2)$
RecombineSequencesNeighborhood	2-opt*	$O(n^2)$
ReverseSegmentNeighborhood	2-opt	$O(n^2/m)$
RelocateSegmentNeighborhood	OR-opt	$O(n^3/m)$
ExchangeSegmentNeighborhood	CROSS	$O(n^4/m^2)$

Relocate Segment Neighborhood: This class implements the Relocate Segment neighborhood known as OR-opt in the literature [10]. This neighborhood allows the relocation of a segment (from 1 element to the whole sequence) from one sequence to another sequence or disjoint segments within the same sequence. It is important to note here that each time we add a new component to HSF nothing prevents us to see if this component can be generalized to a larger family of problems than scheduling. This is typically the case here where this component actually proposes to move a segment of objects within a solution representation of a type list of sequences. The interpretation made in a VRP context is something specific to the problem unrelated to the operator itself therefore in iOpt a relocate segment operator is now available to any scheduling problem and even more problems as long as such a move remains meaningful. This shows a good example of the benefit of generalization when adding new components to iOpt. This allows a faster transfer of solving techniques across families of optimization problems.

Exchange Segment Neighborhood: This class implements the Exchange Segment neighborhood, known as CROSS-exchange in the literature by Taillard E. et al. 1997 [15]. It allows the exchange of a segment of sequence elements with another disjoint segment of sequence elements within the same sequence or with a segment in another sequence.

Recombine Sequences Neighborhood: This class implements the Recombine Segment neighborhood, known as 2-opt* in the literature by Potvin J-Y. and Rousseau, J.M. 1995 [11] allowing the recombination of two sequences by taking the end of one sequence and connecting it to the beginning of another sequence and vice versa.

Reverse Segment Neighborhood: This class implements the reverse segment neighborhood, known in the literature as 2-opt by Lin, S. 1965 [7]. This reverses a segment of sequence elements. It operates only within individual sequences.

As above a generalization of these neighborhoods has been added to iOpt allowing users to easily apply similar techniques to other problems.

4.2   Performance analysis of Neighborhoods

In this section we give a quick performance analysis of the new neighborhoods described above. We ran each individual neighborhood with First Improvement and Fast Local Search (FLS) from a random starting point generated by the *RandomGeneration*[3] class. FLS is a heuristic defined by Voudouris, C. [16] that helps the search to focus only on parts of the search space that have been affected by the latest move. For the VRP, each time a task is considered by the neighborhood but leads to no improvement then the task is "deactivated" and will not be considered at the next iteration of the neighborhood. On the other hand, each time a move is performed, all the tasks affected by this move (i.e. visits allocated to one of the newly modified tours of the vehicles) will be "reactivated". We reach a local

---

[3] This is a search component already provided in iOpt generating a random starting point satisfying all the problem constraints.

optimum when all the tasks are deactivated. Each algorithm is run 10 times using different random seeds thus generating 10 different random starting points for each neighborhood and the iterative process is stopped when we reach the first local optimum[4]. In addition, each new neighborhood is tried with the Insert and Swap move neighborhoods to see how it performs when combined with a basic set of neighborhood operators. These neighborhoods are combined using a *Probabilistic Composite Neighborhood* (PCN); a built-in component of iOpt that selects at each iteration the neighborhood to get the next move from using specified probabilities. For example, given three neighborhoods with respective probabilities 20%, 50% and 30%, at each iteration a random number is generated in [0..1] and if this random number belongs to [0..0.2], we will get the next move from neighborhood1 and so on for the other neighborhoods (i.e. if the value belongs to [0.2..0.7] or [0.7..1]). Here we used the same probability for each neighborhood and when a neighborhood runs out of neighbors, PCN carries on with the remaining non-empty neighborhoods. The idea here is to balance the search effort over several neighborhoods considering more often the most promising neighborhoods (with higher probabilities). We performed the experiments on each of the problems R105, RC206, R202 from benchmark problems of Solomon (1987) [13]. Results for the solution quality of local minima produced are shown in Table 2.

Table 20-2. Solution quality produced by neighborhoods using first improvement and FLS with first improvement to first local minimum (average of 10 runs)

Algorithm	R105		rc206		r202	
	Vehicles	Distance	Vehicles	Distance	Vehicles	Distance
Insert	18.6	1616.0	5.3	1400.3	5.2	1295.8
Swap	20.0	1672.3	5.3	1580.1	5.3	1402.8
Reverse	20.0	1841.7	5.3	1685.5	5.3	1465.3
Recombine	20.0	1580.7	5.3	1466.4	5.3	1308.4
Relocate	17.7	1590.4	5.2	1245.0	5.3	1236.6
Exchange	20.0	1527.9	5.3	1199.9	5.3	1173.5
InsertSwap(IS)	18.3	1547.1	5.3	1403.8	5.2	1266.9
IS+Reverse	17.8	1528.3	5.3	1349.2	5.2	1281.3
IS+Recombine	17.7	1475.3	5.2	1193.9	5.3	1161.7
IS+Relocate	17.9	1531.5	5.2	1248.8	5.2	1251.8

In terms of solution quality, IS+Recombine in general produces results sensibly better than the other neighborhood combinations. But what is more relevant here is how easy any combination of local search iterative processes can be built and fairly compared. In this table, we could easily replace a first improvement strategy by any other strategy: best improvement, best move

---

[4] Note in the case of a Composite Neighborhood, a local optimum is reached if all its sub-neighborhoods reach a local optimum during the same iteration.

with tabu, threshold strategy (simulated annealing) and connect them to any combination of 1 up to 5 different neighborhoods. More importantly, iOpt provides a fairer comparison between techniques as it is based on the same implementation (invariant library, PMF and HSF). As a researcher, it is well known in the meta-heuristic community that it is very often difficult to compare different algorithms because each algorithm performance may vary depending on the random choice generators used, internal data structures, programming language used, machine processing power and very specific tricks or implementation details that the author of the research paper could not mention. Using a toolkit like iOpt allows the user to analyze the actual benefit of a technique independently of its implementation[5]. In addition, because any algorithmic component, algorithm and problem model can be saved in Java or XML files, they can be very easily exchanged between iOpt users allowing them to run any new technique on their own machine.

Table 20-3. Moves per second: performed and evaluated by neighborhoods to first local minimum (average of 10 runs)

Moves per second	r105		rc206		r202	
Algorithm	Performed	Evaluated	Performed	Evaluated	Performed	Evaluated
Insert	6.04	11912	3.3	5556	2.8	5332
Swap	10.3	11788	3.6	7083	3.8	5817
Reverse	0	5374	6.7	1806	6.1	1562
Recombine	3.5	3224	0.7	1004	0.8	1039
Relocate	1.1	6101	0.2	2054	0.2	2119
Exchange	1.1	3503	0	1334	0	1385
InsertSwap(IS)	3.92	9678	1.7	4980	2.1	5461
IS+Reverse	3.35	9167	1.3	3726	1.4	4428
IS+Recombine	1.82	4704	0.5	1762	0.8	2056
IS+Relocate	1.8	6270	0.6	2434	0.6	3097

Table 3 provides the number of moves performed and evaluated per second to the first local minima. iOpt automatically provides to the user a list of statistics during the search process such as the number of moves evaluated, performed, filtered (cf. section on move filtering below), etc.

Furthermore, since any stopping condition can be set on any of those parameters, a better comparison between techniques based on a criterion other than real time can be easily defined. For example, one may want to compare a set of techniques based on a maximum number of moves performed instead of the first local optimum found to identify if the technique makes a good use of the information collected during move evaluations.

---

[5] Even if obviously the fact of using an invariant model has some impact on algorithm performance and behavior, a survey of algorithms is based on a similar ground.

4.3 Start point generator components

In this section, we report the new start point generators added to iOpt/iSchedule for specific Scheduling problems.

Seed and Fill Generation: This class implements a generalization of Solomon's I1 heuristic [13] using the actual cost of insertion rather than "Solomon's C1 & C2" functions (in this class, c1 & c2 functions are defined to be the change in objective function resulting from the insertion). The idea is to choose a seed visit which is far away from the depot and then pack as many visits into that vehicle's route as possible. As an option, the user may also choose to seed a vehicle for the visit with the earliest start time, or the narrowest time window. The idea is to consider the hardest to schedule first.

Solomon's Insertion Heuristic Generation: This class extends the *SeedAndFillGeneration* class by overriding the Solomon's C1 and C2 functions for choosing visits to allocate to vehicles.

Re-optimize after Route Generation: This class extends Solomon's Insertion Heuristic by attempting to re-optimize the current set of routes after each new route is created, by applying Fast Local Search with Insert and Swap moves to the currently allocated vehicles visits only (only activating parts of the sub-neighborhood after a task is added to the Schedule). Bräysy (2001) [1] applies a similar technique re-optimizing individual routes using OR-opt moves.

With regard to start point generators, we can note again that some of those components have been made more generic than the original version with in some case the possibility to make use of iOpt component architecture where a technique like "re-optimize after route generation" initially designed to be used with OR-opt moves becomes a technique that can be now combined with any iOpt neighborhood leading to new possibilities.

4.4 Guided Local Search components

Guided Local Search (Voudouris, 1997 [16]) is known in the literature to provide good results on VRP instances. In this section, we describe search components based on Guided Local Search added to iSchedule for solving Vehicle Routing Problems.

GLS Route Edge: This component implements a Guided Local Search for minimizing the set-up time (travel time) of the sequences of tasks (vehicle routes). Features are associated with each pair of visits and the initial cost for each feature is the travel time between these two visits. The route edge is present in the current solution if the two visits appear to be next to each other in a vehicle route.

GLS Route Removal: The idea of this search component is to periodically penalize individual sequences of tasks (routes of vehicles), in the hope of completely removing a sequence (route), and thus reducing the number of resources (vehicles) required. Only one route at a time is penalized, and it is done in such a way that as tasks are removed from the route, the penalty on the route is reduced, thus encouraging tasks to be gradually reallocated elsewhere. This is done by multiplying the penalty on a route by the number of visits in that route. Unallocated routes have a penalty imposed on them, such that if a task is added to them, they will increase the objective value, to prevent the shuffling of tasks from one resource to an unallocated resource.

GLS Unallocated Tasks: The idea of this search component is to increase the cost of unallocated tasks to encourage them to be allocated, even if this means increasing the "original" objective function. In this way, a task may be swapped with another task, and a sequence of such swaps may lead to a point where the number of unallocated tasks can be reduced.

GLS Aspiration: This component covers the case where GLS penalties prevent a move from being made which improves the original objective function, thus missing a potential new best solution. This component forces such a move to be made.

4.5   Tabu Search components

Many algorithms fall under the umbrella of Tabu Search. Any heuristic search method which uses some kind of memory can be cast into the Tabu Search framework. In this section we describe a memory based algorithm, which is particularly good for solving Vehicle Routing Problems, but may also be applicable to other scheduling problems due to its generic implementation.

Resource Allocation List Memory: This method is specific to Scheduling problems. It may be particularly useful for problems where any resource can be used for a particular task (this is a generalization of the Vehicle Routing technique of Taillard et al. 1997 [14]). It works by storing sets of tasks allocated per resource in previous solutions in memory and selecting a compatible subset of these sets of tasks to generate a new starting point, then applying a local search to the starting point. The improved solution is then broken down into resource allocation lists and placed into memory. Specifically the algorithm works as described in Table 4:

Table 20-4. Resource Allocation List Memory (generalization of Taillard's algorithm)

1. Generate N solutions using start point generator + local search and store each resource list of allocated tasks in R (where the number of tasks is greater than 1).

2. Generate a new solution:
 T = R
 While (tasks remain unallocated and resource allocation lists exist in T)
 – Select and add a resource allocation list to the current solution, r from T probabilistically weighted according to the best objective value of the solution(s) it has been previously present in.
 – Remove all resource allocation lists from T having tasks in r
 End while

3. Apply the local search to new solution
4. Add all generated resource allocation lists in the new solution to the set R of resource allocation lists
5. Go back to step 2 until stopping condition is satisfied

Position Tabu: The component set moving (Inserting, Swapping or a Composite Move involving either of the previous) a position tabu for a number of iterations after it has been moved. For VRP problems, this means preventing a task from being moved again for a certain number of iterations.

Adjacent Visit Tabu: This component makes edges, which have been removed or added to a vehicle's tour, tabu for a certain number of iterations.

4.6   Move filtering components

When a filter component is added to an algorithm, it is called by HSF before a move is evaluated to check whether this move is worth being evaluated via invariant propagation or not. This technique is particularly efficient on benchmark instances where the number of constraint types is limited and facilitates the development of such components.

VRP Move Filter: This component executes a quick test to check if a move generated by a neighborhood is worth considering by the neighborhood search. In the case of VRP, this component will check compatibilities between the tasks to be moved and the resources (quick time windows checking or if each task is allowed to be allocated to the destination resource) as well as estimate the impact on travel distances to avoid using invariant propagation for obvious non-improving moves (i.e. a move leading to a travel time increase will be discarded).

4.7   Evolutionary algorithms components

Although pure genetic algorithms may be poor for solving combinatorial optimization problems such as scheduling, combining a Genetic Algorithm with a local search or a more complex metaheuristic such as tabu search, may produce a good way of diversifying the search. For this reason, we have added some basic Crossover components for iSchedule to combine scheduling solutions.

4.7.1   Crossover components

Schedule Uniform Resource Crossover: This operator executes a uniform resource based crossover combining two parent schedule sequences solutions to generate a child scheduling solution. It generates a child solution that gets the complete route of a particular resource from parent 1 (minus any task which is already allocated) if the probability test is satisfied[6], otherwise it takes the allocated tasks for that resource from parent 2 (minus any task which is already allocated). Any tasks left unallocated at the end, are allocated greedily, picking the best insertion point for each task (if it is possible to allocate the task). The idea with this operator is that some good resource-task list pairings will be combined into a child that might lead to a better solution.

Schedule Uniform Task Crossover: This operator executes a uniform crossover combining two parent schedule sequences solutions to generate one child schedule sequences solution. This operator goes through one resource at a time, allocating a task from either parent1 (with probability 0.5) or parent2 to the same resource in the child solution. Any unallocated tasks left over at the end are allocated (if possible) greedily, inserting them at the best (according to the objective function) point. The idea is to merge characteristics of good resource-task lists hopefully leading to a better combination.

Route Crossover: This operator executes a route based crossover combining 2 parent schedule sequences solutions to generate a child solution. This is done by making a list of routes in both parent solutions and picking N routes (the maximum number of routes in the parents) at random, allocating them to the child solution (excluding any visits already allocated in previous routes).

---

[6] A probability test is performed by randomly generating a real number within [0..1[(using the random function of package java.util.Math for example) and comparing this value with the acceptance probability. Given an acceptance probability equals to 0.5 if the generated value is smaller than 0.5 then the test is positive otherwise it is negative.

Any tasks left unallocated at the end, are allocated greedily, picking the best insertion point for each task. The idea with this operator is that some good routes will be preserved from the parent solution and combined into the child solution.

4.8 Algorithms for vehicle routing

Due to the amount of time required to run a whole metaheuristic solver on a problem in order to gain meaningful results in comparison to so called "hard-coded" algorithms for VRP, we chose to run our preliminary experiments on a small subset of Solomon's benchmark problems. These problems were: r105, rc101, rc206, r202, c103 and c204. The first two algorithms we ran are described in the next subsections.

4.8.1   FLS for VRP

After the experiments on the neighborhoods, we have identified a good local search component (subtree of search components) that we use with both algorithms. It is based on FLS combined with a first improvement neighborhood search for searching the neighborhoods. The single neighborhoods are combined using a Probabilistic Composite Neighborhood (see section 4.2) and these single neighborhoods are:
- InsertMoveNeighborhood
- SwapMoveNeighborhood
- RecombineSequencesNeighborhood
- ReverseSegmentNeighborhood
- RelocateSegmentNeighborhood with maximum segment size = 3.

4.8.2   GLSFLS for VRP

The first algorithm we ran was a version of GLS (GLS has previously been applied by Kilby et al., (GreenTrip project 1999) [6], although our version is enhanced by the *RouteRemovalGLS* component) combined with FLS. Start points were generated using the *ReoptimizeAfterRouteGeneration* component. GLS was built using components combined using the following components:
- RouteEdgeGLS: penalizing route edges
- RouteRemovalGLS: penalizing a route to try reducing the number of routes in the solution
- GLS Aspiration: used to ignore penalties if a move was found which improved the best solution so far (even if that move increased the augmented objective function, due to the penalties).

4.8.3   RouteMemoryFLS for VRP

The motivation for selecting the second algorithm is because it uses a disruptive idea of storing routes as described in the Resource Allocation List Memory component (cf. Section 4.5). This algorithm generates 20 starting solutions using *RandomGeneration* component. Each of these start points is then improved using FLS (as described above). The algorithm then generates new solutions by selecting a non-intersecting (in terms of tasks) subset of the routes stored in memory and then greedily inserting any other tasks remaining unallocated. This new solution is then improved using FLS and its routes stored in memory (similar to Taillard et al. 1997 [14]). The maximum number of routes stored in memory is limited to a maximum of 300 (keeping the best 300 without repetitions, using the best objective value of solutions that each route has appeared in).

4.8.4   Results

In the tables below we list the best solution costs per problem found after allowing each algorithm to run for 3 hours on a PC with the following specifications: Pentium4 1.8Ghz, 768M and Windows XP Pro. Our objective here is to identify if the *RouteMemory* based technique has some positive impact on the results (within the iOpt environment obviously). It is important to note these results have been generated without the use of any move filtering technique.

Table 20-5. Preliminary results for complete algorithms GLSFLS and RouteMemoryFLS after 3 hours running time without move filtering

	GLSFLS		RouteMemoryFLS		Best known results	
Problem	Vehicles	Distance	Vehicles	Distance	Vehicles	Distance
r105	14	1441.28	17	1455.471	14	1377.11
rc101	15	1675.508	18	1720.773	14	1696.94
rc206	3	1219.055	4	1119.519	3	1146.32
r202	4	1112.785	5	1083.968	3	1191.7
c103	10	828.9369	11	895.4877	10	828.06
c204	3	694.8435	3	618.5498	3	590.6

These results are encouraging, since we gain some good quality solutions near to the best known solutions found by hard-wired heuristics in some cases (e.g. c103). However, these results should be treated with caution as we have only run the algorithms on a limited subset of Solomon's problems. We really need to run on all problems in order to evaluate the algorithms properly, and also if we are to compare with hard-wired algorithms, we need to make our algorithm faster (this we do by using our move filtering mechanism), so we can be competitive with them. In any case, we have shown that the new neighborhoods can significantly improve the quality of solution produced by the iSchedule framework on problems such as Vehicle Routing. Combined with the new start point generators and meta-heuristics, this has made a significant improvement to the quality of solutions which can be found within a reasonable time period.

5.   VRP EXPERIMENTS

In this section, we give a short analysis of the performance of iOpt and iSchedule on the Vehicle Routing Problem. We ran our best algorithm GLSFLS (see Fig. 1 as viewed in the iOpt Tool Suite) with the move filtering described above allowing 3 hours maximum run time on Solomon's benchmarks. It is important to note that the speed of our GLSFLS has been significantly increased by our move filtering component up to 10 times faster allowing a speed greater than 100,000 moves evaluated per second for GLSFLS. In the literature similar component are very common and usually allow a hard-wired algorithm to run at a speed of at least 250,000 evaluated moves. FLSGLS speed becomes therefore more competitive for a toolkit such as iOpt written in 100% pure Java software.

Figure 20-1. Tree of search components for GLSFLS

In the table below, we show the results of these algorithms in comparison to those published for GreenTrip GLS (another GLS based algorithm developed during the Green Trip project which later formed the basis of the ILOG dispatcher product in [6]) and the current best known results in the literature [5]).

Table 20-6. Comparison of results between iOpt, GreenTrip [6] and the best known results

Algorithm	iOpt/iSchedule GLSFLS		GreenTrip GLS		Best known results	
Problem Class	Mean #Vehicles	Mean #Distance	Mean #Vehicles	Mean #Distance	Mean #Vehicles	Mean #Distance
C1	10.00	828.84	10.00	830.75	10.00	828.38
R1	12.08	1242.69	12.67	1200.33	11.92	1209.89
RC1	11.88	1403.99	12.12	1388.15	11.50	1384.16
C2	3.00	591.31	3.00	592.24	3.00	589.86
R2	2.91	968.63	3.00	966.56	2.73	951.91
RC2	3.38	1144.99	3.38	1133.42	3.25	1119.35
Average	7.21	1030.07	7.36	1018.58	7.07	1013.92

From the results shown in Table 5, we can see that our best iOpt/iSchedule algorithm (GLSFLS) for vehicle routing is really competitive with the best in the literature. It should also be noted, that our algorithm on average performs significantly better with regard to the number of vehicles (7.21 as opposed to 7.36) used and distances (which is a lesser objective) compared to GreenTrip GLS. In addition, the GreenTrip GLS initially starts with a given number of vehicles (determined using instances knowledge) and reduces this number while a solution with all the tasks allocated is found whereas FLSGLS specifically tries to reduce the number of vehicles used during the search as well as minimizing travel distance. This makes it a more robust solution with no need to identify an initial maximum number of vehicles.

Overall, according to the survey from Gendreau M. and Bräysy O. (2003) [5][7], our GLSFLS is ranked 12th out of the best 21 systems listed in their paper and more impressively would have obtained the best results out of the systems published before 2001. This definitely confirms that a solution built using a toolkit like iOpt can be really competitive to hard-wired solutions.

---

[7] It is important to note that since our experiments, some new techniques have been developed providing better results on some VRP instances but as far as we know not on those Solomon's instances (Mester, D., and Bräysy, O. (2004) [11] and Prins, C. (2004)[15]).

6. CONCLUSION

In this paper, we have shown that solving a combinatorial optimization problem such as VRP using iOpt\iSchedule has many benefits. First of all, the modeling facility of iOpt\iSchedule allows the user to quickly state his VRP problem with resources, tasks, constraints and an objective function. This model uses underneath an invariant (one-way constraint) network that can efficiently propagate the value changes of decision variables to constraints and objective function. As iOpt is using only value propagation in contrast to domain propagation currently used in constraint programming (as in toolkits like Ilog Solver and Chip), modeling any user-specific relation is therefore more straightforward to implement. For this reason, the problem modeling facility of iOpt is flexible, easy to enrich with new relations and avoids the need for the user to implement a tedious code to maintain the consistency of his network of relations. In the case of a hard-wired solution, the user can only develop a solution for his particular problem without any guarantee that he will be able to easily include any future requirement. Another consequence is nobody except the author will be able to use it making it unavailable to other solving techniques. As iOpt keeps problem models and algorithms separated, modeling a problem can be done independently of the algorithms.

Secondly and more importantly, the Heuristic Search Framework of iOpt proposes a methodology for designing metaheuristics that proposes to break down any heuristic search into algorithmic components that can then be exchanged between algorithms. In this paper, by breaking down the best-in-class solving techniques for VRP (this has also been done for other problems, although for reasons of space we do not report this here), we have shown how the most advanced techniques can be compared more fairly leading to better understanding of the techniques themselves. In addition, by doing this decomposition, new combinations of these efficient techniques are immediately available and can now be investigated. Once this work has been done, it is more likely that nobody else will have to do it again as iOpt allows us to maximize the reuse of the work done by previous researchers/users. Furthermore, iOpt offers even the capability to include any new heuristic technique or operator discovered in the future into HSF opening up more possibilities by immediately being able to create new combinations with existing techniques and/or operators.

To conclude, we believe that the use of a toolkit such as iOpt within a community of users will have a significant impact on real world applications where the latest heuristic techniques coming from the research community will be able to reach the industrial world much faster. On the other hand, the use of iOpt in a community of researchers will help the community focus on the key issues of the techniques without sacrificing too much performance as shown in the case of the VRP problem where our GLSFLS algorithm competes with the best existing hard-wired algorithms.

REFERENCES

[1] Bräysy, O. (2001): "A Reactive Variable Neighborhood Search Algorithm for the Vehicle Routing Problem with Time Windows". In *INFORMS Journal on Computing* 15:4, 347-368

[2] Dorne, R., Voudouris, C., Liret, A. and Mills, P. (2005): "iSchedule: A new framework for solving scheduling problems". To be submitted to *INFORMS Journal on Computing.*

[3] Dorne, R., Ladde, C. and Voudouris, C. (2003): "Heuristic Search Builder: the iOpt's Tool to Visually Build Metaheuristic Algorithms". Presented at *5th Metaheuristics International Conference (MIC' 2003)*, Kyoto, Japan.

[4] Dorne, R., and Voudouris, C. (2001) "HSF: A generic framework to easily design Meta-Heuristic methods". Presented at *4th Metaheuristics International Conference (MIC' 2001)*, Porto, Portugal, 423-428.

[5] Gendreau M. and Bräysy O. (2003): "Metaheuristic Approaches for the Vehicle Routing Problem with Time Windows: A Survey". Presented at *5th Metaheuristics International Conference (MIC' 2003)*, Kyoto, Japan.

[6] Kilby, P., Prosser, P. and Shaw, P. (1999): "Guided Local Search for the Vehicle Routing Problems With Time Windows". In Meta-heuristics: Advances and Trends in Local Search for Optimization, Voss, S., Martello, S., Osman I.H. and Roucairol, C. (eds.), Kluwer Academic Publishers, Boston, 473-486.

[7] Lin, S. (1965): "Computer Solutions of the Traveling Salesman Problem Bell System". In *Technical Journal* 44, 2245- 2269.

[8] Michel L., Van Hentenryck, P. (2000): "Localizer". In *Constraints*, **5**, Issue 1-2, 43-84.

[9] Mester, D. and Bräysy, O. (2004): "Active guided evolution strategies for the large scale vehicle routing problems with time windows", Computers & Operations Research 32:1593-1614.

[10] Or, I. (1976): Traveling Salesman-Type Combinatorial Problems and their Relation to the Logistics of Regional Blood Banking. PhD thesis. Northwestern University, Evanston, Illinois.

[11] Potvin, J.Y. and Rousseau, J.M. (1995): "An Exchange Heuristic for Routing Problems with Time Windows Journal". In the *Operational Research Society* 46, 1433-1446.

[12] Prins, C. (2004): "A simple and effective evolutionary algorithm for the vehicle routing problem", Computers & Operations Research 31:1985–2002.

[13] Solomon, M. (1987): "Algorithms for the Vehicle Routing and Scheduling Problems with Time Window Constraints", *Operations Research* **35**, 254-265.

[14] Taillard, E., Badeau, P., Gendreau, Michel, Guertin, F. and Potvin, J-Y (1997): "A Tabu Search Heuristic for the Vehicle Routing Problem with Soft Time Windows". In *Transportation Science* 31, 170-186.

[15] Voss, S. and Woodruff, D. (2002): Optimization Software Class Libraries. Kluwer Academic Publishers.

[16] Voudouris, C. (1997): Guided Local Search for Combinatorial Problems. PhD thesis Department of Computer Science, University of Essex, Colchester, UK.